PESTICIDES

PESTICIDES

Preparation and Mode of Action

R. Cremlyn

The Hatfield Polytechnic, Hertfordshire

JOHN WILEY & SONS

Chichester · New York · Brisbane · Toronto

Library of Congress Cataloging in Publication Data:
Cremlyn, Richard James William Campbell-Davys.
    Pesticides: preparation and mode of action.
    Includes index.
    1.   Pesticides.   I.   Title.
SB951.C68   632′.95   77-28590
ISBN 0 471 99631 9 (Cloth)
ISBN 0 471 27669 3 (pbk.)

Typeset in Great Britain by Preface Ltd., Salisbury and printed at Unwin Brothers Ltd., The Gresham Press, Old Woking.

Contents

v

Preface

The book discusses the growth in the application and sophistication of chemical pesticides which has been particularly rapid since the end of World War II.

The physicochemical factors and biochemical reactions important in pesticides are discussed as an introduction to the subsequent chapters dealing with the major chemical groups used to control different kinds of pests.

The greater awareness of the dangers of environmental pollution arising from the widespread use of chemical pesticides is reflected in the increasing emphasis placed on safer and more selective chemicals. In many countries, candidate chemicals have to satisfy stringent legislative criteria before they can be marketed as pesticides; the effects of pesticides on the environment are discussed together with future developments. Areas currently of special interest include systemic fungicides, selective herbicides against grass weeds, chemicals controlling the growth of insects and plants, and microbial insecticides.

It is hoped that the book will prove valuable to all interested in chemical pesticides, and especially to undergraduates in universities, polytechnics, and colleges studying for degrees in agriculture, agricultural chemistry, applied biology, and applied chemistry. It should also prove of use to students taking courses for National Diplomas and Certificates in Agriculture. The author wishes to thank Dr. H. A. Jones of John Wiley and Sons Ltd. for his help during the passage of the book through the press. My thanks are also due to those firms who allowed me to reproduce photographs of some of their pesticides in action, and to Mrs Paula Iddiols for her expert typing of the manuscript.

R. J. Cremlyn

Chapter 1
Introduction

Pesticides or agrochemicals are chemicals designed to combat the attacks of various pests on agricultural and horticultural crops. They fall into three major classes: insecticides, fungicides, and herbicides (or weed killers). There are also rodenticides (for control of vertebrate pests), nematicides (to kill microscopic eelworms), molluscicides (to kill slugs and snails), and acaricides (to kill mites).

Pesticides may also be divided into two main types, namely contact or non-systemic pesticides, and systemic pesticides. Contact or surface pesticides do not appreciably penetrate plant tissues and are consequently not transported, or translocated, within the plant vascular system. The earlier insecticides, fungicides, and herbicides were of this type; their disadvantages are that they are susceptible to the effects of weathering (wind, rain, and sunlight) over long periods and new plant growth will be left unprotected and hence open to attack by insect and fungal pests. The early agricultural fungicides were, therefore, protectant fungicides – in other words they are designed to prevent the development of the fungal spores, but once the fungus has become established and the infection starts to ramify through the plant tissues such non-systemic fungicides possess little eradicant action and usually cannot halt the infection.

It is not surprising that the earlier pesticides were non-systemic in character because these are easier to discover, since phytotoxicity does not present such a great problem as it does in the case of systemic pesticides where the chemical comes into intimate contact with the tissues of the host plant. The danger of the fungicide causing damage to the host plant is especially formidable in the case of systemic fungicides, since here both host and pest are plants and for success the chemical must show selective toxicity to the fungus. The severity of these problems is reflected in the long time which elapsed before commercial systemic fungicides appeared on the market.

The effectiveness of many relatively insoluble surface fungicides depended on the fungal spores dissolving sufficient toxicant from the mixture to be killed, but on the other hand, there must not be enough soluble toxicant present to appreciably damage the host plant.

In contrast, many of the more recent pesticides are systemic in character – these can effectively penetrate the plant cuticle and move through the plant vascular system. Examples are provided by the phenoxyacetic acid selective herbicides (1942); certain organophosphorus insecticides like schradan (1941), and the more recently discovered systemic fungicides like benomyl (1967).

Systemic fungicides are also sometimes termed plant chemotherapeutants and can not only protect the plant from fungal attack, but also cure or inhibit an established infection. They are little affected by weathering and will also confer immunity on all new plant growth.

Pests can be divided into various groups. In the plant kingdom, characterized by the ability of the organism to photosynthesize carbohydrates from air and water with the aid of the green pigment chlorophyll, higher plants growing where man does not want them are termed weeds and are important pests. Of the lower plants, algae are not generally of great importance as pests, although in some circumstances, e.g. in lakes and other slow moving water, excessive algal growth or 'bloom' may cause considerable damage and require treatment with chemicals (algicides).

Fungi or non-photosynthetic plants cannot obtain their nutrients from air and water since they do not have chlorophyll; consequently they feed directly on decaying plant or animal matter (saprophytic fungi) or on living plants or animals (parasitic fungi). There are thousands of different species of fungi mainly found in soil — some, like yeasts, are unicellular while others are composed of a network of branched filaments (hyphae). A number of fungi are serious pests attacking both living crop plants and also crops in storage.

Several bacteria are causal agents of plant diseases, although they are not nearly as important as the phytopathogenic fungi. Bacteria can be observed under the microscope and can be classified according to their shape; thus a spherical bacterium is termed a *coccus* while a rod-shaped one is a *bacillus*.

Viruses, like bacteria and fungi, attack plants and animals and some species cause significant plant diseases. Viruses form a distinct category of living organism because they are not true cells. Unlike bacteria they are too small (100–300 Å) in diameter to be observed with an ordinary microscope, but they can be revealed under the electron microscope — each virus consists of a single strand of DNA or RNA surrounded by a protective coat of protein.

Several higher animals (vertebrates) are important pests, e.g. mice, rats, and rabbits; another group of pests is represented by the true insects (arthropods) which are invertebrates. The latter possess three pairs of legs and the adult body has three parts; the arachnids (mites and ticks) differ from true insects in having no distinct division of the body into three parts; also they usually have four pairs of legs. In the lower orders of animals, certain nematodes, parasitic worms often with unsegmented bodies, are important crop pests.

Historical Aspects

Ever since the dawn of civilization man has continually endeavoured to improve his living conditions; in his efforts to produce adequate supplies of food man has been opposed by the ravages wrought by insect pests and crop diseases. The blasting mentioned by the prophet Amos (760 B.C.) was the same cereal rust disease which is still responsible for enormous losses. The father of botany, Theophrastus (300 B.C.), described many plant diseases known today such as scorch, rot, scab, and rust. There are also several references in the Old Testament to the plagues of

Egypt for which the locust was chiefly responsible and even today locusts cause vast food losses in the Near East and Africa[1,2].

The major pests inhibiting the growth of agricultural crops are insects, fungi, and weeds, and the idea of combating these pests by the use of chemicals is not new.

Sulphur was known to avert diseases, as well as insects, before 1000 B.C. and its use as a fumigant was mentioned by Homer. Pliny (79 A.D.) advocated the use of arsenic as an insecticide and by the sixteenth century, the Chinese were applying moderate amounts of arsenic compounds as insecticides. In the seventeenth century the first naturally occurring insecticide, nicotine — from extracts of tobacco leaves — was used to control the plum curculio and the lace bug. Hamberg proposed mercuric chloride as a wood preservative (1705) and a hundred years later Prévost described the inhibition of smut spores by copper sulphate[3].

These early discoveries may probably be ascribed to a mixture of acute observation following trial and error application, together with strong undertones of superstition.

Fossil evidence preserved in rocks indicates that the agents of plant disease were operating long before the appearance of man on the earth. The tales of woe about the ravages of blights, mildews, and plagues are prominent in the earliest written records. There are many such references in the Bible, when plant diseases and plagues were seen to be visited upon man by God as a punishment for his sin. Although rather contradictory to the Christian concept of a loving God, this idea lingered on in subsequent centuries. It was certainly held in Hungary when during the eleventh and thirteenth centuries the cereal crops were ravaged by disease attacks. Those disasters, not emanating from the fury of God, were considered to be due to the actions of evil spirits — witches, goblins, and hobgoblins — which opposed the good forces in the world. Such demonic forces were held responsible, right up to the eighteenth century, for damage to agricultural crops caused by insect pests and diseases, and magical prescriptions to control pests abound in the early agricultural literature. It was not until the middle of the nineteenth century that systematic scientific methods began to be applied to the problem of controlling agricultrual pests.

About 1850 two important natural insecticides were introduced: rotenone from the roots of the derris plant, and pyrethrum from the flower heads of a species of chrysanthemum. These are still widely used insecticides (see Chapter 4). About this time too, soap was used to kill aphids, and sulphur as a fungicide on peach trees. A mixture of sulphur with lime to soften it, later called lime sulphur, was first suggested by Weighton (1814), and in 1902 it was observed that lime sulphur was effective against apple scab[4]. Also a treatise by Forsyth (1841) described a combined wash composed of tobacco, sulphur, and unslaked lime to control insects and fungi. During the nineteenth century new inorganic materials were introduced for combating insect pests; for instance, an investigation into the use of new arsenic compounds led in 1867 to the introduction of an impure copper arsenite (Paris Green) for control of Colorado beetle in the state of Mississippi, and in 1892 lead arsenate was used for control of gipsy moth. By 1900 Paris Green was used so extensively as an insecticide that it caused the introduction of the first State legislation governing the use of insecticides in the United States of America[5]. The Irish Potato Famine of 1845—49 provides an illustration of what can occur when a staple food crop is stricken by a disease against which there is no known defence. The potato crop was virtually destroyed by severe attacks of the fungal disease known as potato late blight resulting in the deaths of more than a million people (some 12% of the population) from starvation and the emigration of a million and a half people, chiefly to the United States of America[6,7].

A large number of fantastic theories were propounded to account for the epidemic and a distinguished chemist of the time, Dr. Lyon Playfair, was consulted. He did not however realise that the causal agent of the disease was a fungus *Phytophthora infestans*, whose spores were capable of very rapid reproduction so that a whole field of potatoes could be destroyed virtually overnight. Some ten years after the potato famine evidence was accumulated, largely from the researches of the Rev. M. J. Berkeley in England and Anton de Bary in Germany, which indicated that the potato disease was due to a parasitic fungus, though another decade had elapsed before this idea was widely accepted. One valuable chemical treatment for the control of pathogenic fungi, like potato blight and vine mildew, was discovered accidentally by Millardet in 1882[7]. A local custom of the farmers in the Bordeaux district of France was to daub the roadside vines with a mixture of copper sulphate and lime in order to discourage pilfering of the crop. At this time the crops of the French vineyards were being destroyed by the downy mildew disease and Millardet observed that although the vines away from the road were heavily infested with mildew, those alongside the road which had been treated with the mixture were relatively free from the disease. Millardet subsequently carried out further experiments which established the effectiveness of Bordeaux mixture (copper sulphate, lime, and water) against vine mildew. The mixture was widely applied, the disease was arrested and Millardet became a national hero.

This success stimulated the search for other chemical pesticides and the succeeding years witnessed the successful introduction of new materials containing copper, mercury, or sulphur. In addition, during this period, the manufacture of equipment for effectively applying these materials to the crop was begun.

Just as the search for fungicides to control phytopathogenic fungi was stimulated by the threat of famine, so the development of industrial fungicides was stimulated by the growth of railways and the consequent need to protect wooden railroad ties (some 3200 ties per mile of track) from rotting. This posed a tremendous challenge and meant that a successful chemical treatment would be guaranteed a profitable return and numerous patents were issued covering proprietary products containing creosote and salts of copper, mercury, and zinc for this purpose[7].

Many of the well-known poisons have been applied at one time or another for the control of insects and other pests. They were sometimes quite successful, although the hazards to the operators were great. Cyanide, generally as hydrogen cyanide gas, has been used as a fumigant in buildings to kill bedbugs and wood-boring beetles; also in California from 1886 onwards against scale insects on citrus trees. Tents were placed over the trees and hydrogen cyanide generated inside. Initially this treatment proved a considerable success but after a time failures became apparent. These resulted from the development of resistant strains of the insect — this was the first reported example of resistance to an insecticide. In 1897 formaldehyde was introduced as a fumigant and in 1913 organomercurials were first used as fungicidal seed dressings against cereal smut and bunt diseases[7].

In 1896 a French farmer applying Bordeaux mixture to his vines noticed that it caused the leaves of yellow charlock in the vicinity to turn black. This fortuitous observation is probably the origin of the idea of selective herbicides. A little later it was discovered that by spraying a solution of iron sulphate on to a mixture of cereal and dicotyledonous weeds, only the weeds were killed. During the next decade many other inorganic compounds, such as copper sulphate, ammonium sulphate, and sulphuric acid, were found to exhibit selective herbicidal action at suitable concentrations[8].

In 1912 W. C. Piver developed calcium arsenate as a replacement for Paris Green and lead arsenate, which soon became important for controlling the boll weevil on cotton in the United States of America. By the early 1920's the extensive application of arsenical insecticides caused widespread public dismay because treated fruits and vegetables were sometimes shown to contain poisonous residues. This stimulated the search for other less dangerous pesticides, and led to the introduction of organic compounds, such as tar, petroleum oils, and dinitro-*o*-cresol. The latter compound eventually replaced tar oil for control of aphid eggs, and in 1933 was patented as a selective herbicide (Sinox) against weeds in cereal crops. Unfortunately this is a very poisonous substance which was first used as an insecticide in 1892 to control the Nun moth, an important pest of forest trees[9].

The 1930's really represent the beginning of the modern era of synthetic organic pesticides — important examples included the introduction of alkyl thiocyanate insecticides (1930); salicylanilide (Shirlan) (1931), the first organic fungicide; dithiocarbamate fungicides (1934), valuable as foliar sprays for the control of a range of pathogenic fungi such as the scabs and rots of fruit and potato blight; 2,4-dinitro 6-(1'-methyl-n-heptyl)phenyl crotonate or dinocap (1946) and chloranil (tetrachloro-1,4-benzoquinone, 1938) were two other protectant fungicides, the former was especially valuable against powdery mildews. Other organic compounds used during this period were azobenzene, ethylene dibromide, ethylene oxide, methyl bromide, and carbon disulphide as fumigants; phenothiazine, *p*-dichloro-benzene, naphthalene, and thiodiphenylamine as insecticides.

In 1939 Dr. Paul Müller discovered the powerful insecticidal properties of *d*ichloro*d*iphenyl*t*richloroethane or DDT, and following successful initial field tests in Switzerland against Colorado potato beetle, it was manufactured in 1943 and soon became the most widely used single insecticide in the world (see Chapter 5). Present fears about the long-term deleterious effects of DDT and other organochlorine insecticides on the ecosphere (see Chapter 14) must not allow us to forget our tremendous debt to DDT. This compound controls louse-borne typhus and is equally effective against malaria-carrying mosquitoes; its use certainly materially helped the Western powers to win World War II since it permitted military operations to be conducted in the tropics where otherwise the danger of epidemics would have been too great[9]. Following the success of DDT, several useful insecticidal analogues such as methoxychlor were discovered and a number of different types of organochlorine compounds were also found to be potent contact insecticides[10].

Benzene hexachloride (or correctly hexachlorocyclohexane) was first prepared by the English chemist Michael Faraday in 1825, although its insecticidal properties were not recognized until 1942. From about 1945 several insecticidal chlorinated hydrocarbon cyclodiene compounds were introduced, though they did not come into widespread use until the middle 1950's. Common examples included aldrin, dieldrin, heptachlor, and endrin.

The organophosphorus compounds[11] represent another extremely important class of organic insecticides. Their early development stemmed from wartime research on nerve gases for use in chemical warfare by Dr. Gerhard Schrader and his team in Germany. Early examples included the powerful insecticides schradan (octamethylpyrophosphoramide), which acts as a systemic insecticide against aphids and red spider, and the contact insecticide parathion (O, O'-diethyl p-nitrophenyl phosphorothionate) which is remarkably effective against aphids, red spider, and eelworms. Unfortunately both these compounds are highly poisonous to mammals and later research in this field has been increasingly directed towards the discovery of more selective and less poisonous insecticides. Malathion (1950) was the first example of a wide-spectrum organophosphorus insecticide combined with a very low mammalian toxicity and more recently other safe compounds such as the selective aphicide menazon (1961) have been developed (see Chapter 6). An important advantage of organophosphorus insecticides is that they are generally rapidly degraded after application to non-toxic materials; consequently they are not persistent, like organochlorine insecticides, and therefore do not tend to accumulate in the environment and along food chains (see Chapter 14).

A closely related group of insecticides are the carbamate esters first discovered by the Geigy company in Switzerland in 1947, although the most generally effective member of the group carbaryl or Sevin (N-methyl α-naphthylcarbamate) was not introduced until nearly a decade later. This is becoming of increasing importance as a possible replacement for DDT (see Chapter 6, p. 97).

In 1943 Templeman and Sexton working for Imperial Chemical Industries in England independently discovered the herbicidal activity of the phenoxyacetic acids. Two well-known examples are 2-methyl-4-chloro-(MCPA) and 2,4-dichloro-(2,4-D) phenoxyacetic acid. These compounds are translocated in plants and are extremely valuable for the selective control of broad-leaved weeds in cereal crops. They are very safe to use, and these two compounds in fact are the most widely used pesticides in Britain (see Chapter 8).

In 1951 Kittleson working for the Standard Oil Company in America introduced an important new fungicide called captan (or N-trichloromethylthio-tetrahydrophthalimide). This had outstanding properties as a protectant fungicide against a wide spectrum of pathogenic fungi on fruit and vegetable crops. Subsequently a number of other N-trichloromethylthio compounds have been marketed as foliage fungicides (see Chapter 7, p. 114).

The bipyridylium herbicides diquat and paraquat were introduced by Imperial

Chemical Industries Ltd in 1958. These are very quick-acting herbicides which are absorbed by plants and translocated causing desiccation of the foliage. These herbicides are strongly adsorbed by clay constituents in soil, so they are effectively deactivated as soon as they come into contact with soil. They are useful total weed killers and will rapidly kill off all top growth. Paraquat is used in 'chemical ploughing' in which the weeds are killed by spraying with paraquat, followed by immediate reseeding — this method is specially valuable in areas where there is danger of soil erosion.

The idea of the internal treatment of plants with chemicals (plant chemotherapy) is not new and goes back at least to the twelfth century when various substances such as spices, colouring matters, and medicines were inserted into the boreholes of fruit trees to endeavour to improve the fruit[12]. Some rather macabre experiments were carried out by Leonardo de Vinci in the fifteenth century in which arsenic was injected into fruit trees to make the fruit poisonous. The study of plant pathology developed rapidly during the eighteenth century and experiments on the movement of substances, such as dyes and mineral salts, were carried out. Certain plant diseases were shown to result from nutrient deficiency, such as chlorosis from lack of iron, and attempts were made to cure the plants by injection of mineral salts. In the early 1900's toxic compounds, such as potassium cyanide, were injected into plants in an effort to kill insect pests. An examination was also made of the effects of injecting dyes and disinfectants into plum trees infected with silver leaf disease, and later studies[12] showed that 8-quinolinol sulphate was an effective chemotherapeutant against the disease. In America in the 1920's injection of sweet chestnut trees was studied as a means of controlling blight; lithium salts had some inhibiting effect but injections of thymol were much more effective. Some early workers also tried root applications of chemicals for the control of phytopathogenic fungi; thus Massee (1903) claimed to reduce cucumber mildew by root treatment with aqueous copper sulphate, and Spinks (1913) found that lithium salts inhibited the development of powdery mildew on wheat and barley.

However, little significant progress in plant chemotherapy occurred until the late 1930's, because by this time the limitations of the conventional surface fungicides were obvious. Also many new organic compounds were becoming available, and outstanding success had been achieved in the field of human chemotherapy. In 1935 the important range of sulphonamide drugs was introduced, and in 1938 Hassebrauk demonstrated that root treatment with sulphanilamide protected wheat seedlings from attack by rust spores.

In 1940 Chain and Flory showed that penicillin was highly effective against bacterial infections in man. This stimulated the search for other medically useful antibiotics and chloramphenicol, 'Aureomycin', and streptomycin were soon discovered, and by 1952 streptomycin was being used for systemic control of certain fungal pathogens and bacterial diseases of plants (see Chapter 7, p. 123).

World War II not only enhanced the development and commercial production of antibiotics, but it also provided the basis for Schrader's work on organophosphorus compounds, several of which proved highly efficient systemic insecticides[11]. The very valuable phenoxyacetic acid selective herbicides were also introduced, so that by the 1950's a range of commercial systemic insecticides and herbicides was available. But it was not until the late 1960's that effective systemic fungicides appeared on the market[13], and their development represents the most recent breakthrough in the field of plant chemotherapy.

The major classes of systemic fungicides developed from 1966 are oxathiins, benzimidazoles, thiophanates, and pyrimidines. Other effective systemic fungicides at present in use include antibiotics, morpholines, and organophosphorus compounds[13] (see Chapter 7, p. 127).

Ever since man had a home it was invaded by rats and mice which also raided his food stores. The rat is one of man's most formidable enemies. It does significant damage to the fabric of buildings and is the carrier of some of man's dreaded diseases such as Weil's disease and bubonic plague, the black death of the Middle Ages which in 1348—49 killed a quarter of the population of Europe, and between 1896 and 1917 was responsible for nearly 10 million deaths.

Chemicals that control rats are termed rodenticides (see Chapter 10). The first really effective compound, warfarin, was developed by the Wisconsin Alumni Research Foundation in 1944. It is an anticoagulant, which has been used in human medicine. Rats and mice are killed by internal haemorrhage and they will eat it in a suitable bait without becoming poison-shy[14]. However, in Britain a resistant strain of rats has appeared which are immune to normal doses of warfarin and these resistant or 'super' rats have increased considerably. Luckily almost simultaneously, a new rodenticide, norbormide or Raticate, has been discovered (1964) as a result of research for a chemical effective against rheumatoid arthritis.

The large-scale pesticide industry really dates from the end of World War II with the commercial introduction of the phenoxyacetic acid selective herbicides and the synthetic organochlorine and organophosphorus insecticides. It has been estimated[15] that the exports of the major pesticide manufacturing countries in 1949 were worth the following amounts:

U.S.A.	£28 million
U.K.	£3 million
France	£1 million
Switzerland	£0.6 million

In the period from 1949 to 1965 the United Kingdom exports had expanded to £10 million and Western Germany emerged as a major pesticide exporting country. The total value of all pesticides applied to world crops in 1949 was approximately £200 million which clearly emphasizes the point that most of the pesticides are used in the manufacturing countries and not exported. The relatively low proportion exported may be partly due to the fact that many underdeveloped countries import chemical intermediates from industrialized countries and then carry out the final stages of production themselves in which case the chemical intermediates do not count as pesticide imports. Such relatively primitive countries like to have political control of pesticide manufacture. Thus in India benzene hexachloride (Chapter 5, p. 60) is the most widely used insecticide because it can be produced simply in crude form. The annual production of this compound in

India in 1965 was some 30,000 tonnes which is about four times as much as was produced in the United States of America, though it represented only 20% of the total American production of all organochlorine insecticides[15]. The Pesticide Reviews of the American Department of Agriculture give the value of American pesticide production in 1965 as some £200 million so that a reasonable estimate of total world pesticide production would be £600 million. British pesticide production was nearly £23 million, composed of £12 million of herbicides, £6 million of insecticides, £3.6 million of fungicides, and £1 million of miscellaneous other pesticides.

In temperate countries herbicides are the major type of pesticide used. Thus in Britain current sales to farmers are 66% herbicides, 20% fungicides, 10% insecticides, and 4% miscellaneous agrochemicals. In contrast, in tropical and subtropical areas a much larger relative amount is spent on insecticides.

The rapid growth in the application of pesticides in industrial countries like those of Western Europe and the United States of America has been stimulated by the high cost and stortage of agricultural labour. A recent assessment[16] of the current world usage of pesticides based on their user cash value suggested about 45% in the United States of America, 25% in Western Europe, 12% in Japan, leaving some 18% for the rest of the world including the communist countries. It therefore appears that not more than 10% of the market is in the underdeveloped countries. However it is precisely in these countries that there is the greatest need for plant protection chemicals. They contain 49% of the world population and 46% of the total world cultivated land area and suffer the greatest losses of crops due to pests. On a world scale pests destroy about one third of the annual crop during growth, harvesting, and storage, but in the underdeveloped countries, e.g. India, Africa, and Latin America, losses are some 40% of everything produced[17].

In these countries, even after the harvest has been gathered, there are appreciable losses in storage due to insects, rats, mites, and fungi. Losses in the tropics are much greater than in temperate countries; in West Africa estimated losses in storage are 25% and in India some 8%.

Properly constructed stores can keep out rats, and fungi and mites can be inhibited by controlling grain moisture content at $< 13\%$. Insects are more difficult to control by physical methods and there is a need for the further development of safe insecticides for use on stored crops[16].

Clearly there is a tremendous potential for expansion in pesticide sales to such countries although their poor economic state clearly discourages increased expenditure on pesticides, and Roberts[16] concludes that while extrapolation of existing trends suggests a steady growth in the use of pesticides, it is unlikely that there will be any marked change in their geographical distribution, although by the end of the 1980's perhaps some 16% of pesticides will be used in the developing countries.

The increasing use of pesticides in the underdeveloped countries is absolutely vital if such countries are to obtain the greater supplies of food necessary to feed their large populations adequately. Several examples highlight the value of pesticides in reducing crop losses. Thus in Ghana, which is the world's premier cocoa exporting country, the application of insecticides has almost trebled the yields by effectively controlling the damage to the crop by the capsid bug, and in Pakistan extensive use of insecticides on the sugar crop increased the yield by 30%[17]. The United Nations Food and Agriculture Organization (FAO) has estimated[17] that without the use of pesticides some 50% of the total cotton production in developing countries would be destroyed by pests. It is becoming increasingly evident to the world food authorities that pesticides may be the single most important factor in improving food production in the underdeveloped countries.

Herbicides are likely to remain the major class of pesticides used in developed countries while insecticides take pride of place in developing countries. Current world consumption is 43% herbicides, 32% insecticides, and 19% fungicides, leaving 3% for growth regulators and 3% for miscellaneous agrochemicals.

Over the last twenty years there has been substantial growth in the number of pesticides available, and from 1950 to 1968 the user cash value of pesticides has increased from £200 million to £1230 million – a sixfold increase of which the North American continent accounted for 50%[18]. It is estimated[16] that a steady increase in the use of crop protection chemicals will continue, so that a level 1.3–1.6 times the present usage will be reached by the early 1980's and possibly 2.4–4.2 times the present consumption by the end of the 1980's. Further estimation indicates that a threefold increase in food production will be needed by the end of this century to feed the world's expanding population and this will probably require a fivefold increase in the use of pesticides[19,20].

In 1976 Braunholtz[21] has estimated world sales of crop protection chemicals to be worth some £3,600m of which North America accounted for nearly 40% and Western Europe 25%. It was expected that future growth rates are likely to be 4–5% per annum into the early 1980's.

In countries like the United States of America the development of a pesticide from discovery to marketing takes some ten years and many of the pesticides discovered in the 1940's and 50's are still widely used. There is a clear need for the introduction of new products, especially more systemic nematicides, unique fungicides and compounds to combat resistant fungi, also new plant growth regulators and plant nutrition control agents (see Chapter 15). There is real danger that excessive emphasis on potential environmental hazards, especially in the United States of America, may cause the unjustifiable elimination of many valuable pesticides and stifle the development of essential new products due to over-regulation. These and other factors, such as escalating development costs imply a slower rate of introduction of new chemicals for pest control.

There will be increasing emphasis on integrated biological and chemical pest control methods to reduce the use of chemicals which will mean less dangers of environmental pollution, and less opportunity for the emergence of resistant strains of pests. In developed countries, by the end of the 1980's there should be progressive shift from animal to crop production, resulting from increased introduction of vegetable proteins. Improvements in the technology of texturized vegetable meat analogues and their expanding use as meat substitutes will mean that soya beans and leguminous crops will become more important.

References

1. Cremlyn, R. J. W., *Pest Articles and News Summaries*, **17**, (3), 291 (1971).
2. Hassall, K. A., *World Crop Protection – Vol. 2 Pesticides*, Iliffe Books Ltd., London, 1969, p. 1.
3. Horsfall, J. G., *Principles of Fungicidal Action*, Chronica Botanica Co., Waltham, U.S.A., 1956.
4. Whitten, J. L., *That We May Live*, Van Nostrand, Princeton, U.S.A., 1966, p. 26.
5. De Ong, E. R., *Chemistry and Uses of Pesticides*, 2nd edn., Reinhold Publ. Corp., New York, 1956.
6. Large, E. C., *The Advance of the Fungi*, H. Holt, New York, 1940, p. 13.
7. McCallan, S. E. A., 'History of fungicides' in *Fungicides, An Advanced Treatise* (Ed. Torgeson, D. C.), Academic Press, New York, 1967, p. 1.
8. Martin, H., *The Scientific Principles of Crop Protection*, 6th edn., Arnold, London, 1973, p. 268.
9. Mellanby, K., *Pesticides and Pollution*, Collins, London, 1970, p. 111.
10. O'Brian, R. D., *Insecticides: Action and Metabolism*, Academic Press, New York, 1967, p. 182.
11. Fest, C. and Schmidt, K. – J., *The Chemistry of Organophosphorus Insecticides*, Springer-Verlag, Berlin, 1973.
12. Wain, R. L. and Carter, G. A., 'Historical aspects' in *Systemic Fungicides* (Ed. Marsh, R. W.), 2nd edn., Longman, London, 1977, p. 6.
13. Cremlyn, R. J. W., *Internat. Pest Control*, **15** (2), 8 (1973).
14. McMillen, W., *Bugs or People*, Appleton-Century, New York, 1965, p. 128.
15. Hartley, G. S. and West, T. F., *Chemicals for Pest Control*, Pergamon Press, Oxford, 1969, p. 16.
16. Roberts, E. H., *Proc. 8th Brit. Insectic. and Fungic. Conf.*, Brighton, 3, 891 (1975).
17. *Pesticides in the Modern World* – A symposium prepared by members of the Co-operative Programme of Agro-Allied Industries with FAO and other United Nations Organisations, Newgate Press, London, 1972.
18. Waitt, A. W., *Pestic. Sci.*, 6, 199 (1975).
19. Goring, C. A. I., *Proc. 8th Brit. Insectic. and Fungic. Conf.*, Brighton, 3, 915, 1975.
20. Kohn, G. K., 'The pesticide industry' in *Handbook of Industrial Chemistry* (Ed. Kent, J. A.), 7th edn., Van Nostrand-Reinhold, New York, 1974, p. 619.
21. Braunholtz, J., *Chem. & Ind.*, 1977 (23), 929.

Chapter 2
Physicochemical Factors

As mentioned in the introduction (Chapter 1), chemicals have been used for pest control for centuries but the large-scale pesticide industry really only dates from the end of World War II. The valuable new chemicals developed over the last 25 years or so have arisen as a result of effective collaboration between teams of chemists and biologists[1].

Organic chemists first decide on some novel groups of compounds which they feel may have potential as pesticides; then they devise successful synthetic routes to obtain some candidate members of the selected groups. The chemicals, after purification and characterization by spectroscopic and analytical data, are submitted to the biologists for evaluation for potential pesticidal activity. In the initial primary screens a fairly high dosage of up to 500 p.p.m. may be used to pick out all compounds having some activity on the test organism. If activity is found the tests are repeated on that organism at lower dose rates, for instance at 200, 100, and 10 p.p.m. A series of further increasingly stringent tests are then applied, each stage causing candidate compounds to drop out. These screens are so good at eliminating compounds that the chances of a newly synthesized compound being a marketable pesticide are only about 1 in 7500, but with effective screening techniques this random screening can give a reasonable return[2]. In fact nearly all the present pesticides have been discovered as a result of more or less random screening of the chemicals available to the company[3(a)]. The precise screening methods vary from company to company but generally the candidate chemicals are tested (a) against a range of insects of economic importance, both by spraying and orally by incorporating the chemical in the insects' food, (b) against a range of plant pathogenic fungi growing in nutrient agar plates (*in vitro* tests) and also against fungi growing on plants (*in vivo* tests), (c) against 10—15 different species of weeds growing in the greenhouse, and both soil and foliar applications are investigated, (d) against certain other pests, such as slugs, eelworms, rats, mice, and mites.

As an illustration the Plant Protection Division of Imperial Chemical Industries Ltd. at Jealott's Hill Research Station in Berkshire England, annually screen some 8000 chemicals against disease organisms, pest insect species, weeds[4], and for possible useful effects on plant growth. If a chemical shows promise, the range of test organisms is widened and chemists set to work to obtain structural modifications ('molecular roulette') of the original chemical in an effort to enhance activity. The examination of a number of structural analogues hopefully will enable a structure–activity relationship to be elucidated so leading both to a better pesticide and possibly throwing light on the mode of biocidal action. Only very few candidate chemicals survive the rigorous laboratory screening procedures. Those remaining are then formulated and applied in various ways to find out how to obtain the optimum results under field conditions. These experiments are initiated on a relatively small scale in the greenhouse and continue using small-scale plots and ultimately progress to large-scale field trials in Britain and at overseas field stations.

The chemicals must finally be tested under the actual conditions in which they are to be used in practice by Government Research Stations and farmers. During the various stages of development there is a steady decline in the number of chemicals under test as those which have proved unsuitable for various reasons are rejected. While development proceeds, efficient and economic methods of large-scale synthesis have to be devised by chemists; the production of large amounts of the material may present problems not apparent during the small-scale preparation.

Fairly early in the development of the pesticide, the toxicological properties have to be evaluated – this must cover not only determination of the acute toxicity, but also any possible longer-term effects on the environment (see Chapter 14). The acute toxicity is determined by testing the candidate chemical against various mammals, generally rats and mice. The toxicity is usually recorded as the LD_{50} value – this is the dose required to kill 50% of the population of test animals and is expressed as mg/kg of the body weight of the animal. There should be at least 10 animals in the experiment and administration of the chemical can be oral (in the animal's food), by intravenous injection, or to the skin (dermal). Dermal toxicities are often slightly less than the oral values while intravenous toxicities are higher than the oral figures. In this book, the average values of the oral toxicities (LD_{50} values) against rats are quoted to give some indication of the mammalian toxicities of the different types of pesticides. The smaller the LD_{50} value, the more toxic the chemical so that toxicities of chemicals can be graded by the LD_{50} values as follows:

	LD_{50} (mg/kg)
1. Extremely toxic	1
2. Highly toxic	1–50
3. Moderately toxic	50–500
4. Slightly toxic	500–5000
5. Practically non-toxic	5000–15,000
6. Relatively harmless	>15,000

In this respect it is vital to understand the mode of pesticidal action, and the chemical must have minimum ecological effects if it is to be a commercially viable pesticide in the modern pollution-conscious society. It usually takes a period of at least six years from the date of the initial discovery of useful biocidal properties of a chemical to the marketing stage and, with the increasing stringency of tests which a candidate chemical has to pass, this gestation period is growing. Any candidate pesticide for the British market must be cleared at several stages by the Pesticides

Safety Precautions Scheme and the Ministry of Agriculture, Fisheries and Food must be satisfied that it does the specified job before it may be included in the Ministry's annual list of approved products[5]. Most countries in the world have comparable requirements; in the United States of America they are particularly tight with reference to any residues left after using the chemical. If the mode of action and toxicological studies are satisfactory large-scale field trials are initiated.

At this stage the study of different types of formulation of the chemical have to be undertaken. Getting the right type of formulation is vital in obtaining the optimum results[3(b)] for the pesticide, indeed it is true that a pesticide is only as good as its formulation. The majority of pesticides are applied as dusts or sprays; they may also sometimes be used as granules, vapours, or seed dressings.

The formulator tries to bring the active ingredient into a convenient form for application, either directly as powders, granules, or self-dispensing packs, e.g. aerosols, or sometimes after mixing with water or other readily available liquid diluent. The product must be formulated so that it is as effective as possible, but it must also be stable and reasonably safe in storage and transport. The active ingredients of most pesticides are relatively insoluble in water, but are fairly soluble in organic solvents like petroleum and xylene. These are, however, insoluble in water and so if the pesticide is dissolved in a suitable organic solvent and the solution diluted with water the organic layer quickly separates out from the water in the spray vessel[6].

This problem can be overcome by addition of emulsifiers, which are surface-active agents (surfactants), to the solution of the pesticide enabling it to form a stable emulsion on mixture with water in the spray tank and the emulsion can then be sprayed onto the crop. Many water-insoluble pesticides are marketed in the form of thick self-emulsifiable concentrates consisting of a solution of the pesticide in an organic solvent, e.g. petroleum or other hydrocarbon oil containing oil-soluble surfactants. The formation of emulsifiable oils has been aided by the development of non-ionic emulsifying agents, e.g. polyglycol ethers, and polyethylene oxides such as (1), (where $n \simeq 12$):

$$H_{17}C_9 - \langle\!\!\langle\bigcirc\rangle\!\!\rangle - OCH_2CH_2(OCH_2CH_2)_nOH$$

(1)

Such non-polar compounds have the advantage of being much more soluble in oils than ionic surfactants, and their effect is not greatly altered by the presence of saline impurities so that they do not form insoluble scums in hard water which was a major trouble when soaps were used. An effective self-emulsifying oil often requires the use of several different surfactants, for instance two or more non-ionic agents with widely differing numbers of ethylene oxide groups in their chains, or more generally a mixture of anionic and non-polar surfactants. Cationic surfactants are also used in certain cases, e.g. with some fungicides which are themselves cationic surfactants. In this case, use of anionic surfactants would lead to loss of surfactant properties and formation of an insoluble gum — for this reason pesticides should not be mixed with each other until it is clearly established that they are compatible.

If the active ingredient is soluble in water, the chemical can be dissolved in water to give an aqueous solution containing a suitable concentration of the active ingredient which can be sprayed directly onto the crop. The chemical may also be used as a wettable powder. The powder is prepared by grinding the material in a ball mill. However, few organic compounds produce a free-flowing powder without first mixing them with an inert inorganic mineral diluent, e.g. a talc or clay, and addition of dispersing and wetting agents is also needed to give a fine suspension when the powder is mixed with water in the spray tank, which must not settle out on the bottom of the tank for at least half an hour, otherwise frequent agitation will be needed to prevent formation of sticky powders that tend to cake on storage. In certain instances, especially with inorganic compounds like copper fungicides, the mineral diluent is not necessary since these chemicals have no tendency to stickiness. Clay mineral diluents contain strongly acidic centres which may catalyse the decomposition of certain pesticides, whereas other chemicals may be sensitive to alkaline diluents. A knowledge of the chemical reactivity of the active ingredient, especially towards acidic or basic hydrolysis, is valuable information to aid selection of the mineral diluent to be used in the formulation of a wettable powder. The emulsion spray separates into two phases on impact; the water running off the plant surface while the pesticide solution remains on the surface. With non-systemic pesticides, it is essential to obtain a high coverage of the surface by the spray and this is often enhanced by the addition of various wetters and spreaders; further the persistence of the dried deposit is sometimes improved by the use of chemical additives termed 'stickers'. Currently approximately 75% of all pesticides are applied as sprays, chiefly as water emulsions made from emulsifiable concentrates.

Pesticides may also be applied as dusts, although this method of application is not generally as effective as spraying. The pesticide powders are made by adding a solution of the active ingredient in a suitable solvent to a finely ground carrier. The resultant dust concentrate may be further diluted with a solid diluent. The particles must be the optimum size; if too small, they will cling together in the air blast and the coverage of the target surface will be poor. On the other hand, if too coarse the dust will not penetrate into the interior of plants and the active ingredient may separate from the particles of the diluent, this does not occur if the diameters of the particles are all less than 20 μm. The pesticide powder must not cake in humid climates or be so hard that damage is caused to the dusting machinery. The electrostatic charge is also important. This is acquired by friction as the dust passes through the blower, since the majority of leaf surfaces are negative positively charged dusts adhere better. A dust must be more dilute than a wettable powder generally about 2–5% of the active ingredient, because dusting machines usually cannot produce an even discharge at less than about 16.8 kg/ha (15 lb/acre). In spite of their ease in handling, formulation, and application, dusts are the least effective of the pesticide formulations because they have a relatively poor rate of deposition on the target. For instance in aerial dusting, some 10–40% of the pesticide reaches the crop while the remainder drifts away from the target. In similar conditions, aerial spraying using a water-emulsion formulation deposits 50–80% of the pesticide on the target.

Several insecticides are effective as seed dressings, especially against wheat bulb fly on cereals[7] and, in this context, the stable pyrethroid biopermethrin (Chapter 4, p. 44) appears very promising. With fungicides application by seed treatment has become increasingly important with the development of a number of commercial systemic fungicides, e.g. carboxin (see Chapter 7, p. 130). Various seed dressing techniques include the use of adhesive dust, wet slurries, and solutions[8], which were originally devised to deal with broad-spectrum surface fungicides, particularly organomercurials, which are largely used as seed dressings on cereals. The techniques are, however, equally applicable to systemic fungicides and when used as seed sterilants, as well as protectants during germination, seed dressings provide a most economic and convenient method of application, since they avoid the high labour costs inherent in spraying or dusting operations. The majority of fungicides are however applied as foliar sprays or dusts; ultimately the efficiency depends on the amount of the active toxicant that reaches the site of action which is sometimes not the formulated compound but a metabolite, e.g. benomyl (see Chapter 7, p. 129). Gray showed[8] that the antifungal action of streptomycin was increased by the addition of glycerol, which acts as a humectant to the spray mixture. This slowed up the rate of drying on the foliage allowing more time for the compound to be absorbed by the foliage. Surfactants are often added to fungicides to improve their efficiency by increasing their wetting properties and the solubility of the active compound as illustrated by the effect of the surfactant Tween 20 on triarimol[7].

Small-scale application of pesticides can be conveniently carried out from aerosols – these give very fine spray drops by discharging a liquid contained in a pressure vessel above its boiling point through a very fine orifice. The droplets are further broken up by the propellant boiling when the liquid jet comes into contact with the air. This method of application is however expensive and is only used in small gardens, glasshouses, and in the home where convenience outweighs the cost of aerosol cans. Some pesticides are sufficiently stable to be applied as smokes from specially manufactured smoke generators. The compound is mixed with an oxidant (e.g. sodium chlorate) and a combustible material (e.g. a carbohydrate) giving rise to a hot non-inflammable gas when it comes out from the orifice of the smoke generator[3(b),6]. Such generators are used in glasshouses, dwelling houses, and stores and the insecticides Gammexane HCH, DDT, and parathion and the fungicides Karathane and captan are often applied this way. A recent method of formulation is as granules; the pesticide is incorporated into porous clay pellets of 0.5–1.5 mm diameter which are coated with the pesticide by spraying them with a solution of chemical to give the desired concentration ranging from 2 to 20% of the active ingredient. This type of formulation has the advantage that the pesticide can be applied in winds of up to 20 m.p.h. without any appreciable drifting, in contrast to application as dusts or sprays; also the granules can often force their way through the foliage to reach the ground.

Encapsulation is a new formulation method in which the active ingredient is wrapped in very small beads of polyvinyl chloride or other materials, which break down at different rates so giving a slow release of the pesticide which can increase the life of, say a highly volatile, insecticide from a few minutes to several days. The majority of modern agricultural pesticides are effective at doses of 1 kg/ha (0.89 lb/acre) or less; consequently in order to achieve a reasonably uniform coverage the size of the particles used in dusting must be small (< 20 μm diameter), and similarly when spraying small droplets must be produced. This demands for dusting the use of an inert mineral diluent and for spraying dilution with water. Care must however be taken that too much water is avoided since it tends to increase run off. Generally the amount of water is so arranged that run off just occurs at the required applied dosage. A great variety of machinery has been developed for pesticide application, such as hydraulic sprays with either horizontal or vertical spray attachments and large air-blast machines used in orchards and plantations[9]. (Plates 3 and 4.) When spraying, the greater the volume of water employed and the lower the pressure applied the larger the spray droplets, and the coarser spray will be easier to direct to the target surface than a fine one and will be much less subject to drifting. Thus under these conditions weeds growing between rows of crops can be safely sprayed without drift onto the crop plants, whereas a fine spray can easily be blown hundreds of yards in a slight breeze and is obviously unsuitable for spraying herbicides in a restricted area. However there is a continuous move towards low volume spraying for economic reasons — this has been made feasible by improvements in spraying equipment. In the Sudan, DDT formulations have been effectively sprayed from the air onto cotton in very fine droplets at less than 1 gal/acre (11.2 l/ha). A typical low volume, low pressure farm sprayer tank has a capacity of 90—220 l and gives an output of some 1360 l/h. Such a sprayer operated from a tractor power take-off would spray at 225 l/ha when moving at some 5 km/h[9]. Water is used as the propellant and high pressures are needed to project the spray to the tops of standard trees in an orchard; this can be achieved by using powerful air blast sprayers in which the pesticide is mixed with a comparatively small volume of water. The solution or suspension is atomized by a spinning disc and the spray blown up to the trees. The same principle is used in modern knapsack sprayers, which can apply liquids at concentrations of a few kg/ha, and are extensively used by small farmers. They have the advantage of being less heavy than the older sprayers and are consequently much more pleasant to use. In many instances, controlled droplet application (CDA) is more important in pesticide application than ultra-low volume application (ULVA). With modern spray equipment the optimum droplet size can be selected for the particular target species, e.g. insect, fungus, or weed.

For a drift-free herbicide, the size of the spray droplets applied should be $\simeq$300 μm at 15 l/ha, whereas efficient mosquito control is achieved with 20 μm droplets at 5 ml/ha[10]. It is generally necessary to achieve the maximum retention of the spray droplets[11]. In many cases this will depend on the nature of the plant surface; however, generally retention is favoured by small droplets and low volumes of the spray liquid, thus for cereals not more than 450 l/ha (40 gal/acre) should be applied. The fight against pests indoors and in stored products often requires different pesticide formulations. Fumigation is widely used for controlling insect pests in ships holds, warehouses, silos, and rolling stock (see Chapter 9). The fumigant is usually released from pressure cylinders. Fumigation is also applied for soil sterilization and control of soil pests like nematodes (Chapter 11).

Physical factors are of considerable importance in determining the effectiveness of pesticides. Thus with surface-protectant fungicides like Bordeaux mixture (Chapter 7) the activity depends very much on the particle size used in the spray and the tenacity of the dried deposit on the leaves of the treated crop[12]. In many pesticides the achievement of the optimum balance between oil and water solubility is an important factor. Increasing the length of an alkyl chain aids oil or lipid solubility while depressing aqueous solubility. In a series of 2-alkylimidazolines it was found[12] that fungitoxicity increased with the number of carbon atoms in the alkyl side-chain up to 17 carbon atoms and afterwards activity decreased. With N-dodecylguanidine (dodine), on the other hand, the optimum fungitoxicity occurred with a side-chain of 12 carbon atoms[13].

Pesticides can be broadly classified into physical or chemical toxicants; in the former there is no clearly identifiable toxophore and the toxic effect depends on the physical properties of the whole molecule. For instance, oils such as kerosene are physical toxicants[14] – they kill plants by disorganizing the cells by causing them to lose water, and they flood the breathing pores of insects causing rapid asphyxiation. Oils also wet insect cuticles and consequently they become trapped by the surface-tension forces of water. These suffocating and entangling effects are utilized in the method of killing mosquito larvae by spreading a thin film of oil over the surface of water in which mosquitoes breed[2]. Oil washes have been used for a long time to control insects and spider mites in orchards, and this method of controlling mites is becoming increasingly important as the mites are rapidly developing resistance to chemical acaricides (Chapter 5, p. 56). The effectiveness of the treatment is enhanced by adding some polyisobutene to the oil which leaves a permanent sticky deposit on the trees and leaves which immobilizes the mites.

If pesticides are to be active they must reach the ultimate site of action within the target organism. Thus even surface fungicides, like Bordeaux mixture, must be able to penetrate the fungal spore; similarly contact insecticides have to penetrate the insect cuticle, and contact herbicides the plant cuticle when they impinge on it. The requirements if the pesticides are to be systemic in action are much more stringent because in addition they must have the capacity to be absorbed by the roots or leaves or seeds of plants and be translocated to other parts of the plant. In this way the whole plant, including new growth, is protected from fungal attack, or an established fungal infestation. With systemic insecticides, the total plant is rendered poisonous to any insect that eats or sucks it but insects that just alight on the leaves are not hurt, so systemic insecticides are likely to be more selective in their toxicity than contact insecticides. Phytotoxicity is a much more difficult problem to overcome with systemic pesticides because they are brought into intimate contact with the host plant.

Translocation of chemicals in plants may be considered in three stages[15]:

(a) *Entry into the free space within the tissues.* Water and solutes after penetrating the leaf cuticle pass into free space defined as that part of the plant in contact by diffusion with the external environment. The cuticle controls the loss of water from the plant and aqueous solutions of many organic molecules can be transported across the cuticle, although the majority of chemicals enter by the root tips. Diffusion of an aqueous spray from the leaves into the free space will only occur while a liquid film remains on the leaf surface so in cases where the speed of drying of the spray is a limiting factor uptake may be enhanced by addition of a humectant, e.g. glycerol.

(b) *Movement in the xylem.* The xylem vessels provide a system of water pathways communicating with the environment by free diffusion. The bulk of the movement of water and soluble minerals from the roots to the leaves is *via* the non-living xylem, and this process does not involve the expenditure of metabolic energy.

(c) *Movement in the phloem.* This is within the living parts of the cell and does require metabolic energy. Chemicals which have arrived at the leaves in the xylem are then distributed to the growing tissues of the plant *via* the phloem.

Movements (b) and (c) are distinct within the plant. The former occurs passively in the transpiration stream but requires evaporation at the surface and so can be reversed by immersion of the plant leaf in water so checking evaporation at the immersed surface. The latter is dependent on the metabolic activity and can be prevented by treatments inhibiting metabolism or immobilizing the phloem.

Movement in the phloem has been demonstrated for some chemicals, e.g. 2,4-D, while other compounds like dalapon and maleic hydrazide move more freely and transfer from the phloem to the xylem enabling them to be widely distributed in plants. Certain chemicals move from the leaves to the roots via the phloem and may be released into the surrounding soil. Such downward translocation may be valuable for control of soil fungi as illustrated by pyroxychlor (Chapter 15, p. 225). This effect was demonstrated with the hormone herbicides 2,3,6-trichloro- and 2,3,4,6-tetrachlorobenzoic acids (Chapter 8, p. 146), when sufficient chemical was exuded into the soil to produce malformations in adjacent plants.

Most of the early systemic fungicides showed typical xylem movement. These included several antibiotics, e.g. griseofulvin and streptomycin. Experiments with the first generation of commercial systemic fungicides, (Chapter 7, p. 127), such as benomyl demonstrated that these are mainly transported in the xylem, but the pyrimidine systemic fungicides (e.g. ethirimol and dimethirimol) (Chapter 7, p. 132) appear to move unchanged in the phloem.

The majority of systemic organophosphorus insecticides are also transported in the xylem. If the compound is to move in the essentially aqueous plant sap, it must have sufficient aqueous solubility or be converted to such a compound after metabolism in plant tissue[16(a)]. On the other hand, if a compound is to achieve penetration via the foliage, it must be transported across the waxy leaf cuticle demanding appreciable lipid solubility. To function as a systemic pesticide by foliar application the candidate compound must therefore have a reasonable lipophilic—hydrophilic balance. If the compound is too lipophilic, it will remain held in the cuticular waxes and if too hydrophilic, will never penetrate the cuticle. Thus in the series of O, S-dimethyl N-n-alkylphosphoramidates a parabolic relationship was established between systemic movement and the logarithm of the octanol-water partition coefficient, (P), or relative hydrophobic constant (π). π was defined as the difference between the logarithm of the partition coefficient of the parent compound (P_0) and the derivative (P) so that $\pi = \log P/P_0$. This implies the existence of an optimum lipophilic—hydrophilic value for maximum systemic translocation in plants, and in this series π was calculated to be 1.19 The N-n-propyl derivative, with a value of π of 1.31, nearest to the optimum value, showed the greatest systemic movement in a cotton leaf petiole. Some of the organophosphorus systemic insecticides (Chapter 6, p. 74) are translocated unchanged (e.g. mevinphos), while in other cases it is the active metabolite that is translocated in the plant (e.g. schradan)[16(a)].

Soil has been formed by the gradual breaking down of the rocks of the Earth's crust over many millions of years and different rocks produce soils with different characteristics[17]. When a pesticide is applied to the soil, the architecture of the soil becomes an important factor in its effectiveness. Soil is a dynamic system containing many inorganic and organic compounds which are constantly being chemically and microbiologically transformed. It is a very heterogeneous system containing solid, liquid, and gaseous components. The solid phase is mainly present in a finely divided form with a large surface area which is important for understanding the behaviour of chemicals in the soil system. The texture of the soil depends on the size of the particles it contains; sand particles have diameters of from 2 to 0.02 mm, while silt ranges from 0.02 to 0.002 mm diameter, and clay particles have diameters of less than 0.002 mm. They have a surface area corresponding to 2.3 m^2/g, and play an important role in the dynamics of pesticides in soil[17].

Clay minerals are composed of thin molecular layers held together by attractive forces forming an assemblage of layers. In the kaolinite type, the packing of the layers is so strong that no water molecules or ions can penetrate between them. On the other hand, in the montmorillonite type the individual layers are much more loosely packed, so that water molecules and ions can penetrate between the layers. The lattice of colloidal clay minerals is negatively charged, and the charge is neutralized by positively charged ions from the soil solution which are electrostatically attracted to the surface of the clay minerals. The cations are not however fixed and they can be exchanged by other cations, so montmorillonite has a much greater cation exchange capacity than kaolinite. Soil also contains organic colloids (humic substances) which are mainly negatively charged and possess a large surface area. All chemicals in soil are exposed to the physical and electrostatic attractive forces exerted by the soil colloids. Pesticides are adsorbed onto these surfaces making it harder for them to be taken up by plants and microorganisms and partially protecting them from chemical and enzymic attack. Greenhouse experiments have shown a good correlation between pesticidal activity and such factors as the amount of clay and humus colloids present in the soil, soil humidity, and pH value. However under field conditions, the correlations are not nearly as good probably due to the effects of variation in climatic conditions, e.g. the intensity of sunlight and temperature, so great care is needed in the interpretation of the results of field tests. Certainly the activity of a pesticide is often markedly influenced by the moisture content of the soil at the time of application.

In conclusion as the famous soil scientist Sir John Russell pointed out 'in an apparently solid clod of earth only about half is usually solid matter, the other half is simply empty except for the air and water it contains'. Thus a good silt loam is only half solid material by volume, because when the soil is in good physical condition the individual particles are packed together in aggregates and the pore spaces between the aggregates are partly filled with air and water[17].

Soil generally contains some 1–5% by weight of organic matter, but in peat soils this may reach some 95% of the dry weight of the soil. Bacteria are far the most important and numerous of the soil microorganisms – one spoonful of soil will contain billions of bacteria! Beneficial bacteria play an important role in breaking down organic matter to humus which is vital for the maintenance of good soil structure; other bacteria such as *Azotobacter* and *Clostridium* can convert atmospheric nitrogen to nitrates which subsequently can be utilized by plants. Other soil microorganisms are fungi; some are beneficial and some harmful to plants. The latter are the parasitic fungi which attack plant tissues, e.g. the species responsible for club root in brassicas. *Actinomycetes* attack beneficial bacteria and some are parasitic to plants, like that causing scab on potatoes.

The physicochemical properties of pesticides themselves are of great importance. For instance, water solubility which may vary widely, thus the solubility of the herbicide simazine is only 5 p.p.m. whereas diuron is 42 p.p.m., and dalapon 500,000 p.p.m. This is an important factor in distribution processes in the soil: high solubility results in easier passage into the soil solution and penetration below the surface layers, but the material will also often be readily leached out from the soil by heavy rainfall.

The basic strength or alkalinity of a pesticide measures the ability of the compound to become positively charged by adsorption of hydrogen ions, as is illustrated by the herbicidal triazines[18] (2):

(2)   (3)

(where R,R' = low alkyl (C_2–C_4) radicals and X = Cl, OCH_3, or SCH_3) The hydrogen ions in the soil are attracted by the negatively charged centres of the triazine (2), converting the originally neutral molecule into the cation (3). The greater the alkalinity of a pesticide, the larger its role in cation exchange reactions in the soil. Many other herbicides, e.g. ureas, contain nitrogen atoms which permit the adsorption of hydrogen ions with formation of cationic structures:

$$\geqslant N: \ + H_2O \ \rightleftharpoons \ \geqslant \overset{+}{N}\!-\!H \ + OH^-$$

Extreme cases are provided by compounds like paraquat (Chapter 8, p. 159) that are themselves cations and so very readily exchange with cations in the soil colloids. These exchange reactions result in the herbicide becoming very strongly bound to the soil so that it is no longer available to be absorbed by the plant roots which explains why paraquat and diquat are deactivated as soon as they come into contact with the soil. They act only through the leaves of plants, and when applied to soil they do not appreciably affect emergent seeds, hence the use of paraquat in 'chemical ploughing'. On the other hand, when bipyridylium herbicides are introduced into a nutrient solution without adsorptive colloids they are absorbed by the roots of plants growing in the solution and exert their full herbicidal activity.

Another extreme case is presented by herbicides which are strong acids (4), e.g. 2,4-D, MCPA, and dalapon, which evolve hydrogen ions, and are converted to anions (5). This causes them to repelled by the negatively charged soil colloids and consequently they are not adsorbed onto their surfaces and so such compounds behave completely differently in soil from alkaline pesticides:

$$RCO_2H \ \rightleftharpoons \ RCOO^- + H^+$$
$$(4) \qquad\qquad (5)$$

Another important physical property is volatility[19] related to vapour pressure, which is very considerable with such compounds as 2,4-D, and carbamates, whereas ureas and triazines have only slight volatility. With certain pesticides the volatility is so high that they must be mixed immediately into the soil, otherwise they are vaporized into the surrounding atmosphere before they have a chance of controlling the pest.

High volatility is an essential feature for the effective action of soil fumigants (Chapter 9) and several nematicides (Chapter 11). Other important physicochemical features of pesticides are the molecular shape and size; their importance is well illustrated by the organochlorine insecticides such as DDT and HCH (Chapter 5, p. 54). The distribution of the active ingredient of a pesticide in the soil obviously depends on its chemical nature and mode of formulation. When a compound exists as a finely divided suspension with only comparatively little of the solid actually dissolved, quick distribution into the soil solution depends on the size of the suspended particles. Distribution continues into the soil colloids (soil particles), the biophase (plants and soil microorganisms), and the gaseous phase (soil air).

Application of a pesticide to the soil at an average dose of 5 kg/ha wil not saturate the surface area of the soil colloids in the top 3 cm of soil. Consequently pesticide molecules will be continually withdrawn from the soil solution via adsorption onto the soil colloids. This dynamic process results in further solid material from the applied suspension passing into the soil solution. The process of dissolution can therefore be accelerated by adsorption so even very insoluble materials can be dissolved in a reasonable time, and the compound becomes distributed between the adsorbed, solution, and gaseous phases. There is, of course, no final equilibrium attained since the soil system is not static but is in a continual state of flux. The pesticide concentration changes constantly as a result of plant absorption, and biochemical and chemical degradation. With cationic materials, like paraquat, nearly all the compound is firmly adsorbed onto the soil colloids so that practically none of the chemical remains in the soil solution. On the other hand, with acidic pesticides, like 2,4-D, there is very little adsorption and the chemicals are concentrated in the soil solution and gaseous phases. Chemicals can finally be lost into the atmosphere by evaporation from the gaseous soil phase, and this process may represent a serious loss of the active ingredient with some pesticides, such as 2,4-D.

The selectivity of some herbicides, like the triazines, is a result of their low aqueous solubility combined with fairly high degree of adsorption onto soil colloids[18]. Such chemicals do not therefore penetrate more than the top 15 cm of soil, so that deep-rooted plants, e.g. fruit bushes, are not affected by them although shallow-rooted weeds are killed.

Leaching of pesticides from the soil is obviously an important factor, which again will be determined partly by adsorption. This is also decisive in determining the distribution of the material into the biophase because plants and micro-organisms cannot directly take up compounds adsorbed onto soil colloids. The process of adsorption onto soil colloids is in equilibrium with the removal (desorption) of the compound into the soil solution:

$$\text{Pesticide in soil solution} \quad \underset{\text{desorption}}{\overset{\text{adsorption}}{\rightleftharpoons}} \quad \text{Pesticide adsorbed onto soil colloids}$$

If any element in the equilibrium changes, Le Chatelier's Principle is obeyed and so the equilibrium shifts in such a way as to oppose the change. Thus if the water content in the soil increases as a result of precipitation, then the concentration of the chemical in the soil solution is maintained by increased desorption from the soil colloids. Similarly, when the concentration in the soil is reduced through absorption by plant roots or degradation increased desorption maintains the equilibrium. On the other hand, if the pesticide concentration in the soil increases, the balance is now maintained by increased adsorption onto the soil colloids. Soil organic matter can absorb much more pesticide than the clay mineral of similar surface area, and with different clay minerals, adsorption is proportional to the surface areas. In the case of organic substances the specific nature of the adsorbent is important; for instance, bog moss adsorbs very much more triazine herbicide than Wisconsin peat (Table 1).

The titrable acidity of soil measures the quantity of hydrogen ions in the soil solution and pesticides like triazines with basic centres can absorb positive hydrogen ions which then become firmly bound to the negatively charged soil colloids and consequently adsorption of basic compounds increases with the soil acidity, and hence the herbicidal activity of triazines is less in acidic than in alkaline soils.

Table 1.

Soil components	Surface area (m^2/g)	$K_d = \dfrac{\text{adsorption (mg/kg)}}{\text{solution (mg/l)}}$	
		Simazine	Atrazine
Organic matter:			
Bog moss	500–800	82.3	91.8
Wisconsin peat	500–800	21.5	21.5
Clay minerals:			
Montmorillonite	600–800	12.2	5.3
Illite	65–100	8.5	4.3
Kaolinite	7–30	0	0

In contrast, products with acidic groups like 2,4,5-T and dalapon are relatively easily leached down from the surface layers of the soil by water and are therefore valuable for controlling deeply rooted weeds and shrubs. Leaching is measured by applying the compound to a column of soil and noting the quantity of the pesticide eluted from the column with known volumes of water. The degree of leaching is dependent on the aqueous solubility of the compound, its chemical nature, and the pH value of the soil. Leaching will be favoured by a small adsorption capacity of the soil sample (i.e. one containing relatively few soil colloids such as clay and humus), and by high temperature and rainfall.

The adsorption of a pesticide or its distribution into the biophase (plants and miroorganisms) depends on the absorption capacities of the biophase, and on the nature of the soil. A soil of high adsorption capacity can lead to total inactivation of the pesticide because it never penetrates to the pest[16(b)]. Thus simazine applied to grasses at 10–20 kg/ha showed no herbicidal action when the grasses were growing in soil containing activated charcoal; whereas in similar experiments using soil without charcoal, the grasses were completely killed. Application rates must therefore be adjusted to take account of the particular soil conditions. Pesticides may be chemically and biologically degraded in soils, plants, and pests. The optimum life of a pesticide depends on the nature of both the crop and the pest; thus relatively short-acting herbicides are ideal for vegetables otherwise subsequent crops may be injured by herbicidal residues in the treated soil. On the other hand, long-lasting herbicides, like simazine, are valuable for controlling weeds in sugar cane and fruit bushes. The majority of pesticides are biologically attacked, although the organochlorine soil insecticides such as aldrin, and dieldrin, tend to resist biodegradation and are therefore extremely persistent[16(b)]. Comparatively few pesticides suffer purely chemical attack, although in some instances this can be catalysed by surface-active soil colloids. Thus the hydrolysis of the chlorotriazines, e.g. simazine, is much faster in the presence of clay minerals, but in contrast biochemical degradation can be hindered by the compounds being adsorbed onto clay minerals. The relative active life of certain pesticides can be reversed by different soil and climatic conditions; among the herbicides methoxy- and chlorotriazines, ureas, and picloram usually have comparatively long-term activity in soil, whereas carbamates, phenoxyalkanecarboxylic acids, and chloroaliphatic acids, like dalapon, have only short-term activity.

Soil colloids clearly play an important role in the fate of pesticides in soil; some organophosphorus systemic insecticides, e.g. mevinphos, phorate, and schradan (Chapter 6) gave better long-term control of aphids when applied to sandy soil than in clay loam. The amount of organic matter in the soil appears to be a major factor restricting the absorption of phorate from the soil by plant roots, and generally in affecting the persistence of organophosphorus insecticides[16(b)]; thus diazinon though stable to neutral hydrolysis nevertheless is easily hydrolysed in the presence of clay minerals. Metallic ions in soils often interact with organophosphorus insecticides. The cupric ion is a very effective catalyst for the degradation of some organophosphorus esters, such as diazinon and Dursban.

Organophosphorus insecticides are generally much less persistent than organochlorines and usually are degraded in soil within 2—4 weeks of application, e.g. diazinon, dichlorvos, dimethoate, malathion, parathion, phorate, and mevinphos.

The amount of pesticide introduced into soil is relatively small and therefore will have little effect on the physicochemical state of the soil. In a French vineyard, where simazine had been used for weed control for ten years, a careful chemical, physical, and microbiological examination of the soil showed no recognizable changes; there was no accumulation of simazine and the yield and quality of the grapes were not impaired.

References

1. *Pesticides in the Modern World* — A symposium by the Co-operative Programme of Agro-Allied Industries with FAO and other U.N. Organisations, Newgate Press, London, 1972, p. 30.
2. Fletcher, W. W., *The Pest War*, Blackwell, Oxford, 1974, p. 36.
3. Hartley, G. S. and West, T. F., *Chemicals for Pest Control*, Pergamon Press, Oxford, 1969, (a) p. 5, (b) p. 248.
4. *Pesticides*, I.C.I. Educational Publications, I.C.I. Ltd., 1975.
5. Approved Products for Farmers and Growers, Ministry of Agriculture, Fisheries, and Food, 1978.
6. Woods, A., *Pest Control: A Survey*, McGraw-Hill, London, 1974, p. 64.
7. Griffiths, D. C., Scott, G. C., Maskell, F. E., Roberts, P. F., and Jeffs, K. A., *Proc. Brit. Insectic. and Fungic. Conf.*, Brighton, 1, 213, (1975).
8. Evans, E., 'Methods of application' in *Systemic Fungicides* (Ed. Marsh, R. W.), 2nd edn., Longman, London, 1977, p. 198.
9. *Farm Sprayers and their Use*, Ministry of Agriculture, Fisheries, and Food, Bulletin No. 182, HMSO, London, 1961.
10. Bals, E. J., *Proc. Brit. Insectic. and Fungic. Conf.*, Brighton, I, 153, (1975).
11. Evans, E. and Rickard, M. H., *Proc. Brit. Insectic. and Fungic. Conf.*, Brighton, 3, 853, (1975).
12. Cremlyn, R. J., *Internat. Pest Control*, 5, 10 (1963).
13. *Pesticide Manual*, British Crop Protection Council, 4th edn., 1974, p. 232.
14. Martin, H., *The Scientific Priniciples of Crop Protection*, 6th edn., Arnold, London, 1973, p. 203.

15. Crowdy, S. H., 'Translocation' in *Systemic Fungicides* (Ed. Marsh, R. W.), 2nd edn., Longman, London, 1977, p. 92.
16. Eto, M., *Organophosphorus Insecticides*, CRC Press, Cleveland, Ohio, U.S.A., 1974, (a) p. 275, (b) p. 192.
17. Berger, K. C., *Sun Soil, and Survival*, Univ. of Oklahoma Press, U.S.A., 1972.
18. Esser, H. O., Dupuis, G., Ebert, E., Mareo, G. J., and Vogel, C., '*s*-Triazines' in *Herbicides — Chemistry, Degradation, and Mode of Action* (Eds. Kearney, P. C. and Kaufman, D. D.), 2nd edn., Vol. I, Dekker, New York, 1975, p. 129.
19. Plimmer, J. R., 'Volatility' in *Herbicides — Chemistry, Degradation, and Mode of Action* (Eds. Kearney, P. C. and Kaufman, D. D.), 2nd edn., Vol. 2, Dekker, New York, 1976, 891.

Chapter 3
Important Biochemical Reactions in Pesticides

All living organisms are composed of one or more cells, which are therefore the basic units of life[1,2]. Many simple organisms contain just a single cell; such unicellular organisms include several species of algae. In contrast, more complex organisms, e.g. larger plants and animals, are made up of many cells (multicellular) in order to maintain adequate exchanges of energy and materials with the environment.

The cell is composed of a colloidal suspension of substances known as the *cytoplasm*; to preserve the structural identity of the cell it must be in a state of equilibrium with the surrounding environment. The energy needed for cellular activities comes from the respiration of glucose which requires oxygen, this reaches the cell by diffusion across the cell surface, but if the diameter of the cell exceeds approximately 1 mm oxygen will be unable to diffuse to the centre of the cell. This sets an upper limit to cell size, and in fact the majority of cells are considerably less than 1 mm in diameter.

The cell is separated from its environment by the *plasma* (or cell) membrane (Figure 3.1) which controls the exchange of materials between the interior of the cell and the external environment. The plasma membrane is essentially lipid in nature so that only those molecules possessing a reasonable lipid solubility will be able to permeate through the membrane. In the cytoplasm, cellular activities are located in membrane-bounded structures called *organelles* (small bodies), the most conspicuous of which is the *nucleus* which is usually spherical. The major proportion (> 95%) of cellular DNA occurs in the nucleus and this directs the synthesis of more DNA and of the messenger RNA which controls protein synthesis. Nuclear DNA is bound up with protein molecules and is organized into strands folded into tubular structures known as chromosomes (coloured bodies) — the name deriving from the fact that these structures are readily stained with certain types of dyes. In cells that are on the point of dividing the chromosomes are coiled into short thick segments, while at other times they unwind into longer thinner structures. The spherical *nucleolus* (Figure 3.1) is often associated with a chromosome in the central portion of the nucleus and is composed of DNA, RNA, and protein; probably the DNA in the nucleolus consists of a large number of identical segments arranged end to end, and each segment probably directs the synthesis of one of the ribosomal RNA molecules used to form

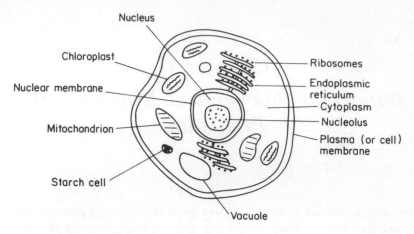

Figure 3.1   Diagram of a plant cell.

ribosomes, while other DNA segments in the nucleolus may direct the synthesis of ribosomal protein. The nucleolus is therefore almost certainly responsible for ribosome synthesis.

A *ribosome* provides a surface for protein synthesis and is a small particle found in the cytoplasm consisting of two spherical subunits each containing molecules of protein and ribosomal RNA (r-RNA), the latter probably being synthesized in the nucleolus. In some cells ribosomes appear to be scattered throughout the cytoplasm while in others the ribosomes are attached to membranes within the cytoplasm providing a large surface area for metabolic reactions, known as the *endoplasmic reticulum* (Figure 3.1).

Generally two organelles are involved in the production of cellular energy: *mitochondria* contain different arrays of enzymes and the electron carriers and are the site of the chemical oxidation (respiration) of glucose and the production of energy-rich phosphorus compounds, e.g. ATP. This is the powerhouse of the cell providing energy for the various cellular reactions, while *chloroplasts* are the site of photosynthesis, the conversion of light energy into chemical energy. These organelles (Figure 3.1) both contain membranes to which are bound enzymes, and DNA, the latter enabling them to reproduce themselves within the cell. Chloroplasts also contain free ribosomes which are probably associated with protein synthesis; the chemical products of oxidation and photosynthesis are transported *via* the various membranes for utilization by the rest of the cell. These organelles, when isolated from the cell, can continue to function; thus isolated mitochondria will oxidize glucose and produce high-energy phosphates; likewise isolated chloroplasts will carry out photosynthesis.

Plant cells generally contain one or more large *vacuoles* (Figure 3.1) which are regions within the cytoplasm bounded by a single membrane containing a complex solution of inorganic and organic molecules, and also functioning as stores for enzymes. The cytoplasm of plant cells often contains starch granules and chloroplasts are unique to plant cells. Some animals possess another kind of organelle known as a *lysosome*. This is a droplet bounded by a membrane containing several hydrolytic enzymes which can split the major cellular components, for instance nucleic acids, proteins, and polysaccharides. In higher animals lysosomes are important for the utilization of dead tissue and for degrading those cellular parts in need of frequent renewal.

The cell walls of plant cells are rigid, whereas an animal cell is bounded by a flexible cell wall. The cell walls or membranes control the exit and entry of substances between the cytoplasm and the external environment. The concentration of solute molecules within cells is usually higher than their external concentration, so that water tends to diffuse into the cell by the process termed osmosis which causes the cell to expand and press against the cell wall so reducing the inflow of water and when these two pressures become equal water movement ceases when the cell no longer expands.

In large multicellular organisms, each individual cell need no longer have all the capacities necessary for the survival of the whole organism so the cells often become specialized performing particular functions, e.g. a red blood cell is uniquely developed to transport oxygen. Such specialized cells are often assembled together to form a tissue and several different tissues may be further grouped together into large functional units known as organs, like the kidney, responsible for excretion and water balance in animals, and the leaf which carries out photosynthesis and gas exchange in plants.

The major chemical constituents of cells are carbohydrates, lipids, proteins (some of which function as enzymes), and nucleic acids. The intact cell also contains simple inorganic cations, such as sodium, potassium, calcium, and magnesium and anions like chloride and phosphate. In unicellular organisms the cell absorbs directly the nutrients it requires from the external environment, but in more complex animals and plants, the food is first digested or metabolized to simpler low molecular weight compounds, e.g. amino acids (from proteins), fatty acids (from lipids), and glucose (from starch and sucrose), which are then carried to the cell in the blood stream or the plant sap.

Living organisms, apart from certain bacteria, derive the energy needed for their vital life processes from organic compounds synthesized by another organism, or they utilize the energy from light.

Respiration

Organisms, obtaining their energy requirements from the respiration of preformed organic molecules obtained from their food, include all animals, the majority of bacteria, fungi, and the non-green cells of plants[3(a)]. The energy derived from the respiration (biological oxidation) of foods such as carbohydrates, fats, and proteins is finally trapped in energy-rich phosphate bonds of adenosine-5′-triphosphate (ATP) which can then supply the energy needed for biochemical processes. The oxidation of glucose by burning in air can be expressed as:

$$\underset{\text{glucose}}{C_6H_{12}O_6} + 6\,O_2 \longrightarrow 6\,H_2O + 6\,CO_2 + \text{heat energy}$$

However the biological oxidation process is much more complex and occurs in four main stages (Scheme 3.1):

(a) *Formation of acetyl coenzyme A (acetyl CoA).* This can be derived from breakdown of carbohydrates, fats, or proteins, but pesticides apparently only interfere directly with the route from glucose. Initially the foods are metabolized to simpler molecules: proteins are converted to α-amino acids; carbohydrates, like starch, to glucose; and lipids to fatty acids and glycerol.

The conversion of glucose into pyruvate involves the operation of a sequence of nine enzymic reactions occurring in the presence of oxygen. The net result of the process of respiration is the transformation of each molecule of glucose into approximately thirty molecules of ATP together with pyruvate. Finally, in the presence of oxygen, pyruvate is converted to acetyl CoA:

$$CH_3COCO_2H + NAD + CoASH \longrightarrow CH_3COSCoA + NADH_2 + CO_2$$

| pyruvate | nicotinamide adenine dinucleotide | coenzyme A | acetyl CoA | reduced form of NAD |

The conversion of glucose to pyruvate is termed glycolysis, and the enzymes controlling the process are in the cytoplasm of the cell (Figure 3.1).

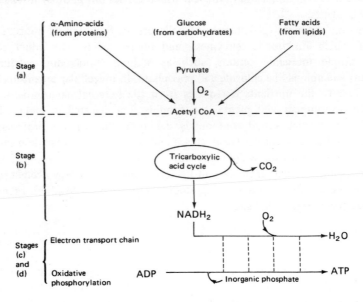

Scheme 3.1 An outline of the biochemical reactions involved in respiration.

(b) *The tricarboxylic acid cycle.* This involves progressive oxidation of acetyl CoA and the transference of the reducing power chiefly associated with the production in the cycle of the reduced form of nicotinamide adenine dinucleotide ($NADH_2$) to the electron transport chain.

(c) *The electron transport chain*. This oxidizes reduced compounds, such as $NADH_2$, with liberation of energy. Oxidation may be defined as loss of electrons and as a consequence of the operation of the electron transport chain electrons pass from a relatively high level in $NADH_2$ to progressively lower levels in the electron carriers composing the chain until they combine with oxygen and protons to form water.

Reducing power is transferred from acetyl CoA to NAD *via* the tricarboxylic acid cycle. The electron transport chain is fed by the transfer of hydrogen atoms from acetyl CoA, the electrons from $NADH_2$, and from the reduced succinate dehydrogenase flavoprotein. Succinate dehydrogenase and cytochrome oxidase are two enzymes involved in the cycle; the enzymes and the electron carriers are bound to the mitochondrial membrane. Cytochrome oxidase occurs in very low concentration and the activity of uncouplers may be measured by reference to the concentration of this protein.

(d) *Oxidative phosphorylation* (Scheme 3.1). This couples the energy released in the electron transport chain to affect the phosphorylation of adenosine-5'-diphosphate (ADP) to adenosine-5'-triphosphate (ATP), the energy being trapped in the phosphate bonds. The precise nature of the coupling between electron transport and oxidative phosphorylation is not known. Oxidative phosphorylation may be measured by suspending mitochondria in an oxygen electrode which measures the concentration of oxygen in solution.

Photosynthesis

Plants also break down glucose to release water and carbon dioxide and the liberated energy is trapped in ATP by biochemical mechanisms similar to those already described for animals.

The important difference between plants and animals is the origin of their glucose: green plants and other photosynthetic organisms such as algae and certain bacteria, have the ability to utilize light energy which is absorbed by the green pigment chlorophyll to synthesize glucose from carbon dioxide and water[3(b), 4].

$$6 CO_2 + 6 H_2O \xrightarrow[\text{light}]{\text{chlorophyll}} \underset{\text{glucose}}{C_6H_{12}O_6} + 6 O_2$$

This process is known as photosynthesis.

The light energy is absorbed by the chlorophyll molecules arranged in trapping units so that one chlorophyll molecule receives all the energy from the assembly. The energy is converted to chemical energy by reduction of an acceptor (A) in the plant which in turn produces an oxidized chlorophyll molecule which must be reduced again before it can become the focus of the light absorption system. The net result is the reduction of A by a donor molecule and the cyclic oxidation—reduction mechanism powered by light energy will result in an overall gain in chemical energy by the system as a whole.

The chemical energy is then used by the plant to reduce carbon dioxide to carbohydrates.

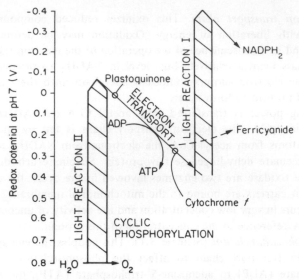

Figure 3.2   The scheme of reactions in photosynthesis.

There are apparently two light reactions in photosynthesis and both are of the general type shown in Figure 3.2. The light reactions I and II are coupled in series by the photosynthetic electron transport path as illustrated by Figure 3.2.

This resembles the mitochondrial electron transport chain in respiration, and also involves quinone and cytochrome components; it is coupled to the photophosphorylation of ADP → ATP.

Electrons are energized in light reaction II in which the donor molecule of Figure 3.2 is H_2O, while A is unknown. Removal of electrons from water results in the production of oxygen, and the passage of electrons down the photosynthetic electron transport chain is coupled to photophosphorylation, but it can be inhibited and uncoupled by a number of compounds although these may not be identical to those that interfere with the mitochondrial electron system. Electrons, after transfer down the photosynthetic electron transport chain, have lost energy and are reactivated in light reaction I, which forms an unknown reduced acceptor which is more electronegative than NADP and can therefore reduce it to $NADPH_2$[5]. The electron transfer *via* light reactions I and II results in the production of oxygen, ATP, and $NADPH_2$ (Figure 3.2).

Photosynthesis is localized in subcellular organelles called chloroplasts; these can be isolated by centrifugation of macerated leaves, and have been extensively used in studies on the mode of action of herbicides. Herbicides that owe their activity to interference with photosynthesis generally inhibit photosynthetic electron transport by preventing light reaction II. Cell-free preparations of green plants or isolated chloroplasts can catalyse the photolysis of water in the presence of an electron acceptor (A) such as ferricyanide:

$$2\,A \;+\; 2\,H_2O \xrightarrow{\text{light, chloroplasts}} 2\,H_2A \;+\; O_2$$

This is known as the Hill reaction[6(a)] and may be measured by determination of the evolution of oxygen with an oxygen electrode, or by a spectrophotometric method which determines the amount of H_2A formed. The Hill reaction with ferricyanide is apparently powered solely by light reaction II and the majority of herbicides, such as ureas and triazines, that act by inhibition of photosynthesis probably have light reaction II as their primary site of action. On the other hand, the bipyridylium herbicides function by utilizing photosynthesis for the production of stabilized free radicals which in the presence of oxygen give rise to toxic entities which kill the plant (see Chapter 8).

Transmission of Nervous Impulses

The nervous system is characteristic of mammals and insects and carries signals from the various receptor sites (e.g. eyes, ears, and nose) to the brain[4]. Other specific receptor cells record senses such as temperature, taste, and pain. The external environment also transmits information to the nervous system by means of discrete electrical impulses along a long fibre of neurons (or nerve cells) known as the axon, eventually reaching the brain, so that the appropriate response can be made to the received stimuli. The brain and spinal cord together are termed the central nervous system (CNS) and this sends out signals and communicates with the rest of the body by the peripheral nervous system[3(c), 6(b)]. Neurons utilize the electrical charges carried by ions and the activity of the nervous system ultimately depends on the neurons capacity to maintain an unequal distribution of sodium and potassium ions on each side of the cell membrane. Under resting conditions, the electrical potential inside the membrane is negative with respect to the outside, and the concentration of sodium ions (Na^+) inside the nerve cell is low, relative to the outside, whereas for the potassium ions (K^+) the reverse is true. This situation arises from the fact that Na^+ ions are actively transported out of the cell, while K^+ ions move into it. The unequal distribution of ions on the two sides of the cell membrane gives rise to an electrical potential.

The transmission of the nervous impulse is therefore an electrical process in which the current is carried by ions. When the axon meets another neuron there is a junction called a synapse generally some 20–30 nm wide. Nerve impulses are

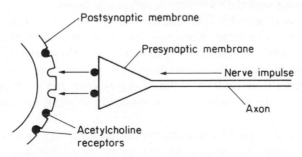

Figure 3.3   Schematic diagram of nerve synapse.

transmitted at the synapse by the release of a chemical transmitter, generally acetylcholine, although other neurotransmitters such as L-glutamate and γ-amino-butyric acid are also involved in some synapses.

When a nerve impulse arrives at the presynaptic membrane, acetylcholine is simultaneously released from the presynaptic cells and the chemical transmitter diffuses across the synaptic cleft to the postsynaptic membrane, where it binds to the acetylcholine receptor sites. The liberated acetylcholine must not persist in the synapse too long, otherwise there would be a continuous chain of nerve impulses. The transmitter is generally eliminated by combination with the enzyme acetyl-cholinesterase present in the postsynaptic membrane (Figure 3.3).

Acetylcholinesterase catalyses the hydrolysis of acetylcholine to choline which does not act as a transmitter of nerve impulses.

$$(CH_3)_3\overset{+}{N}CH_2CH_2OCOCH_3 \underset{}{\overset{\text{acetylcholinesterase}\ (+H_2O)}{\rightleftharpoons}} (CH_3)_3\overset{+}{N}CH_2CH_2OH + CH_3CO_2H$$

acetylcholine   choline

The combination of acetylcholine with the receptor causes the postsynaptic cell to pass an impulse; subsequently acetylcholinesterase hydrolyses the acetylcholine so that stimulation of the receptor ceases, and the synapse is then available for release of a new transmitter.

The organophosphorus and carbamate insecticides (Chapter 6) owe their insecticidal properties to phosphorylation or carbamoylation of the enzyme acetylcholinesterase. This poisons the enzyme so that it cannot catalyse the hydrolysis of acetylcholine to choline; consequently there is an accumulation of acetylcholine at the synapse which permits the continuous transmission of nerve impulses, and effective nervous coordination breaks down − the insect or mammal suffers convulsions and finally death.

Mammals contain several receptors having the capacity of binding to acetyl-choline, and O'Brian has demonstrated[7] that the receptor isolated from fly heads has different binding properties from a vertebrate receptor. The receptor on the postsynaptic membrane may not be acetylcholinesterase since enzyme inhibitors do not block the receptor and may even enhance its activity.

Nicotine (Chapter 4, p. 39) almost certainly owes its insecticidal properties to its ability to combine with the acetylcholine receptor in insects[4].

A comparison of acetylcholine receptor and the enzyme acetylcholinesterase as sites of action for insecticides shows that few compounds are considered to act on the receptor, while very many of them, e.g. organophosphates and carbamates, react with the enzyme. The observation[7] that insect receptors differ in their binding properties from those in mammals suggests that the receptors may offer a useful target for the development of compounds showing selective toxicity to insects.

DDT and other organochlorine insecticides probably act by interference with axonal transmission by binding to the nerve membrane and upsetting the sodium–potassium ion balance (Chapter 5, p. 55). Pyrethroids affect both the peripheral and central nervous systems of insects, and the convulsion of the insect appears to be initiated by loss of potassium (Chapter 4, p. 46).

Transport and Biological Processes in Plants

In order to understand what happens when a pesticide is applied to a plant either directly by foliar spray, seed dressing, or *via* the soil, it is necessary to know the basis of plant biochemistry.

The plant root system absorbs water and minerals from the soil solution; the centre portion of the root contains two main types of conducting tissues, the xylem and phloem[6(c)]. The xylem is responsible for the movement of water throughout the plant since it is continuous from the root tip to the leaf veins, and contains an unbroken column of water. Water and minerals are carried up from the roots to the rest of the plant by the transpiration stream and the water movement is largely caused by the evaporation of water from the leaf surfaces which reduces the pressure in the leaf veins and consequently more water is sucked up from the roots. The suction pressure is substantial and may be 10–20 times the normal atmospheric pressure and so the water is easily transported upwards against the force of gravity.

The overall flow in the xylem is controlled by the opening and closing of small pores in the leaf surface (stomata) permitting movement of gases between the leaf cells and the surrounding air[6(c)]. In contrast organic compounds produced by photosynthesis are transported in the phloem. This is a more complex conducting tissue than the xylem, since it also permits downward movement of certain chemicals from the leaves to the roots[4]. Such chemicals could prove valuable for the control of soil pathogens and there is evidence that some synthetic organic compounds, e.g. pyroxychlor (Chapter 15, p. 225), act by downward translocation (see Chapter 2, p. 19). Sugars move downwards and upwards in the phloem and the phloem sap contains approximately 25% of carbohydrates, chiefly sucrose, together with small amounts of amino acids.

Transport in the phloem system is governed by osmotic pressure; the leaf cells with high concentrations of sugars possess larger osmotic pressures than the non-photosynthetic leaf tissues so the pressure difference forces the sugars downwards towards the roots so providing nourishment for the whole plant. The plant skeleton is composed mainly of thick tough walls of xylem tissue which are necessary to withstand the suction forces in the xylem.

The characteristic shapes of different plants are controlled by hereditary instructions contained in the genetic code of the plant DNA and these are relatively independent of environmental factors. Hereditary instructions however also provide physiological responses that can modify plant growth, e.g. decide when lateral buds become active or how the plant responds to alterations in such external factors as temperature, light, and nutrients. The physiological changes in plants are controlled by chemical hormones (Chapter 8, p. 166) and by mimicking such compounds useful herbicides and growth regulators can be produced, for instance the phenoxyacetic acid selective herbicides (Chapter 8, p. 142).

Protein and Nucleic Acid Biosynthesis

Ribonucleic acid (RNA) and deoxyribonucleic acid (DNA) have been recognized as the genetic material for some 20 years. The nature and distribution of RNA and DNA molecules defines the organism and direct and control the synthesis of all the compounds required by the organism, such as proteins[6(d)]. Proteinoid enzymes are involved in and direct cell growth and maintenance, including processing chemicals from the environment and oxidation reactions (respiration) that provide cellular energy.

The genetic material must be capable of reproduction or replication passing on the essential genetic information to daughter cells during replication so ensuring that the information in the progeny is identical to that of the parent. Chemically, nucleic acids consist of polymers made up of a phosphate group, a sugar (either D-ribose (RNA) or 2-deoxy-D-ribose (DNA)) and four nitrogenous bases (adenine, uracil, cytosine, and guanine (RNA) or adenine, thymine, cytosine, and guanine (DNA))[2].

The genetic information contained in DNA is coded by a linear sequence of two purine (adenine and guanine) and two pyrimidine (cytosine and thymine) bases arranged in two interwoven strands forming part of a double helix. The precise linear structure of RNA or DNA is defined by the arrangement of the four bases; only certain pairs of bases easily form hydrogen bonds with each other, thus guanine (G) bonds with cytosine (C), and adenine (A) with thymine (T) in DNA or uracil (U) in RNA. These specific bonded base pairs C–G, A–T, and A–U are stable and are responsible for the double helical structure of the DNA molecule in which the two nucleic acid strands are held together by hydrogen bonding between the base pairs. The specificity of the base pairs implies that the two DNA strands of each double helix will be different but complementary.

Genetic information is transferred from the nucleic acid to the protein and protein synthesis may be summarized as shown (Scheme 3.2).The information

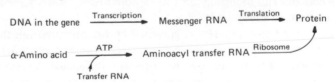

Scheme 3.2   Basis of protein synthesis.

contained in the DNA molecule is coded or transcribed onto a new RNA molecule known as messenger RNA, and is translated into a specific amino acid sequence defining a particular protein. This operation involves a specialized organelle called the ribosome whose sole function is protein synthesis (see p. 28). The specific linear sequence of amino acids is dictated by the messenger RNA to which the ribosomes attach themselves. Protein synthesis from the free α-amino acids in the cytoplasm occurs in three steps:

(a) the α-amino acid reacts with ATP and a specific transfer RNA molecule to give an aminoacyl transfer RNA molecule. Each transfer RNA becomes attached to only one of the twenty α-amino acids usually found in proteins and a specific enzyme is needed to catalyse the formation of each aminoacyl transfer RNA;

(b) the aminoacyl transfer RNA is combined with the messenger RNA by base pairing at a second specific binding site;

(c) protein is synthesized by the ribosome in which the α-amino acid carried by the transfer RNA is joined onto the end of a growing protein chain through formation of a peptide linkage. In this way the ribosome can be visualized as moving down the messenger RNA molecule and 'reading' its message. The synthesis, in spite of its complexity, is quite fast and the peptide chain may grow at the rate of some 25 amino acid units per second.

Some pesticides believed to owe their activity to interference with biosynthetic processes, include several antifungal antibiotics, e.g. streptomycin, blasticidin S, and kasugamycin (Chapter 7, p. 122); probably all act by inhibition of protein biosynthesis. The benzimidazole systemic fungicides (Chapter 7, p. 127), e.g. benomyl, apparently function by inhibition of DNA biosynthesis. The herbicides aminotriazole and 3,4-dichlorobenzyl N-methylcarbamate cause extreme chlorosis of plant leaves and are believed to act by interference with carotenoid biosynthesis. Carotenoids are yellow orange pigments found in all plant tissues but especially in chloroplasts where they protect the plant cell from photooxidations which would otherwise kill the plant.

Lipids, such as fatty acids and steroids, are important constituents of plant cuticular waxes and the herbicidal action of thiolcarbamates (Chapter 8, p. 149) is thought to be due to inhibition of wax formation; also the fungicidae triarimol (Chapter 7, p. 133) may owe its activity to interference with lipid biosynthesis[4].

Many fungal cell walls contain a polymer of N-acylglucosamine known as chitin and interference with chitin formation is thought to be the mode of action of the antifungal antibiotic polyoxin D (Chapter 7, p. 125) and the systemic fungicide Kitazin P (Chapter 7, p. 135).

Cellulose is a vital structural component of plants which is rarely found in animals so its inhibition would appear a good biochemical target for herbicides. However no current commercial herbicide is considered to act by this mechanism, but tris(1-aziridinyl) phosphine oxide (1), used for cross-linking cellulose in fabrics, does have a stunting effect on plants. This may well arise from interference with cellulose biosynthesis since it does reduce cell wall constituents in treated plants[4].

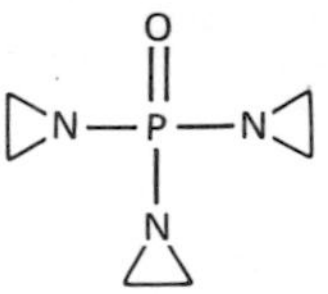

(1)

References

1. Steiner, R. F., *Life Chemistry*, Van Nostrand, New York, 1968, p. 1.
2. Rose, S., *The Chemistry of Life*, Penguin Books, 1970.

38

3. White, A., Handler, P., and Smith, E. L., *Principles of Biochemistry*, 5th edn., McGraw-Hill, New York, 1973, (a) p. 389, (b) p. 514, (c) p. 959.
4. Corbett, J. R., *The Biochemical Mode of Action of Pesticides*, Academic Press, New York, 1974.
5. Hill, R., *Essays in Biochem.*, **1**, 121 (1965).
6. Stephens, G. C. and North, B. B., *Biology*, Wiley, New York, 1974, (a) p. 159, (b) p. 310, (c) p. 195, (d) p. 34.
7. O'Brian, R. D., in *Biochemical Toxicology of Insecticides* (Eds. O'Brian, R. D. and Yamamoto, I.), Academic Press, New York, 1970, p. 1.

Chapter 4
Botanical Insecticides

Plants have evolved over some 400 million years and to combat insect attack they have developed a number of protective mechanisms, such as repellency, and insecticidal action. Thus a large number of different plant species contain natural insecticidal materials; some of these have been used by man as insecticides since very early times although many of them cannot profitably be extracted. However, several of these extracts have provided valuable contact insecticides which possess the advantage that their use does not appear to result in the emergence of resistant insect strains to the same degree as the application of synthetic insecticides (see Chapter 6, p. 101).

Some botanical insecticides survive today; the most important examples, in ascending order of importance, are nicotine, derris (rotenone), and pyrethrum.

Nicotine

The tobacco plant was introduced into Europe about 1560, Sir Walter Raleigh began the practice of smoking tobacco in England in 1585, and as early as 1690 water extracts of tobacco leaves were being used to kill sucking insects on garden plants[1]. The active principle in tobacco extracts was later shown to be the alkaloid nicotine (1), first isolated in 1828 and the structure elucidated in 1893[2,3]. Nicotine occurs in tobacco plants as a salt with citric and malic acids to the extent of 1–8% and may be extracted from the leaves and roots of the plants by treatment with aqueous alkali, followed by steam distillation.

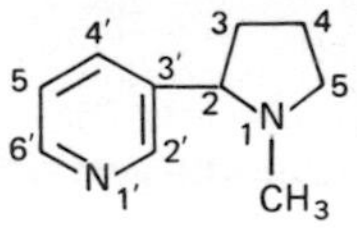

(1) 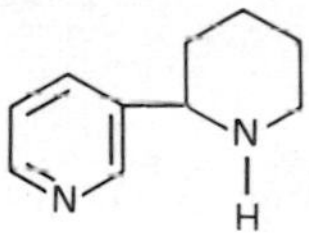   (2) 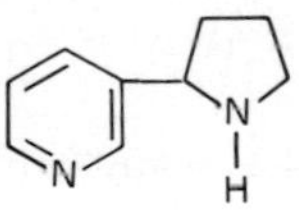   (3)

Natural nicotine is the laevorotatory (−) isomer $[\alpha]_D - 169°$, the optical isomerism arising from the presence of the asymmetric $C_{(2)}$-carbon atom. The dextrorotatory (+) isomer is much less insecticidal so that synthetic nicotine which is the racemic (±) form is only about 50% as active as the natural material[3]. Commercially nicotine is generally used as 'Black Leaf 40' which has been a popular garden spray for a long time. This is a concentrate containing 40% of nicotine sulphate[2,4]. Alkaline activators, like soap and calcium caseinate, are added to liberate the active free nicotine. Nicotine may also be applied as a dust.

Nicotine functions as a non-persistent contact insecticide against aphids, capsids, leaf miner, codling moth, and thrips on a wide variety of crops[5]. However, its use is rapidly declining and it is being replaced by synthetic insecticides, because of its comparatively high mammalian toxicity (LD_{50} (oral) to rats $\simeq$ 50 mg/kg)[1], and its lack of effectiveness in cold weather. The compound is readily absorbed by the skin and any splashes must be washed off immediately[4].

Smith and his coworkers in 1930 showed[2,3] that the bipyridyl derivative neonicotine or anabasine (2) possessed comparable aphicidal activity to that of nicotine (1). This compound, as the laevorotatory (−) isomer, was later isolated from a Turkestan weed *Anabasis apylla*, hence the name anabasine. Aphicidal activity was also shown by nornicotine (3). Subsequent examination of a range of nicotine analogues indicated that high insecticidal activity required the presence of a pyridine nucleus joined between the 2- and 3-positions to a saturated 5- or 6-membered ring. Other essential features included the 3-pyridylmethylamine residue with a highly basic side-chain nitrogen atom at least 4.2 Å away from the pyridine nucleus[3].

Nicotine kills vertebrates because it mimics acetylcholine by combining with the acetylcholine receptor at the neuromuscular junction causing twitching, convulsions, and finally death[4,6]. There is evidence[6] that a similar mode of action accounts for the insecticidal activity where nicotine blocks synapses associated with motor nerves. For activity, it is essential that nicotine (1) and its active analogues, e.g. (2) and (3) should have very similar molecular dimensions to acetylcholine (4) (Figure 4.1).

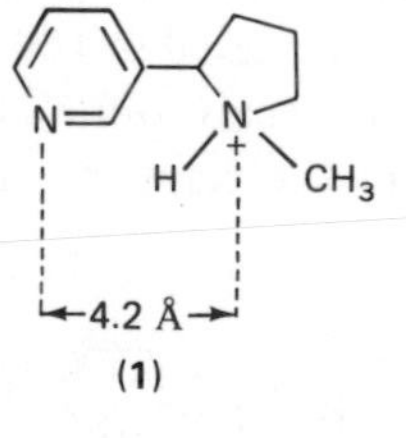

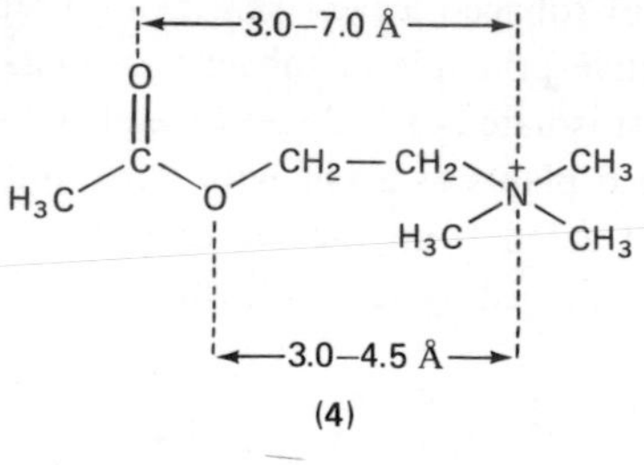

Figure 4.1 Molecular dimensions of nicotine and acetylcholine.

So nicotine penetrates into the synapse and is then converted to the nicotinium ion containing a positive charge on the pyrrole nitrogen atom. This interacts with the acetylcholine receptor just like acetylcholine but, unlike acetylcholine, it is not susceptible to enzymic hydrolysis by acetylcholinesterase. This accounts for the observed toxic symptoms and there is considerable evidence[1,6] that the nicotinium ion is the active entity in nicotine.

Rotenoids

These are a group of insecticidal compounds occurring in the roots of *Derris elliptica* (from the East Indies and Malaya) and a species of *Lonchocarpus* (from South America). Derris has been used as an insecticide for a long time, thus Oxley (1848) recommended it for control of caterpillars[1]. Derris dust is manufactured by grinding up the roots and mixing the powder with a clay diluent[4]. Alternatively the rotenoids can be extracted from the powdered roots with organic solvents[2,3]. The resultant resin by crystallization from ether or carbon tetrachloride gave rotenone (5), a white crystalline, laevorotatory solid whose structure was elucidated in 1932[1,2]:

(5)

The mother liquors, after removal of rotenone, also afforded a number of analogous pentacyclic compounds (rotenoids), but these were not nearly as strongly insecticidal.

Attempts have been made to extract the active principles from derris and lonchocarpus roots to avoid the cost of transporting much inactive material contained in the roots. However, it was found that solvent extraction appeared to accelerate the decomposition to products of low insecticidal potency. This problem could be reduced by hydrogenation of the isopropenyl group in rotenone (5) to dihydrorotenone which, though highly active, was resistant to oxidative degradation.

Rotenoids are toxic to fish and many insects, but are almost harmless to most warm-blooded animals. In 1946 about ten million pounds of dried derris and lonchocarpus roots were imported into the United States of America but this had declined to six million pounds by 1955. Derris was widely used in cattle and sheep dips for the control of ticks and other ectoparasites, but recently has been largely superseded by synthetic insecticides. Now it is primarily used in horticulture against aphids, caterpillars, sawflies, wasps raspberry beetles, and red spider[7]. Rotenone is an extremely safe garden insecticide because it is degraded by light and air and does not leave residues, and has been widely used for more than 50 years (LD_{50} (oral) to rats $\simeq$ 135 mg/kg.

The biochemical mode of insecticidal action appears to involve the inhibition of mitochondrial electron transport, and in isolated mitochondria rotenone (5) inhibits oxidation linked to $NADH_2$, although at low concentrations succinate oxidation was not affected. The inhibition of the electron transport chain appears to arise from the binding of rotenone to a component of the chain, but $NADH_2$ dehydrogenase is not inhibited[6]. The symptoms of insects poisoned by rotenone differ from those produced by insecticides acting on the nervous system, and are characterized by reduction in oxygen consumption, depressed respiration and heartbeat, and eventual paralysis[3].

Pyrethroids

Pyrethrum is a contact insecticide obtained from the flower heads of *Chrysanthemum cinerariaefolium* and has been used as an insecticide since ancient times. The varieties grown in the highlands of Kenya yield the highest proportions of active ingredients; it is also grown commercially in the Caucasus, Iran, Japan, Ecuador and New Guinea[1,2,4]. The production of pyrethrum as an insecticide dates from about 1850 and, unlike nicotine and derris, the use of pyrethrum has increased despite the large-scale introduction of synthetic insecticides. In 1965 the world output of pyrethrum was approximately 20,000 tons with Kenya producing some 10,000 tons.

Pyrethrum owes its importance to the outstanding rapid knockdown action (a few seconds) on flying insects combined with a very low mammalian toxicity due to its ready metabolism to non-toxic products. So, unlike DDT, pyrethrum is not persistent and leaves no toxic residues which may be why this insecticide does not tend to induce the development of resistant insect populations. Pyrethrum is used to control pests in stored foods and against household and industrial pests. Pyrethrum aerosol sprays are excellent home insecticides because of their safety and rapid action[7].

However a major disadvantage of the pyrethrum, especially for use against agricultural pests lies in its lack of persistence due to its instability in the presence of air and light. Insects can often recover from exposure to sublethal doses of pyrethrum which means that the compound must be mixed with small amounts of other insecticides to ensure that the treated insects do not recover.

Pyrethrum is obtained from the dried chrysanthemum flowers by extraction with kerosine or ethylene dichloride and the extract concentrated by vacuum distillation[2]. It contains four main insecticidal components which are collectively termed pyrethrins[8]. These are the esters of two cyclopentenolones (6; $R' = CH=CH_2$ or CH_3) and two cyclopropanecarboxylic acids (7; $R = CH_3$ or $CO_2 CH_3$). The structures of the main pyrethrins are therefore as shown in formula (8).

The alcohols (6; $R' = -CH=CH_2$ and CH_3) are called pyrethrolone and cinerolone, while the carboxylic acids (7; $R = CH_3$ and $CO_2 H$) are chrysanthemic and pyrethric acids. Pyrethrin I is the most active of the natural pyrethrins. The acid components are capable of existing in *cis* and *trans* geometrical isomers due to the presence of the olefinic double bond and each of these isomers can further exist in *dextro-* (+) and *laevo-* (−) rotatory optical isomers. Similarly the alcohols can exist in four stereoisomeric forms. The stereochemistry of the pyrethroids has a vital influence on their insecticidal activity, thus the (−)-*trans*-chrysanthemates are practically inactive as compared with the (+)-*trans*-chrysanthemates, so it is

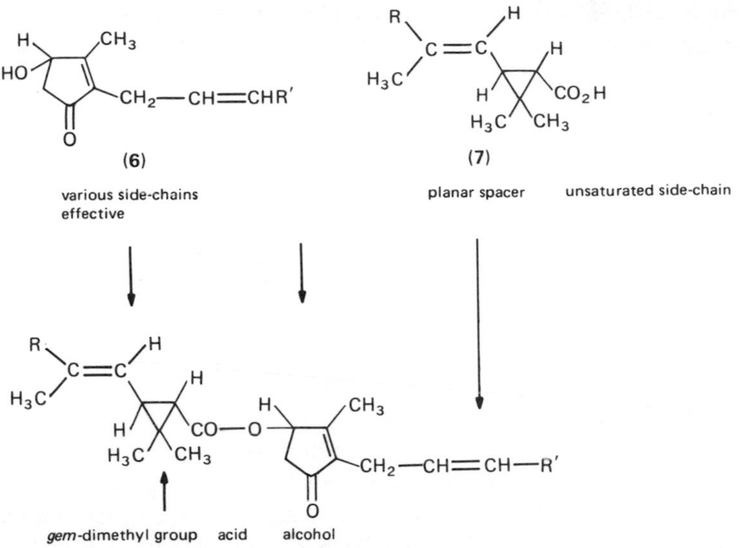

(6)		(7)
various side-chains effective		planar spacer unsaturated side-chain

gem-dimethyl group   acid   alcohol

(8)

Compound	R'	R
pyrethrin I	—CH=CH₂	—CH₃
pyrethrin II	—CH=CH₂	—CO₂CH₃
cinerin I	—CH₃	—CH₃
cinerin II	—CH₃	—CO₂CH₃

fortunate that the natural (+)-*trans*-chrysanthemic acid can now be commercially synthesized.

The synthesis of chrysanthemic acids and of cyclopentenolones[3] opened up the possibility of obtaining synthetic pyrethroids, the first of which was allethrin (9) prepared by esterification of synthetic (±)-chrysanthemic acid (7; R = CH₃) with the alcohol allethrolone (6; R' = H):

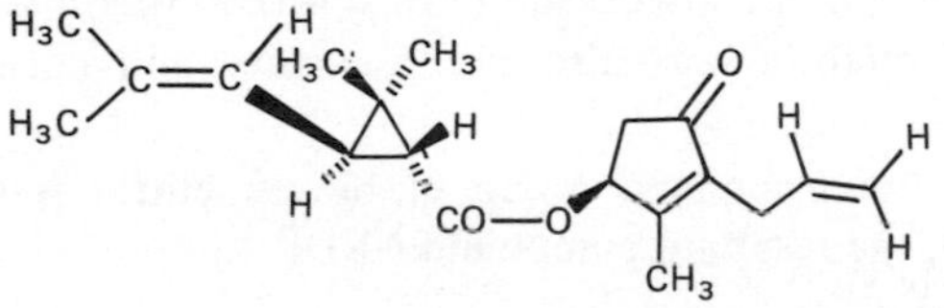

(9)

Allethrin (9) had strong insecticidal activity and removal of the keto group gave another insecticidal synthetic pyrethroid known as bioallethrin (LD_{50} by topical application to houseflies 0.10 μg/insect). Recently there has been considerable exploitation of synthetic pyrethroids both in England and Japan leading to some extremely active molecules. The most active of the natural pyrethrins was pyrethrin I (8) (LD_{50} 0.33 μg/insect) and studies of the metabolism of pyrethrins have shown[9] that the cyclopropane ester linkage appears to resist cleavage and provides an appropriate polar centre to the molecule (8) whose insecticidal activity is probably due to the intact structure not a metabolite. The activity shown by the first important synthetic pyrethroid allethrin (9) indicated that the side-chains in the alcohol component may be modified without loss of activity. Barthel[10] showed that 2,4- and 3,4-dimethylbenzyl esters of chrysanthemic acid (7; R = CH$_3$) were insecticidal and Elliott and coworkers concluded[11] that this implied that the 4-methyl group was equivalent to the methylene group of the natural ester side-chain.

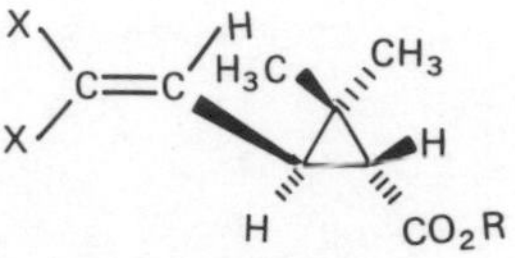

(10)

On this basis, 4-allylbenzyl chrysanthemate (10; X = CH$_3$, R = p-CH$_2$C$_6$H$_4$ CH$_2$CH=CH$_2$), combining the structural features of allethrin (9) and the methylbenzyl chrysanthemates, was synthesized and was shown to be appreciably more active against houseflies (LD$_{50}$ 0.02 μg/insect) than allethrin. Examination of a number of methylbenzyl chrysanthemates showed that 2,3,6- and 2,4,6-trimethyl substitution was particularly effective and so 4-allyl-2,6,-dimethylbenzyl chrysanthemate was prepared which had a broader spectrum of insecticidal activity. This work demonstrated that the cyclopentenone ring of the natural pyrethrins could be replaced by a benzylic system without loss of activity[11]. The benzyl group can also be replaced by other aromatic systems and this approach led to the discovery of 5-benzyl-3-furylmethyl (+)-*trans*-chrysanthemate or bioresmethrin

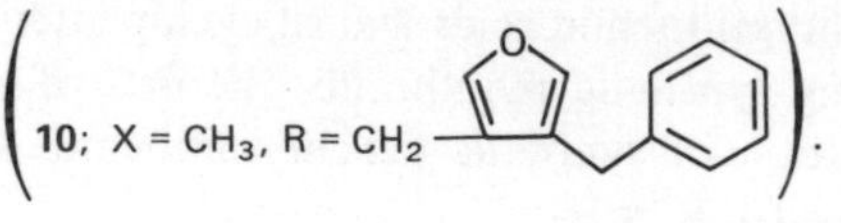

10; X = CH$_3$, R = CH$_2$—

This is an extremely potent insecticide (LD$_{50}$ = 0.005 μg/insect)[9], but although more active, bioresmethrin was also photosensitive and consequently was not persistent.

However, when the isobutenyl group of bioresmethrin was replaced by the dichlorovinyl group, the resultant compound NRDC 134

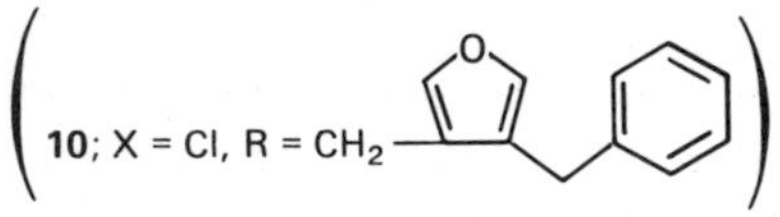

10; X = Cl, R = CH$_2$—

was more toxic to houseflies and mustard beetles than most known insecticides[12].

Chemical and spectroscopic evidence indicated that the furan ring in esters of 5-benzyl-3-furylmethyl alcohol was the probable site of photosensitized oxidative decomposition so, in an attempt to discover more stable pyrethroids, other esters of 2,2-dichlorovinyl cyclopropanecarboxylic acid were synthesized. The ester from 3-phenoxybenzyl alcohol, biopermethrin

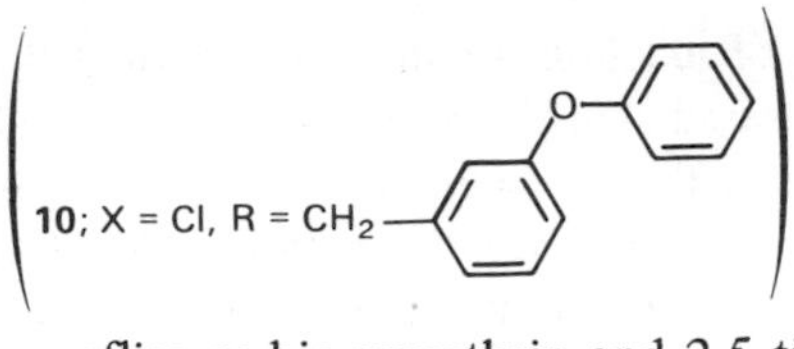

10; X = Cl, R = CH$_2$—

was as good against houseflies as bioresmethrin and 2.5 times as effective against mustard beetles. This compound showed much greater photostability and consequently was a moderately persistent insecticide which should extend its uses in crop protection. It was the first pyrethroid to be effective as a seed treatment against wheat bulb fly[13].

Further work showed[13] that the chlorine atoms of biopermethrin could be replaced by bromine and the introduction of an α-cyano group led to an outstandingly active compound known as decamethrin or NRDC 161.

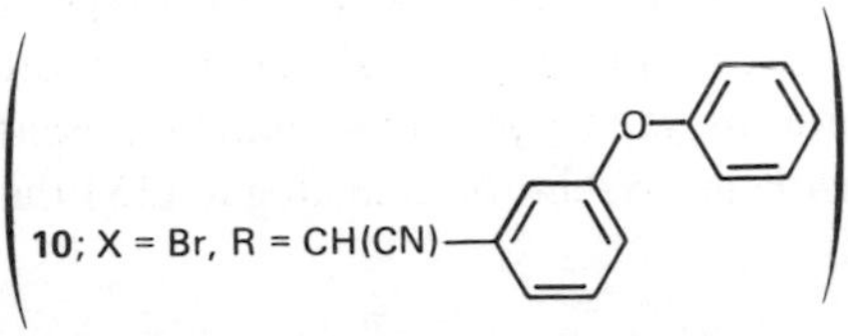

10; X = Br, R = CH(CN)—

LD$_{50}$ to houseflies and mustard beetles was 0.0003 μg/insect ($\equiv$ 0.03 mg/kg) which sets a new standard of insecticidal potency and is some 20 times more active than bioresmethrin. These dihalovinyl pyrethroids appear extremely promising insecticides because of their high activity, combined with reasonable photostability and very low mammalian toxicity[15].

Elliott *et al.*[9,14] have demonstrated that very small alterations in the structure and configuration of pyrethroids can greatly influence their insecticidal potency so the mechanism of toxic action probably involves a very specific interference with the biological system.

Considering pyrethrin I (**8**) and its analogues (p. 42) the isobutenyl side-chain of chrysanthemic acid (**7**; R = CH$_3$) may be substituted by other groups without loss of activity, so the direct interaction of the isobutenyl group at the side of action is unlikely. Measurement of the octanol—water partition coefficients of pyrethroids, including pyrethrin I (**8**), shows[14] that the most active members have a narrow range of polarities and the main function of the acid side-chain is apparently to confer the optimum polarity to the molecule. The side-chain must however contain at least one unsaturated centre, thus the isobutyl (+)-*trans*-chrysanthemates are much less active. The presence of the *gem*-dimethyl group and the carboxy moiety on the cyclopropane ring appears essential for activity. This must be a cyclopropanecarboxylic acid. The alcoholic component must contain an unsaturated side-chain which may be alkenyl, cycloalkenyl, benzyl, or aromatic (e.g. furyl, benzyl, or phenoxy). The whole cyclopentenolone ring can be replaced by structures that maintain the essential stereochemistry between the *gem*-dimethyl group on the cyclopropane ring and the unsaturated centre in the alcohol side-chain, so that for example (+)-*trans*-chrysanthemates of 5-benzyl-3-furylmethyl and 3-phenoxybenzyl alcohols are active[14].

A significant feature of alcohols leading to insecticidally potent pyrethroids is the ability of the unsaturated side-chain on the alcohol to adopt a conformation in which it is not coplanar with the ring. Thus the (+)-*trans*-chrysanthemate of the xanthene alcohol (11) is inactive because rotation about the bond is prevented, whereas it can occur easily in the active esters from, for instance, 3-phenoxybenzyl alcohol as indicated:

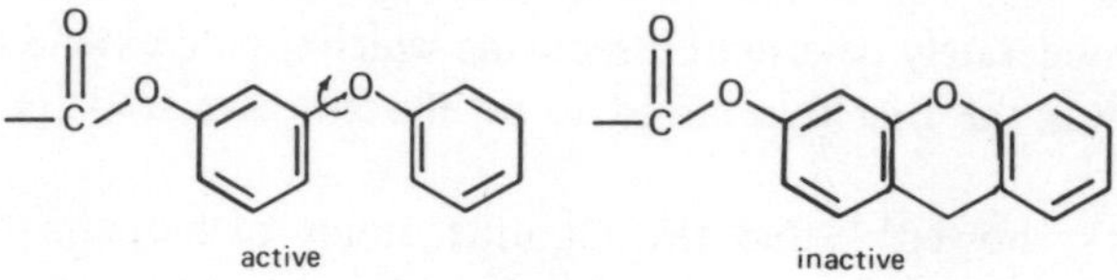

active   inactive

(11)

In substituted 5-benzyl-3-furylmethyl and 3-phenoxybenzyl alcohols nuclear substitution generally reduces the insecticidal activity of the derived chrysanthemates; on the other hand, α-cyano-3-phenoxybenzyl chrysanthemate (12) showed increased activity and with the *cis*-dibromo analogue (13) the two possible optical isomers were obtained:

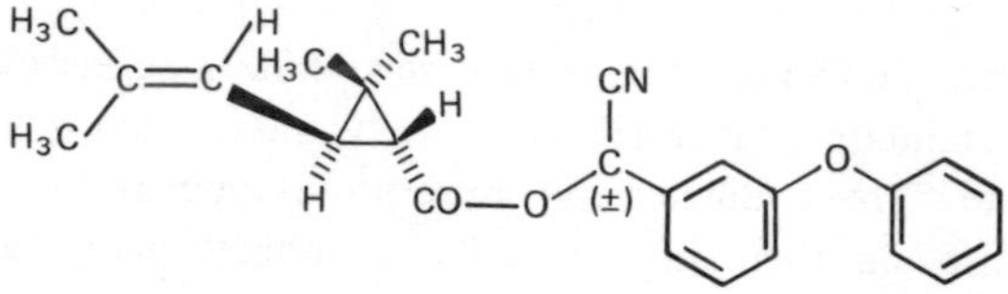

(12)

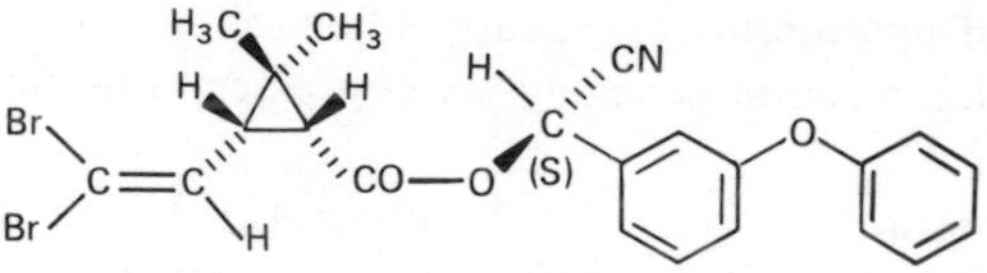

(13)

The crystalline *S*-isomer (13) was much more active than the liquid *R*-isomer. The activity of (13) known as NRDC 161 or decamethrin, was quite exceptional, and again emphasizes the importance of stereochemistry in the activity of pyrethroids and suggests that the mode of action must involve a very specific interaction with a receptor site probably in the central nervous system of the insect[9,14].

Not a great deal is known of how precisely pyrethroids produce their almost instantaneous knockdown effect on flying insects, whereas they generally show only low toxicity to mammals. The primary action must be on the nervous system and histological studies have revealed [3] extensive disruption of nervous tissue. The ability of treated insects to recover from the effects of sublethal doses of pyrethroids is due to rapid enzymic detoxification *in vivo* probably by mixed microsomal oxidases[1].

Pyrethroids effect both the peripheral and central nervous systems of treated insects causing repetitive discharges followed by convulsions of the insect[6].

Application of higher concentrations of pyrethroids resulted in an actual blockage of nerve conduction. The insecticidal activity of pyrethroids displays a negative temperature coefficient — they are more potent at lower temperatures. On the other hand, the repetitive discharge produced by allethrin in the nerve cord of the cockroach showed a positive temperature coefficient, indicating that this effect is unlikely to be responsible for the insecticidal activity. The blocking action of pyrethroids on nerve transmission does, however, exhibit a negative temperature coefficient, probably due to enhanced detoxification at higher temperatures, and this is probably the primary cause of their insecticidal action[6].

The major metabolic pathway of pyrethroids appears to involve oxidation by mixed-function oxidases (MFO); thus with allethrin (9) one of the methyl groups of the isobutenyl side-chain attached to the cyclopropane ring is oxidized to the carboxylic acid (14), the rest of the structure remaining intact:

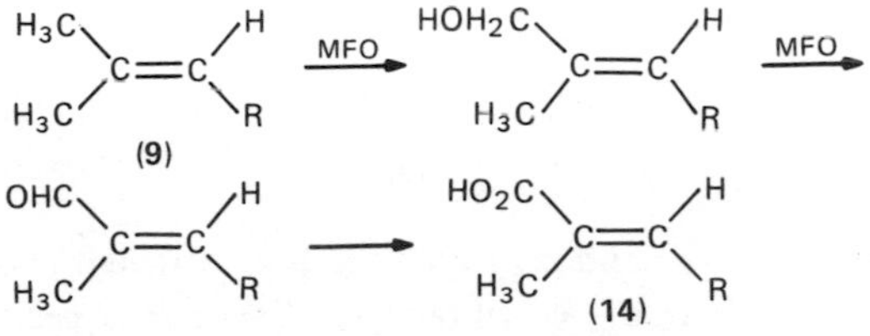

(where R = remainder of the allethrin molecule)

In 1940 Eagleson discovered[3] that by formulating pyrethrum in sesame oil, the insecticidal effectiveness, was increased, although the oil itself was not insecticidal. The synergistic activity of the oil was shown to depend on the presence of compounds containing methylenedioxyphenyl groups, such as sesamin (15), and further work resulted in the discovery of a number of pyrethrum synergists; one of the most common is piperonyl butoxide (16):

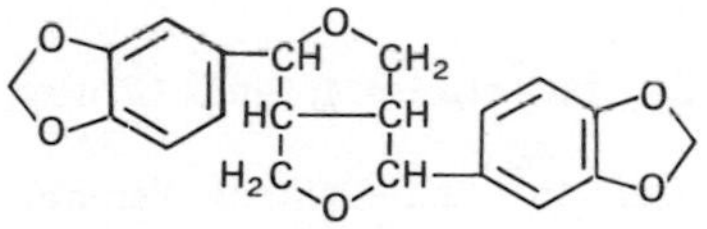

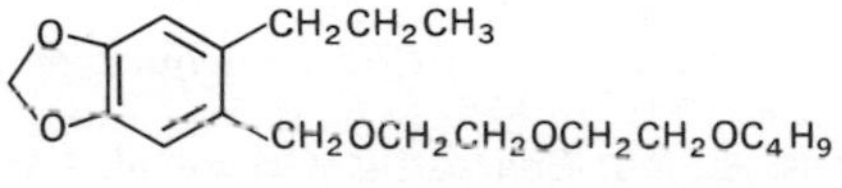

The addition of synergists to pyrethrum preparations allows the amount of active components to be substantially reduced without loss of insecticidal activity; thus when one part of pyrethrum is mixed with two parts of piperonyl butoxide (16), the resultant mixture is as active as seven parts of pyrethrum alone.

The synergists are expensive compounds, but their use is justified with pyrethroids which are both expensive and often not persistent. Pyrethroids are the only insecticides which are commercially formulated with a synergist, because it enhances the effectiveness of the insecticide during its relatively limited life. The mechanism of action of pyrethrum synergists probably involves inhibition of oxidases which otherwise detoxify the active compound. The methylenedioxyphenyl synergists (e.g. **16**) are metabolized by the same oxidation processes as are involved in the breakdown of the pyrethroids, consequently they serve as substitutes for pyrethroids in the enzyme system and inhibit their metabolism[1,3].

A new group of insecticides are related to nereistoxin (**17**), a toxin isolated from a species of marine worm. Examples are cartap (**18**) and the 1,2,3-trithiole (**19**);

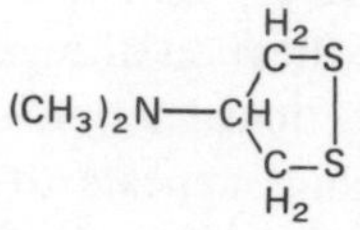

$$
\underset{(17)}{(CH_3)_2N-\underset{\underset{H_2}{C-S}}{\overset{\overset{H_2}{C-S}}{\mid}}CH} \qquad \underset{(18)}{(CH_3)_2N-CH\overset{CH_2SCONH_2}{\underset{CH_2SCONH_2}{\diagdown}}}
$$

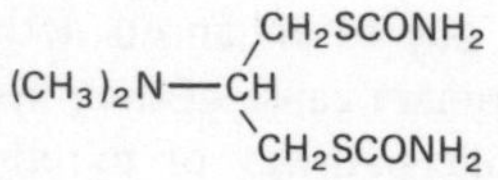

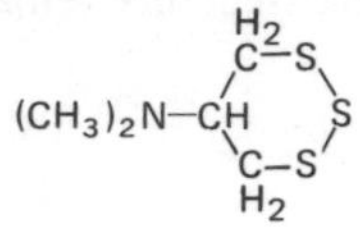

$$
\underset{(19)}{(CH_3)_2N-\underset{\underset{H_2}{C-S}}{\overset{\overset{H_2}{C-S}}{CH}}S}
$$

The latter was developed by Sandoz (1975) and is an efficient contact and stomach poison against a number of economically important pests[15]. Cartap (**18**), generally used as the hydrochloride, is a slow-acting contact insecticide useful against rice stem borers, Colorado and Mexican beetles, caterpillars, and weevils[7]. It has only moderate toxicity to mammals: LD_{50} (oral) to rats 380 mg/kg. Both cartap and nereistoxin, like nicotine (p. 40), act by combination with the acetylcholine receptor in insects, and experiments with cockroaches show[6] that they block synaptic but not axonal nervous transmission.

References

1. *Naturally Occurring Insecticides* (Eds. Jacobson, M. and Crosby, D. G.), Dekker, New York, 1971.
2. Hartley, G. S. and West, T. F., *Chemicals for Pest Control*, Pergamon Press, Oxford, 1969, p. 26.
3. Martin, H., *The Scientific Principles of Crop Protection*, 6th edn., Arnold London, 1973, p. 179.
4. Woods, A., *Pest Control*, McGraw-Hill, London, 1974, p. 82.
5. Approved Products for Farmers and Growers, Ministry of Agriculture, Fisheries, and Food, London, 1978.
6. Corbett, J. R., *The Biochemical Mode of Action of Pesticides*, Academic Press, London & New York, 1974, p. 165.
7. *Pesticide Manual* (Eds. Martin, H. and Worthing, C. R.), 4th edn., British Crop Protection Council, 1974.
8. Elliott, M. and Janes, N. F., 'Chemistry of natural Pyrethrins' in *Pyrethrum, the Natural Insecticide* (Ed. Casida, J. E.), Academic Press, London and New York, 1973, p. 56.
9. Elliott, M., *Bull. Wld. Hlth. Org.*, **44**, 315 (1970).
10. Barthel, W. F., 'Synthetic pyrethroids', *Advances in Pest Control Research*, **4**, 33 (1961).

11. Elliott, M., Janes, N. F., and Graham-Bryce, I. J., *Proc. 8th Brit. Insectic. and Fungic. Conf.*, Brighton, 2, 373 (1975).
12. Elliott, M., Farnham, A. W., Janes, N. F., Needham, P. H., Pulman, D. A., and Stevenson, J. H., *Nature (London)*, 246, 169 (1973).
13. Griffiths, D. C., Scott, G. C., Jeffs, K. A., Maskell, F. F., and Roberts, P. F., *Proc. 8th Brit. Insectic. and Fungic. Conf.*, Brighton, 1, 213 (1975).
14. Elliott, M., Farnham, A. W., Janes, N. F., Needham, P. H., and Pulman, D. A. 'Insecticidally active conformations of pyrethroids', in *Mechanism of Pesticide Action*, A.C.S. Symposium Series, No. 2, 1974, p. 80.
15. Watkins, T. I. and Weighton, D. M., *Repts. on Progr. Appl. Chem.*, 60, 404 (1975).

Chapter 5
Synthetic Insecticides I: Miscellaneous and Organochlorine Compounds

In recent years synthetic insecticides have been gaining ground at the expense of naturally occurring insecticidal products (Chapter 4), apart from pyrethroids whose production has continued to rise inspite of the growth of synthetic compounds.

Far the most important groups of synthetic insecticides are the organochlorine and organophosphorus compounds, though the sales of organochlorine insecticides such as DDT are now declining substantially, in view of the environmental hazards associated with the widespread use of this group of compounds (see Chapter 14, p. 215).

Miscellaneous Compounds

The earliest synthetic contact insecticides were inorganic materials[1 (a)]: the pigment Paris Green, a copper aceto-arsenite of approximate composition $Cu_4(CH_3COO)_2(AsO_2)_2$, was successfully employed in the United States of America (1864) for the control of Colorado beetle on potatoes. Lead arsenate, $PbHAsO_4$, was also used in 1892 against the gipsy moth in forests in the eastern United States. This compound is still one of the most effective agents against codling moth caterpillars in orchards[2,3]. It is almost insoluble in water and is often formulated with organomercurial compounds so providing a mixture which also controls apple and pear scab. In view of the high intrinsic mammalian toxicity of lead, preparations containing calcium arsenates are often preferred because environmental pollution by lead presents a serious problem since lead is a cumulative poison.

Such arsenical compounds are general stomach poisons and any selective toxicity to insects depends on the fact that the insect consumes a much larger amount of vegetable matter in relation to its body weight as compared with man. Moreover the insect consumes the freshly contaminated crop whereas man only eats part of the crop after a suitable interval of time between application of the insecticide and harvesting.

Due to the highly poisonous nature of arsenic and the potential dangers of environmental contamination sodium arsenite has been banned, and the use of inorganic arsenicals has markedly decreased. Much smaller amounts of lead arsenate are now used in orchards because fruit growers rely mainly on organophosphorus and carbamate insecticides (Chapter 6): but where long-term residual insecticidal activity is needed to protect fruit trees against chewing insects lead arsenate is still used to a limited extent[3].

The action of arsenical insecticides is probably due to the production of water-soluble inorganic arsenic ions. Paris Green and sodium arsenite produce arsenite ions, while calcium and lead arsenates give arsenate ions. Both kill insects by inhibition of respiration: arsenate ions (ions of arsenic acid $O=As(OH)_3$) are known[4(a)] to uncouple oxidative phosphorylation, probably because arsenate mimics the phosphate ion and becomes incorporated into key high-energy intermediates which quickly decompose if they contain arsenate rather than the normal phosphate.

Arsenite ions (ions of arsenious acid $As(OH)_3$) act by inhibition of the pyruvate or the α-ketoglutarate dehydrogenase systems, and probably interference with keto acid oxidation is the main site of toxic action.

Dinitrophenols

Dinitrophenols and their derivatives are very versatile pesticides and have been used as insecticides, fungicides, and herbicides. Potassium dinitro-o-cresylate was marketed in Germany in 1892 as a moth-proofing agent. In tests against the eggs of the purple thorn moth, phenol and cresols were found[1(b)] to be toxic, and the activity was progressively increased by the introduction of two nitro groups in the molecule, but further nitration reduced activity. This led to the development of 3,5-dinitro-o-cresol (DNOC), or more correctly 2-methyl-4,6-dinitrophenol (1; R = CH_3), and the corresponding 2-cyclohexyl phenol (1; R = cyclohexyl) formulated in petroleum oil as sprays against the eggs of aphids, winter moth, and red spider mites.

DNOC (1; R = CH_3) is obtained by sulphonation of o-cresol followed by controlled nitration. Such dinitrophenols have high mammalian toxicities and their use has caused environmental damage; they are also so phytotoxic that they cannot be used on foliage, indeed their major use is as rapid-acting herbicides[5] (Chapter 8, p. 141).

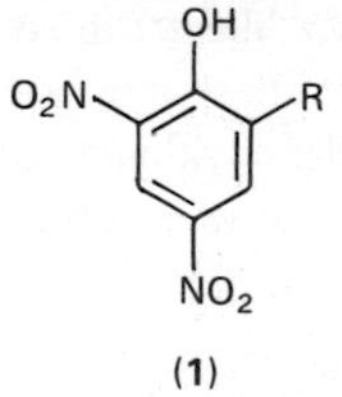

(1) 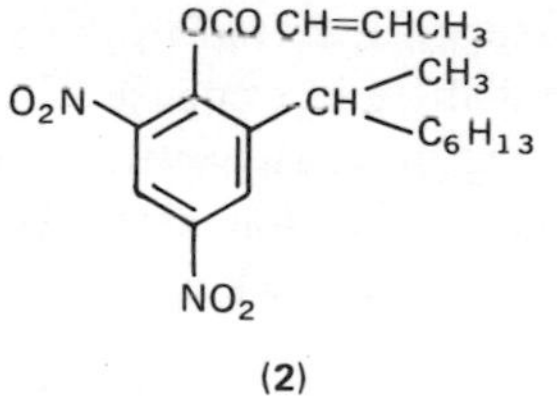 (2) 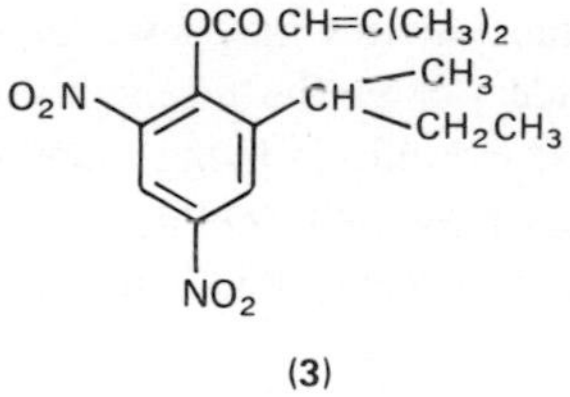 (3)

Certain related dinitrophenols, such as Karathane (2) and binapacryl (3), although less toxic to mammals and insects are useful acaricides and fungicides (Chapter 7, p. 117). Many of the dinitrophenols are highly toxic to all forms of life; they are contact poisons which have the ability to penetrate to the site of action within the cell. The toxicity arises from their interference with respiration through uncoupling oxidative phosphorylation in mitochondria[1(b),4(b)]; in addition some of the herbicidal effects may be due to direct interference with photosynthesis (Chapter 3).

Various esters (e.g. 2 and 3) are used against powdery mildews and mites and these are apparently hydrolysed *in vivo* to the free 2,4-dinitrophenol which is the active toxicant.

Organic Thiocyanates

In a series of alkyl thiocyanates (4) obtained by reaction of alkyl halides with sodium thiocyanate, the insecticidal activity increased with the length of the alkyl chain up to the dodecyl derivative (4; $R = C_{12}H_{25}$) and then declined. The dodecyl compound was the most active because it possessed the optimum oil/water solubility balance for penetration of the insect cuticle (Chapter 2, p. 18). A more useful compound was, however, 2-(2-butoxyethoxy) ethyl thiocyanate or Lethane (4; $R = -CH_2 CH_2 OCH_2 CH_2 OC_4 H_9$) discovered in 1936[1(b),8].

$$R-S-C\equiv N$$

(4)　　　　(5)　　　　(6)

Isobornyl thiocyanate or Thanite (6) obtained from the terpene isoborneol (5) is also a useful derivative; the latter two compounds are the only members of this group currently used as insecticides, mainly in fly sprays. The insecticidal properties of thiocyanates have not been fully exploited probably because of the dramatic successes achieved by the organochlorine insecticides such as DDT.

Now that organochlorine insecticides have been banned in several countries and severely restricted in others, the insecticidal thiocyanates possibly merit further investigation; they are remarkably rapid-acting compounds against flying insects — indeed their knockdown action is almost as quick as that of the pyrethroids (Chapter 4), and they also show[1(b)] ovicidal activity against a number of insect eggs.

Thiocyanates may owe their insecticidal action to the *in vivo* liberation of the cyanide ion in the insect body, and there is evidence[4(c)] that these compounds release cyanide in living mice and houseflies. However, despite this evidence, other factors may well operate in their mode of action, since the release of cyanide ion would not seem to be consistent with their rapid knockdown effect.

Organochlorine Insecticides

The most important member of this group of insecticides is 1,1,1-trichloro-2,2-di-(*p*-chlorophenyl)ethane also termed *d*ichloro*d*iphenyl*t*richloroethane or DDT (7). This compound was first prepared by Zeidler (1874) but its powerful insecticidal properties were not discovered until 1939 by Müller of the Swiss

Geigy Company. DDT is manufactured by condensation of chloral and chlorobenzene in the presence of an excess of concentrated sulphuric acid[1 (b),6,7].

chlorobenzene chloral (7)

The crude product consists of some 80% of the desired p,p'-compound (7) together with approximately 20% of the $o,p,'$-isomer and a trace of the o,o'-isomer. Only the p,p'-isomer has significant insecticidal activity; pure DDT can be obtained as a white powder m.p. 108°C by recrystallization from ethanol. However the increased cost involved in purification is only justified when DDT is used for special purposes such as for aerosol packs when the presence of sticky impurities may block up the fine nozzle through which the solution of the pesticide is discharged[6]. DDT was introduced as an insecticide in 1942 and was manufactured on a large scale during the war. Its application in Naples in 1944 checked, for the first time in medical history, a potentially serious outbreak of typhus in which, during one month, a million people were dusted with the insecticide; and its subsequent use in India and other countries of the far East has substantially decreased the death rate from malaria[9]. The benefits to mankind from the use of organochlorine insecticides have been tremendous and DDT has become the most widely used insecticide in the world. The annual production was more than 100,000 tons in the late 1950s, although since that period the annual production has declined by about 50%. At the time of its discovery, the main advantages of DDT appeared to be its stability, persistence of insecticidal action, cheapness of manufacture, low mammalian toxicity (LD_{50} (oral) rats 300 mg/kg), and wide spectrum of insecticidal activity. DDT kills a wide variety of insects, including domestic insects and mosquitoes, but it is not very effective against mites and does not act nearly as rapidly on flying insects as pyrethrum or thiocyanates[2,7,9].

The discovery of the insecticidal properties of DDT stimulated the search for analogous organochlorine compounds; although many hundreds of compounds were synthesized only comparatively few have been sufficiently active and cheap for commercial exploitation. Some important examples are methoxychlor (8; X = OCH_3, Y = H), Gix (8; X = F, Y = H), and 1,1-dichloro-2,2-di-(p-chlorophenyl) ethane known as DDD or TDE (9).

(8) (9)

None of these showed such a high general activity as DDT. Gix was used as an insecticide by the German Army in World War II, and DDD has a slightly different spectrum of insecticidal activity. All three compounds (8,9) have lower mammalian toxicities than DDT, and methoxychlor does not appear to be concentrated in animal fats — a most valuable asset bearing in mind the widespread environmental pollution arising from the majority of organochlorine insecticides (Chapter 14, p. 215).

DDT was formerly used to control flies in milking sheds but it was found in the milk so it is now banned in most countries for this purpose. However methoxychlor is permitted since no residues of the compound could be detected in the milk[1(b)]. DDD (9) has found use on food crops since it is appreciably less toxic than DDT. Generally for insecticidal potency, a DDT-type molecule (10) must contain p-substituents X which may be either halogens, or short-chain alkyl or alkoxy groups, Y is almost always hydrogen, and Z may be CCl_3, $CHCl_2$, $CH(NO_2)CH_3$ or $C(CH_3)_3$. In a given series (10) with fixed X and Y substituents, successive substitution of the Z substituent by the groups from CCl_3 to $C(CH_3)_3$ is generally accompanied by a progressive decline in insecticidal potency.[8]. The insecticidal activity of DDT and its analogues is greatly influenced by molecular shape and size, and various hypotheses have been proposed to account for the influence of molecular geometry. Thus it has been proposed that for activity, a DDT analogue (10) requires a Z group of sufficient steric size, e.g. trichloromethyl, to inhibit the free rotation of the two planar phenyl rings so that they will be constrained to positions of minimum steric crowding, termed a trihedral configuration. The idea was supported by examination of DDT analogues containing differently sized Z groups; thus Z = t-butyl afforded highly active compounds such as the non-chlorinated p,p'-dimethoxy diphenyl derivative (10; X = OCH_3, Y = H), and when Z = $CH(NO_2)CH_3$, and $CH(NO_2)CH_2CH_3$ the p,p'-dichloro derivatives (10; X = Cl, Y = H) were also insecticidal. On the other hand, there were several cases where molecules with trihedral configurations were inactive.

Another suggestion emphasized[1(b)] the importance of free rotation of the phenyl rings in DDT analogues. If such rotation was impossible, the compound would be inactive as was the case with o,o'-isomer of DDT. The concept was successfully extended to the tetramethyl DDT derivatives; the 2,2',4,4'- and 2,2',5,5'-isomers without free rotation of the phenyl rings were inactive, while the 3,3',,4,4'-isomer in which free rotation is possible was insecticidal, but again there were many exceptions. Both these hypotheses emphasize the importance of molecular topography in structure—activity studies of DDT analogues. The third hypothesis argued that the active DDT-type molecules functioned essentially as physical toxicants. They had the optimum molecular size and shape to penetrate the interstices of the cylindrical lipoprotein molecules so orientated to permit ions and small molecules to pass through, but larger DDT-type molecules would be excluded. When the penetrating molecule possessed regions with strong attractive forces, e.g. chlorine atoms, interaction with the membrane might affect its permeability.

Although such a large amount of DDT has been used since World War II, comparatively little is known about its precise mode of action; the general symptoms of DDT poisoning in insects and vertebrates are violent tremors, loss of movement followed by convulsions, and death, clearly indicating that DDT acts on the nervous system, and produces toxic effects in nervous tissue at much lower concentrations than induce toxic effects in other tissues and enzyme systems[9]. DDT apparently exerts its toxicity by binding to the nerve membrane and interfering with the transmission of nervous impulses, possibly by upsetting the sodium or potassium ion balance across nerve membranes[10,11] (Chapter 3, p. 33).

However, how precisely DDT interacts with the nerve membrane is not known; it also affects other membrane-linked functions such as oxidative phosphorylation in mitochondria and the Hill reaction in chloroplasts. DDT exerts a general effect on membranes but its molecular topography is conducive to special activity on the axonal membrane. DDT forms a complex with lipoprotein, and Holan[12] explained the insecticidal activity of certain diaryl halocyclopropane DDT analogues on the basis of their ability to bind to the lipoprotein interface of the axonal membrane. All the active molecules were regarded as wedges, the base of which was represented in DDT by the two substituted phenyl rings which must contain electron-donor groups. This complexes with the protein of the axonal membrane, while the apex of the wedge is the trichloromethyl group. The size of the apex is critical because it must fit into a pore in the lipid part of the membrane, and for activity the size of the apex corresponded to that of a hydrated sodium ion. The two-point attachment of the wedge to the membrane locks the molecule in position increasing the permeability of nerve membrane to sodium ions so upsetting the ionic basis of normal axonal nerve transmission[4(d),9]. The complex of the insecticide with the membrane probably dissociates more readily at higher temperatures, possibly explaining why the insecticidal activity of DDT displays a negative temperature coefficient. DDT causes repetitive discharge in insect nerves as well as complex changes in the permeability of the nerve membrane to sodium and potassium ions (see Chapter 3, p. 33). The latter effect appears to be the basis of the toxicity of DDT and related organochlorine compounds, the specificity to insects arises from the high permeability of the insect cuticle to the passage of the toxicant while the mammalian skin constitutes an effective barrier.

By 1950, a number of examples of DDT-resistant strains of insects (e.g. houseflies and cabbage root fly) had been reported and the problems raised by them had become serious[13,14]. Resistance to a particular insecticide will be more widespread if the toxicant is persistent, and the insect has a short life cycle, which explains why, given the very large-scale use of DDT and other organochlorine insecticides, there was a rapid emergence of resistant insects (see Chapter 6, p. 101).

Several factors have been proposed[13] to account for the phenomenon of insect resistance (Chapter 6, p. 101), such as morphological characteristics (e.g. the development of thick impermeable insect cuticles), slow rate of uptake of DDT[4(d)], behaviouristic tendencies, and increased detoxification of DDT and other organochlorine insecticides. The biochemical defence mechanisms are almost certainly the most significant[15].

The metabolism of DDT occurs by a number of different pathways, but the most important appears to be the dehydrochlorination of DDT (8) to give the dichloroethylene (DDE, 11)[15,16] by the non-microsomal enzyme DDT-dehydrochlorinase; this enzyme has been isolated in a pure state from DDT-resistant houseflies. In this connexion, it is interesting that dehydrochlorination by boiling alcoholic sodium hydroxide is one of the few chemical reactions of DDT.

DDE (11) is a highly persistent metabolite which is therefore a major environmental pollutant (Chapter 14, p. 218), but has only slight insecticidal activity; in birds and mammals it is further slowly metabolized to the carboxylic acid DDA (12) and this is sufficiently water soluble to be excreted[15,16].

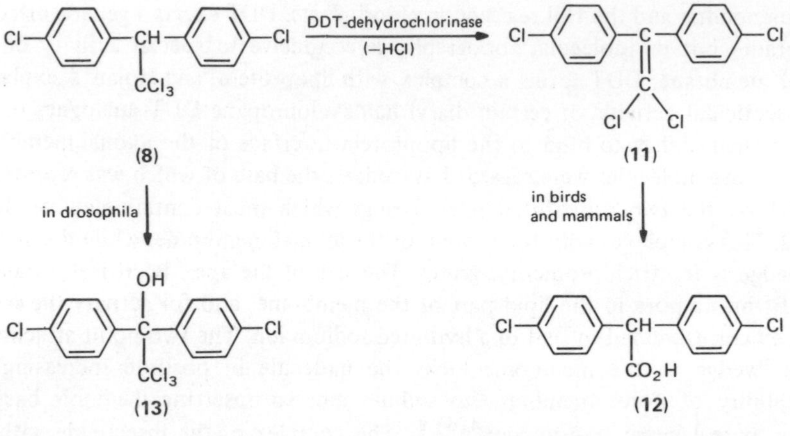

The proposal that metabolic conversion to DDE (11) is the major detoxification mechanism in DDT-resistant insects is supported by experiments in which DDT-resistant houseflies (*Musca domestica*) are shown to become susceptible to a mixture of DDT and certain non-insecticidal structural analogues, such as 1,1-di-(*p*-chlorophenyl)ethanol or 1,1-di-(*p*-chlorophenyl)ethane. Such compounds function as DDT synergists (cf. pyrethrum synergists Chapter 4, p. 47) because they inhibit the *in vivo* activity of the dehydrochlorinase[1(b)].

Sometimes compounds not structurally related to DDT may also show synergistic activity, e.g. *N,N*-di-n-butyl-*p*-chlorobenzenesulphonamide. Another enzymic detoxification pathway appears to operate in DDT-tolerant strains of fruit flies (*Drosophila melanogaster*) and probably in the cockroach. This involves an oxidase-catalysed conversion of the central α-hydrogen atom of DDT to the hydroxyl group giving dicofol (13), a compound used commercially as an acaricide, but without general insecticidal activity.

Mites and ticks (acari) are sufficiently different from insects in their biochemistry for certain compounds to be highly toxic to them (acaricidal) while they are relatively innocuous to true insects. Many acaricides are also fungicidal providing a valuable combination of activity for the control of pests on fruit trees[9]. The low insecticidal activity of many acaricides is also an advantage because most of the natural predators of mites are insects. (Plate 5).

Indeed it has been discovered that several bridged diphenyl derivatives, some of which are closely related to DDT, are valuable acaricides against phytophagous mites and ticks, although they are almost devoid of toxicity to insects[6]. One example, already mentioned is dicofol (13), which was introduced as an acaricide in

1952. Another related compound is chlorobenzilate (14) prepared from p-chloro-benzaldehyde via the benzoin condensation.

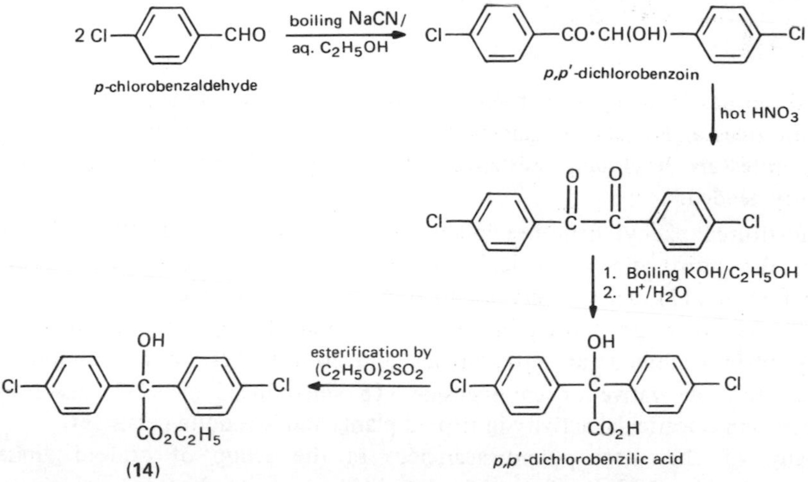

p-chlorobenzaldehyde → p,p'-dichlorobenzoin

A more recent acaricide of this type is bromopropylate or isopropyl 4,4'-dibromobenzilate. The earliest acaricide was azobenzene (1945) which may be used as an aerosol or smoke against red spider in greenhouses. Another early compound was benzyl benzoate, whose acaricidal properties were substantially enhanced by the introduction of two chlorine atoms in the p-positions giving Neotran (15), obtained from methylene chloride and sodium p-chlorophenate:

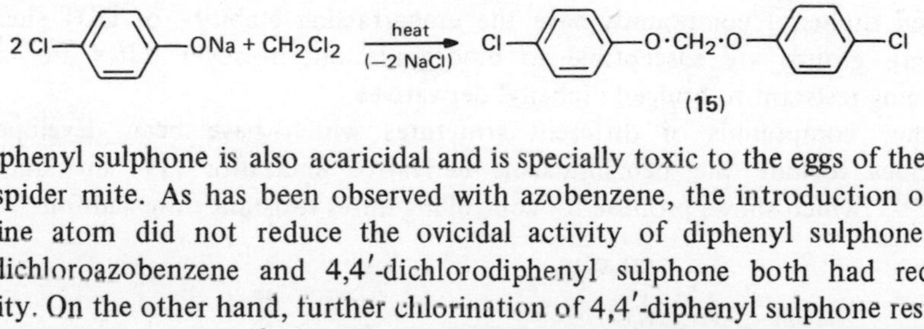

Diphenyl sulphone is also acaricidal and is specially toxic to the eggs of the fruit tree spider mite. As has been observed with azobenzene, the introduction of one chlorine atom did not reduce the ovicidal activity of diphenyl sulphone, but 4,4'-dichloroazobenzene and 4,4'-dichlorodiphenyl sulphone both had reduced toxicity. On the other hand, further chlorination of 4,4'-diphenyl sulphone restored the acaricidal properties[6]; thus both 2,4,5-trichlorodiphenyl sulphone and the 2,4,5,4'-tetrachloro derivative (16) are valuable agricultural acaricides. They are persistent in their action and are less phytotoxic than diphenyl sulphone[6,17]. Tetradifon (16) is obtained from 1,3,4-trichlorobenzene:

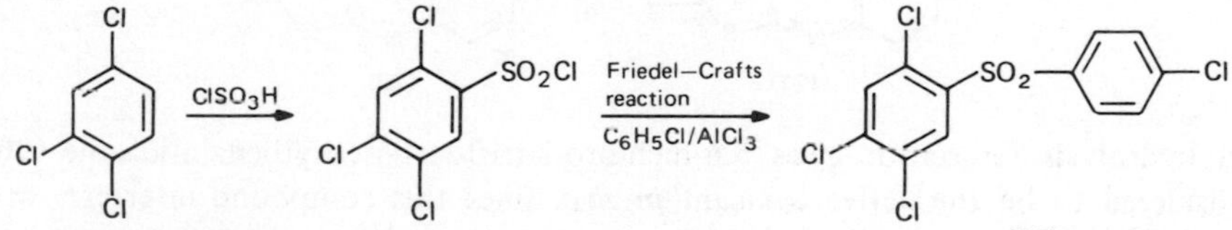

1,3,4-trichlorobenzene

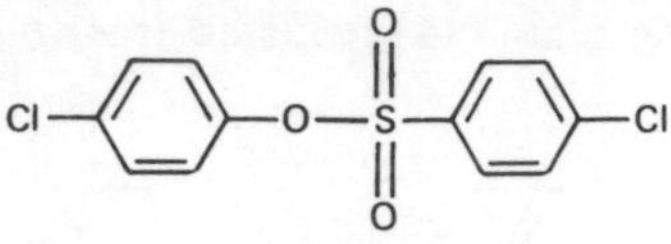

(17) 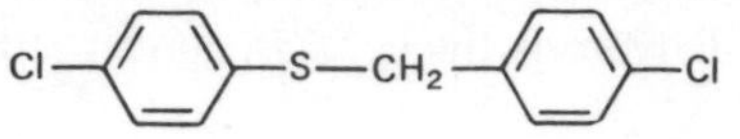   (18)

Tetradifon has for many years been the most important acaricide for the protection of fruit trees and is active against all stages and eggs of phytophagous mites; but many mites are developing resistance to this compound and so new acaricides are urgently needed.

Substituted phenyl benzenesulphonates show acaricidal and insecticidal properties; the most effective ovicide was the 4,4'-dichloro derivative known as chlorofenson (17). This is very useful since with rapidly breeding mites control against the egg stage is often more effective than killing the mites themselves. Benzyl phenyl sulphides are also acaricidal and again the best known example is the 4,4'-dichloro derivative or chlorbenside (18) introduced in 1953; this showed good systemic acaricidal activity in treated plants and was quite persistent.

Many of the most potent acaricides in the group of bridged diphenyl compounds, like DDT itself, contain 4,4'-dichloro atoms. Nothing appears to be known about the biochemical mode of action of these acaricides[4], although this is not really surprising since the study of the physiology of mites is much less developed than that of insects[9]. The wide variety of different possible bridging groups between the two phenyl rings makes it difficult to formulate precise structure—activity relationships[1(c)]. The chemistry of acaricides has been reviewed[18,19]; most of the compounds have selective toxicity to mites, some of them in addition show selectivity towards specific species of mites[1(c)]. None of the bridged diphenyl compounds have the embarrassing stability of DDT since the bridging groups are susceptible to biodegradation; however mites are rapidly becoming resistant to bridged diphenyl derivatives.

Other compounds of different structures which have been developed as acaricides include the benzimidazole derivative fenazaflor (19) introduced in 1966[6,17] which shows promise for controlling mites resistant to tetradifon.

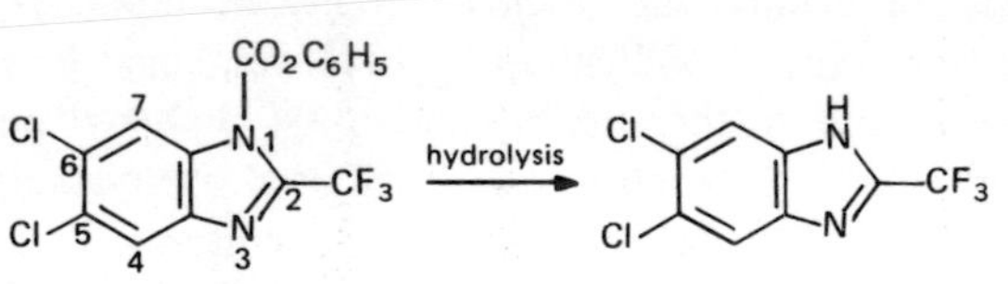

(19)   (20)

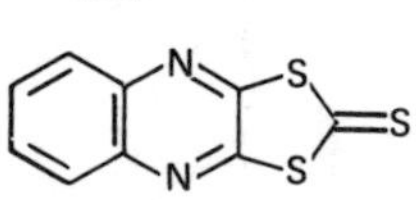

(21) 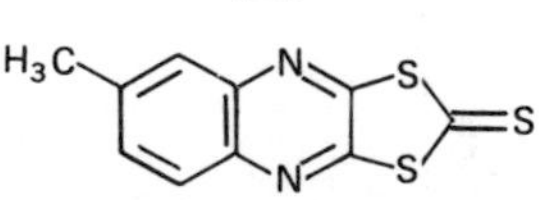   (22)

On hydrolysis fenazaflor gives 5,6-dichloro-2-trifluoromethylbenzimidazole (20), considered to be the active toxicant *in vivo,* since this compound interferes with respiration by uncoupling oxidative phosphorylation[4(b)] in rat liver and mite mitochondria. This is supported by the observation[4(b)] that fenazaflor increased oxygen uptake by mites.

Other aromatic nitrogen heterocylic acaricides are thioquinox (21) and chinomethionat (22). These compounds also possess useful fungitoxicity against powdery mildews; both have low mammalian toxicities and are effective against the summer eggs of spider mites[9]. A number of organophosphorus insecticides show acaricidal properties, e.g. parathion, malathion, and diazinon (Chapter 6), also carbamates (e.g. aldicarb), and dinitrophenols (e.g. DNOC and dinocap).

Some important new types of acaricides include amidines, such as formamidines like chlordimeform (23) which are claimed[19] to function as monoamine oxidase inhibitors. Chlordimeform (1972) acts as a broad-spectrum acaricide effective against adult mites, eggs, and larvae. It is also toxic to cockroaches which arises from its ability to uncouple oxidative phosphorylation (Chapter 3, p. 31). Several other amidines are useful as acaricides and insecticides. An important example is Mitac developed by Boots Ltd. (1973) and prepared from 2,4-xylidine, ethyl orthoformate and methylamine[7]:

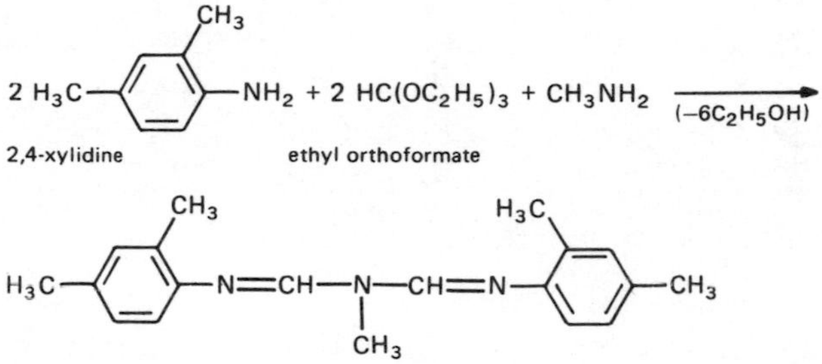

Mitac is active against all stages of a wide range of mites and controls mites which have become resistant to other acaricides[19]. It is also a valuable insecticide against white fly, leaf miners, scale insects, and caterpillars. Mitac provides no danger to the environment and has a low mammalian toxicity: LD_{50} (oral) to rats 800 mg/kg.

Some tin compounds are useful acaricides, the best known example is Cyhexatin (24) which is now one of the leading acaricides; in spite of being rather slow acting and sometimes phytotoxic, it has the important advantage of low toxicity to predatory mites[2,19].

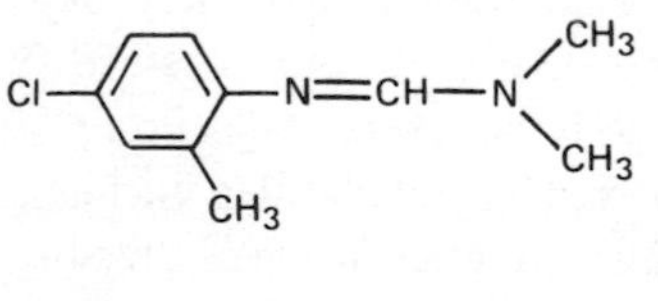

(23) 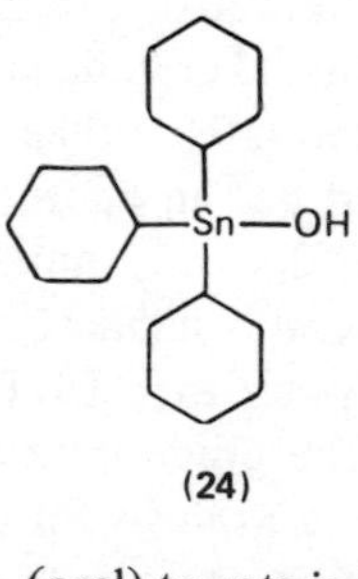   (24)

Cyhexatin has a low mammalian toxicity: LD_{50} (oral) to rats is 540 mg/kg[7].

Hexachlorocyclohexane

This was discovered independently by Imperial Chemical Industries in Britain and in France in 1942[7]. Hexachlorocyclohexane (HCH), or formerly incorrectly termed benzenehexachloride (BHC), is manufactured by treatment of benzene with chlorine under the influence of ultraviolet light without catalysts[7]. Hexachlorocyclohexane can theoretically exist as eight different stereoisomers of which five are actually found in the crude product[1(b)] Assuming for simplicity that the cyclohexane ring is planar then these five isomers may be depicted as follows:

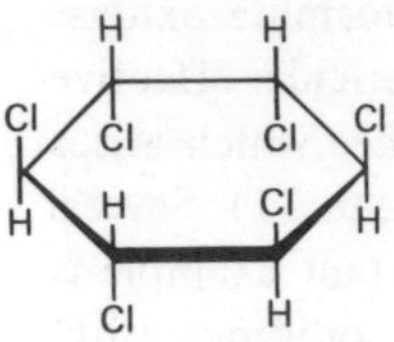

$\alpha-(70\%)$ $\beta-(6\%)$ $\gamma-(13\%)$ $\delta-(6\%)$

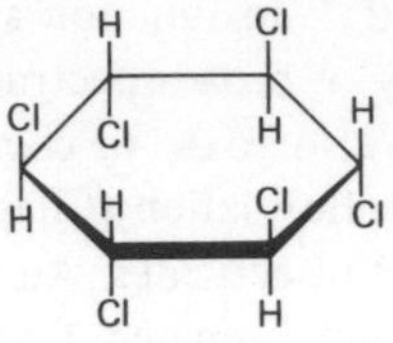

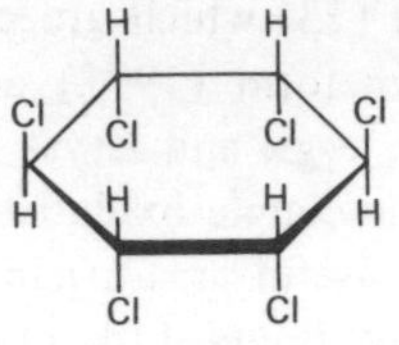

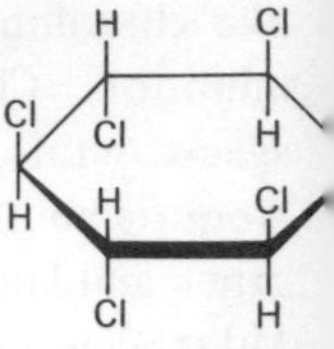

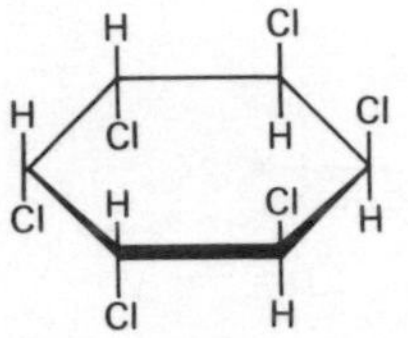

$\epsilon-$(traces only)

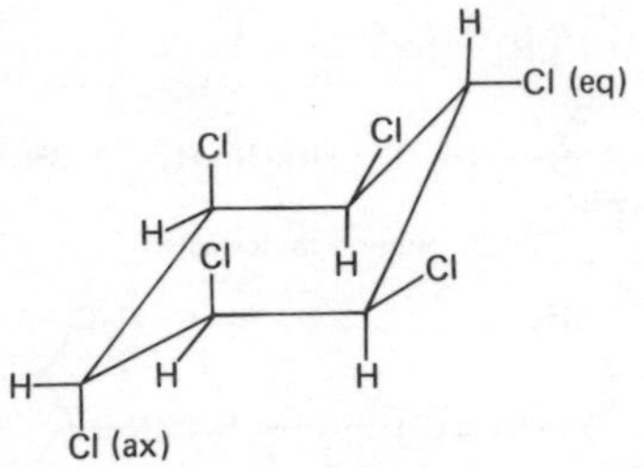

(25)

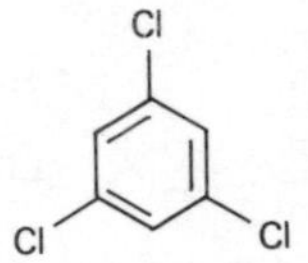

(26)

Only the γ-isomer or lindane has powerful insecticidal properties. In fact the cyclohexane ring has a non-planar chair conformation in which the bonds attached to the six carbon atoms of the ring are either axial (ax) or equatorial (eq). The axial bonds are perpendicular to the general plane of the ring system, while the equatorial bonds lie in the general plane of the ring. On this basis, the true representation of γ-HCH or lindane is depicted in (25).

Since only the γ-isomer is appreciably active, many attempts have been made to increase the proportion of this isomer in the mixture obtained, but without success. The crude product is generally used directly, but can be purified by extraction with hot methanol followed by fractional recrystallization to give substantially pure (99% or more) γ-HCH or lindane[6], the latter name arising from its original isolation by Van der Linden (1912). The pure compound, like DDT, rapidly penetrates the insect cuticle and has an appreciably higher vapour pressure (9 x 10^{-6} mmHg at 20°C) than DDT (2 x 10^{-7} mmHg) so it can exert a significant fumigant action in a dry atmosphere, and is stable to heat so it may volatilize unchanged. Lindane is also more soluble in water than DDT and is an effective seed dressing against the attack of soil insects. The crude mixture of isomers containing some 13% of the active γ-isomer is applied as dusts for control of various soil pests as well as flea beetles and mushroom flies. As sprays lindane is useful against many sucking and biting pests, and as smokes for control of pests in grain stores[2]. The crude material, however, suffers from the disadvantage of an unpleasant musty odour and taste which tends to taint foodstuffs[9]. This apparently is due to the presence of the other isomers as pure γ-HCH or lindane has no smell, but is also considerably more expensive than the crude mixture. The symptoms of insect poisoning superficially resemble those of DDT and γ-HCH is known to be a neurotoxicant; thus a concentration of $10\ \mu M$ increases the frequency of spontaneous discharges in the cockroach nerve cord and extends the synaptic after discharge[4 (d)]. Lindane rapidly penetrates the cuticle of cockroaches and accumulates in the peripheral regions of the central nervous system, and quickly causes tremors, loss of bodily coordination, convulsions, and prostration. Like DDT lindane probably kills insects by bringing about a sodium—potassium imbalance in nerve membranes. However, it is possible that its biochemical site of action is somewhat different from that of DDT because it has been observed[1 (b)] that houseflies exposed to either γ-HCH or the cyclodiene insecticides exhibited a characteristic wing 'fanning' effect which was not observed with DDT. Also DDT-resistant houseflies were generally susceptible to both lindane and the cyclodienes (see Chapter 6, p. 102)[1 (b), 9]. In weakly alkaline media lindane (25) suffers dehydrochlorination to give 1,3,5-trichlorobenzene (26). The metabolism in insects also involves dehydrochlorination; thus one of the initial products of metabolism of lindane in houseflies was the monodehydrochlorinated compound, pentachlorocyclohexene. This was the major metabolite isolated from lindane-resistant houseflies one hour after treatment with the insecticide (25), although pentachlorocyclohexene was not obtained after similar experiments with susceptible flies. Pentachlorocyclohexene is an intermediate metabolite since the amount of this compound isolated from treated flies decreased with time.

Most of the lindane is metabolized in flies to water-soluble compounds which with aqueous alkali gave isomeric dichlorothiophenols (27), probably formed via reaction with a sulphydryl compound (RSH), related to glutathione[15]:

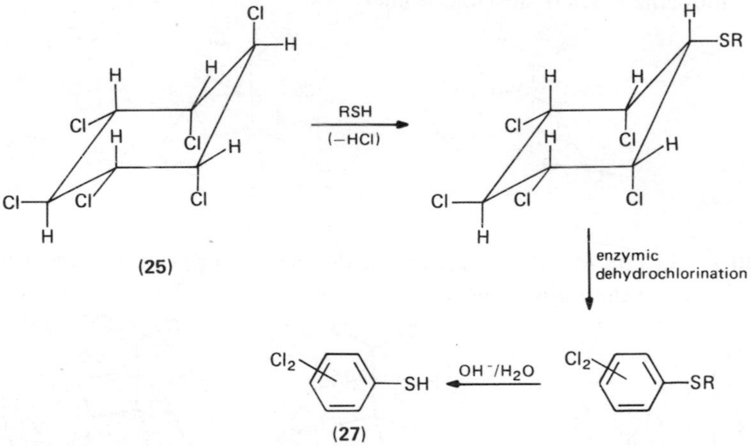

(25)

(27)

The resistance to lindane observed in houseflies seems chiefly due to their enhanced ability to metabolize the insecticide to such non-toxic materials.

The Cyclodiene Group

The insecticidal properties of chlordane (28) were reported in 1945 – this was the first member of a remarkable new group of organochlorine insecticides[8,10]. These compounds are prepared from hexachlorocyclopentadiene (29) by the Diels–Alder reaction; for instance with cyclopentadiene the product is chlordene (30). This is only slightly toxic to insects but subsequent addition of chlorine gave the highly active compounds chlordane (28) and heptachlor (31):

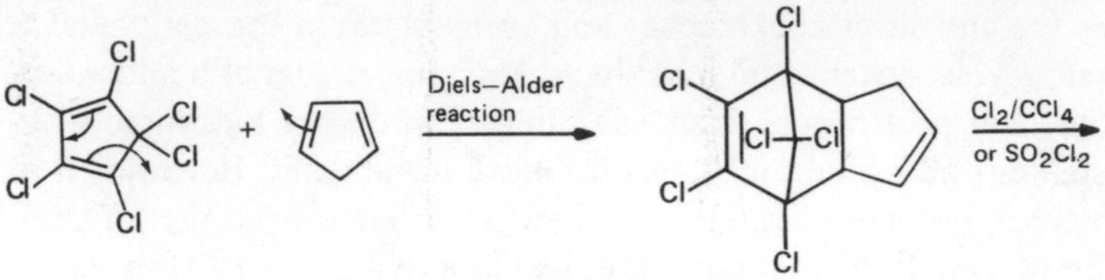

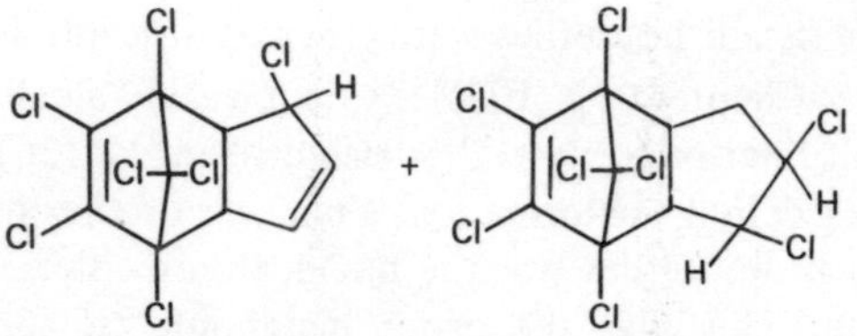

An important feature of the synthesis is that cyclopentadiene is so reactive that it can undergo the Diels–Alder reaction with itself. On the other hand, hexachlorocyclopentadiene (29) cannot undergo a self Diels–Alder and so may be used for the synthesis of a large number of other ring systems.

The stereochemistry of the cyclodienes is complex, for instance chlordene (30) could exist in two possible configurations known as *exo* and *endo* isomers – actually chlordene is solely the *endo* isomer:

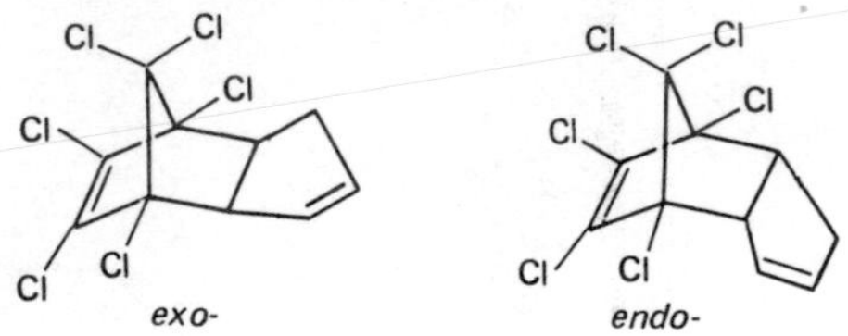

The further addition of chlorine across the double bond gives chlordane which is a mixture of the *cis* and *trans* isomers:

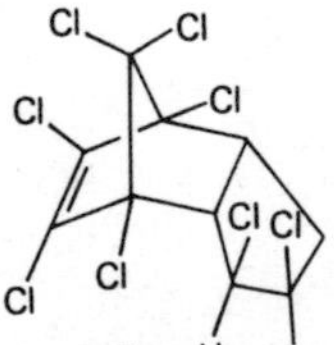

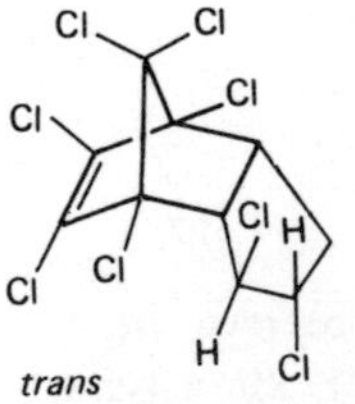

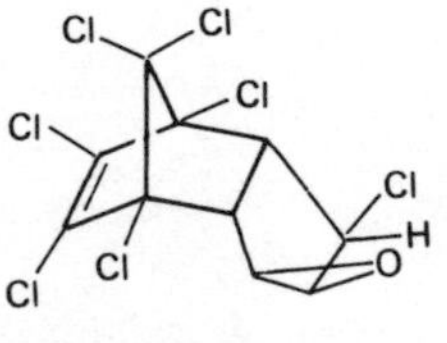

Commercial chlordane also contains some heptachlor (31) which is generally more insecticidal than chlordane so efforts were made to increase the proportion of heptachlor in production. Experiments showed that when the chlorination of chlordene (30) was carried out by sulphuryl chloride in the presence of benzoyl peroxide catalyst, heptachlor (31) was the major compound obtained.

In animal, plant and insect tissue heptachlor is converted to the epoxide (32) which is slightly more insecticidal than the parent compound[19].

The most important cyclodiene insecticides are those containing four fused five-membered rings and are prepared by the addition of various dienophiles to hexachlorocyclopentadiene (29). The simplest Diels—Alder adduct of hexachloro-cyclopentadiene is obtained with vinyl chloride followed by treatment with alcoholic potassium hydroxide:

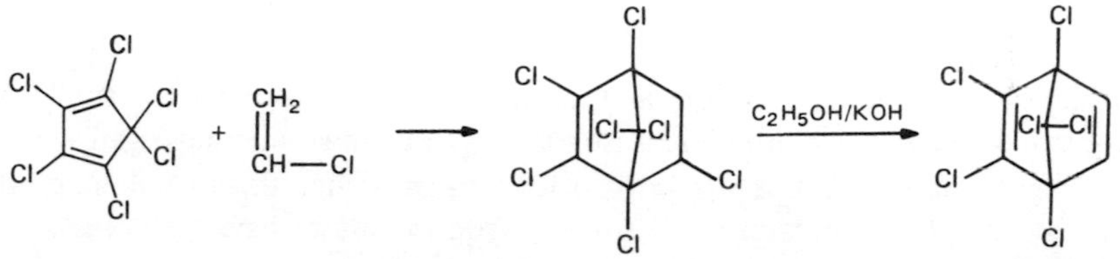

The resultant compound (33) has almost no insecticidal properties, but when reacted with cyclopentadiene it gives isodrin (34) which on peroxidation yields endrin (35):

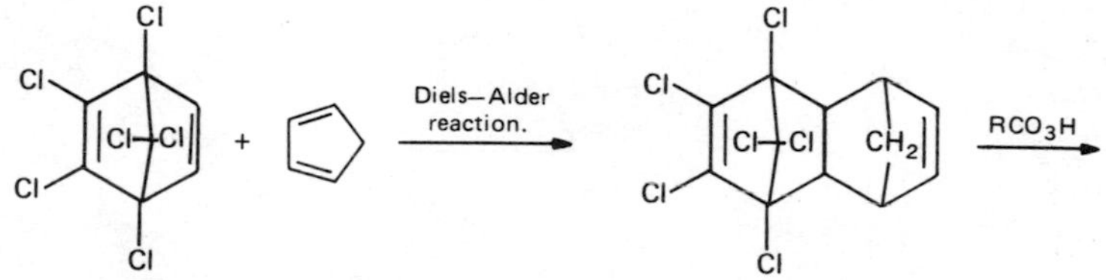

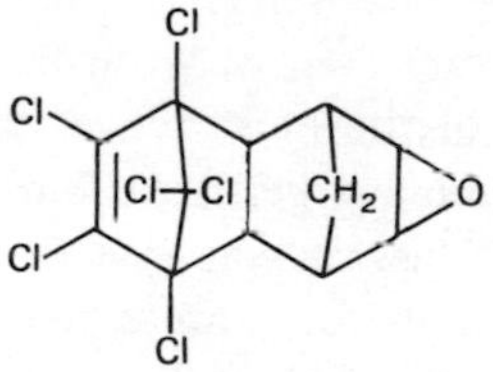

Isodrin is also metabolized to endrin in houseflies and in a variety of animals, e.g. rabbits[15].

Both isodrin (34) and endrin (35) are insecticidal, but only endrin has been commercially exploited. When the Diels—Alder adduct of vinyl chloride and cyclopentadiene is dehydrochlorinated and the resultant product subjected to a second Diels—Alder reaction with hexachlorocyclopentadiene, the product is not isodrin but the isomer known as aldrin which on epoxidation affords dieldrin:

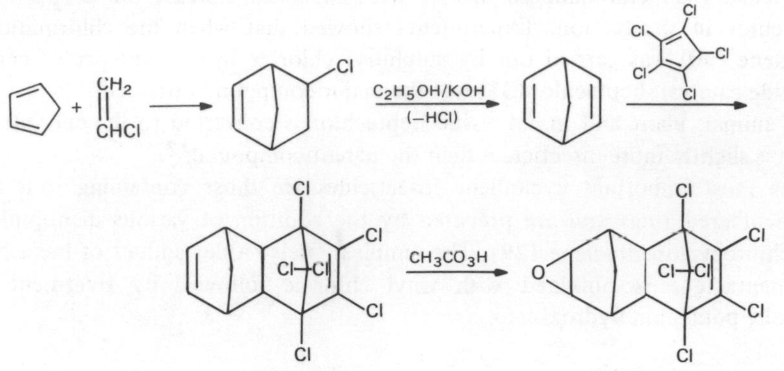

aldrin

dieldrin

Considering the isomers isodrin and aldrin, only the former, by sunlight or treatment with hydrogen bromide, is isomerized to the high-melting caged structure (36). The formation of this derivative (36) demands that the two double bonds must lie in close proximity; therefore isodrin must have the *endo–endo* configuration while aldrin is the *endo–exo* isomer[1] [(b)]:

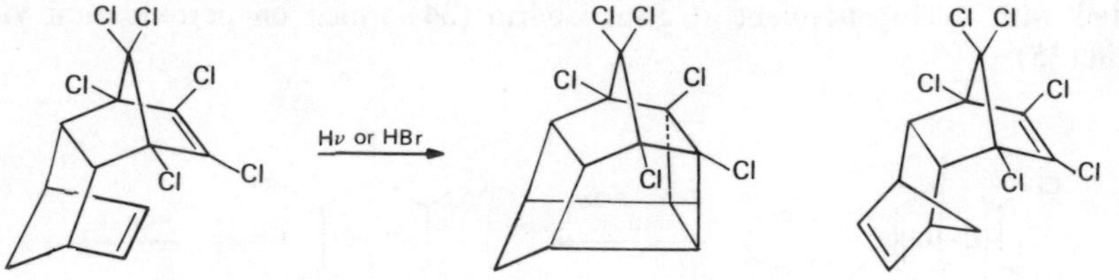

isodrin

(36)

aldrin

Aldrin and dieldrin are the best known members of the cyclodiene group of insecticides and are named after Diels and Alder, the discoverers of the diene synthesis. Both are chemically very stable and do not react even with caustic soda solution. Aldrin, dieldrin, and endrin (35) are some of the most active general contact insecticides – like DDT they are highly lipophilic and persistent but have little systemic action and so are relatively ineffective against sucking insects. They are excellent soil insecticides and are the best compounds for termite control[2]. Dieldrin is remarkably effective against ectoparasites such as blow flies, lice, and ticks and was widely employed in cattle and sheep dips. Dieldrin was also applied for the protection of fabrics from moths and beetles and against carrot and cabbage root flies and as a seed dressing against wheat bulb fly[9]. The epithio derivative of dieldrin containing a sulphur in place of the oxygen atom shows high selective toxicity to houseflies but little activity against other insects. Soloway[20] argued that the selectivity may arise from the ability of most insects, but not houseflies, to detoxify the compound by metabolism to a water-soluble sulphonium salt.

Other related cyclodienes include the cyclic sulphite esters, e.g. endosulfan (37), obtained as follows:

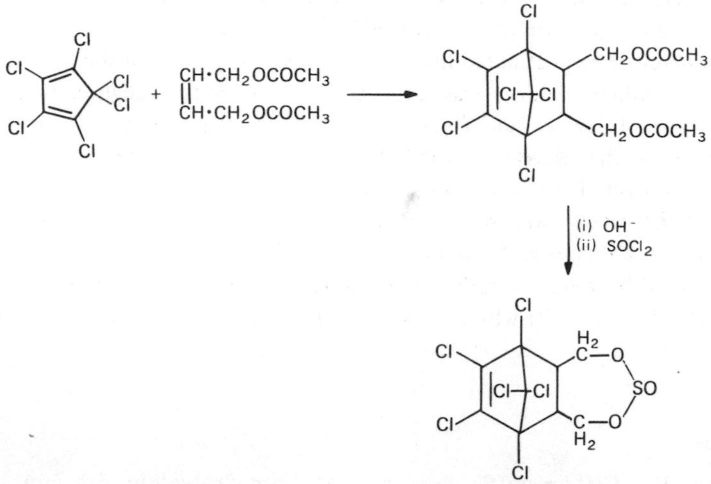

(i) OH⁻
(ii) SOCl₂

(37)

Endosulfan has a similar spectrum of insecticidal activity to aldrin, except that it is also acaricidal[2]. In common with the majority of organochlorine insecticides, the cyclodienes are very persistent lipophilic molecules which are not readily biodegraded and tend to accumulate in the environment (see Chapter 14, p. 215). The cyclodiene insecticides possess significantly higher acute mammalian toxicities as compared with DDT or lindane; thus the LD_{50} (oral) values to rats are: aldrin, dieldrin, and heptachlor 40; endosulfan 35; and endrin 4 mg/kg.

Soloway[20] concluded that active cyclodienes needed to contain two correctly separated electronegative centres which possibly become attached to the biological site of action. The overall molecular topography is also vital; the active compounds

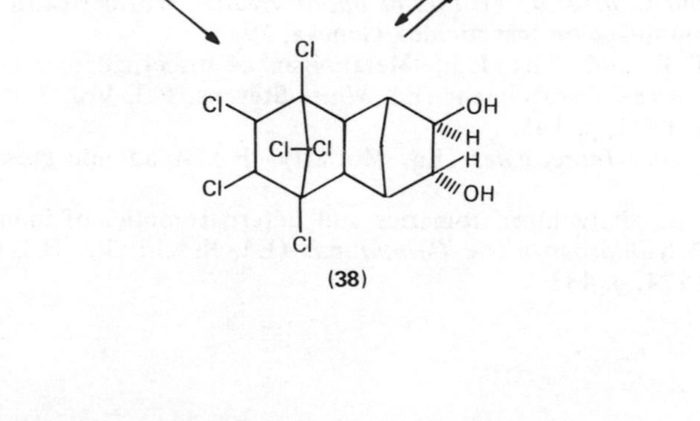

aldrin

metabolism

dieldrin

(38)

have similar molecular shapes and in the epoxides (e.g. dieldrin) the orientation of the oxygen atom appears decisive. Comparatively little is known about the *in vivo* metabolic degradation of the cyclodienes; aldrin is metabolized in insects, soil, and plants to dieldrin *via* aldrin *trans*-diol (38)[4,15,16].

It may therefore be postulated that the *trans*-diol (38) is one of the active forms of dieldrin; thus when dieldrin was applied to a cockroach ganglion there was an inhibition period before an electrophysiological response was observed. But treatment with the *trans*-diol (38) gave a much more rapid response[4]. The symptoms of cyclodiene poisoning clearly show that they act on the nervous system, but the precise site of biochemical action is unknown; thus it is not known whether the cyclodienes and lindane act primarily on the axon or the synapse[4(d)]. Cross-resistance between cyclodienes and lindane suggested that these may act on the same site (see p. 102) which appears to be different from that of DDT and its analogues.

References

1. Martin, H., *The Scientific Principles of Crop Protection*, 6th edn., Arnold, London, 1973, (a) p. 170, (b) p. 217, (c) p. 263.
2. Chemicals for Farmers and Growers, Ministry of Agriculture, Fisheries, and Food, 1978.
3. *Arsenical Pesticides* (Ed. Wollston, E. A.), A.C.S. Symposium Series, No. 7, American Chemical Society, 1975.
4. Corbett, J. R., *The Biochemical Mode of Action of Pesticides*, Academic Press, London and New York, 1974, (a) p. 10, (b) p. 33, (c) p. 27 (d) p. 169.
5. Crafts, A. S. and Robbins, W. W. *Weed Control*, 3rd ed. McGraw-Hill, New York, 1962.
6. Hartley, G. S. and West, T. F., *Chemicals for Pest Control*, Pergamon Press, London, 1969.
7. *Pesticide Manual* (Eds. Martin, H. and Worthing, C. R.), 4th edn., British Crop Protection Council, 1974.
8. Metcalf, R.L., *Organic Insecticides*, Interscience, New York, 1955.
9. Woods, A., *Pest Control*, McGraw-Hill, London, 1974.
10. Hassall, K. A., *World Crop Protection*, – *Vol. 2 Pesticides*, Iliffe Books Ltd, London, 1969. p. 71.
11. Kay, I. T., Snell, B. K., and Tomlin, C. D. S., *Chemicals for Agriculture in Basic Organic Chemistry*, Part 5, Wiley, London, 1975, p. 423.
12. Holan, G., *Bull. Wld. Hlth. Org.*, **44**, 355 (1971).
13. Brown, A. W. A. and Pal, R., *Insecticide Resistance in Anthropods*, World Health Organization, Geneva, 1971.
14. *Ecology and Control of Vectors in Public Health*, World Health Organization, Expert Committee on Insecticides, Geneva, 1975.
15. Fukuto, T. R. and Sims, J. J., 'Metabolism of insecticides and fungicides' in Pesticides in the Environment (Ed. White-Stevens, R.), Vol. 1, Part 1, Dekker, New York, 1971, p. 145.
16. *Organochlorine Insecticides* (Ed. Moriarty, F.), Academic Press, New York, 1975.
17. Green, M. B., 'Polychloroaromatics and heteroaromatics of industrial importance' in *Polychloroaromatic Compounds* (Ed. Suschitzky, H.), Plenum Press, London, 1974, p. 433.

18. Knoles, C. O., Armad, S., and Shrivastava, S. P., 'Chemistry and selectivity of acaricides' in *Insecticides* (Ed. Tahori, A. S.), Vol. 1, Gordon and Breach, New York, 1972, p. 77.
19. Watkins, T. I. and Weighton, D. M., *Repts. Progr. Appl. Chem.*, **60**, 404 (1976).
20. Soloway, S. B., *Advan. Pest Control Res.*, **6**, 85 (1965).

Chapter 6
Synthetic Insecticides II: Organophosphorus and Carbamate Compounds

The organic chemistry of phosphorus goes back to 1820 when Lassaigne first studied the reactions of alcohol with phosphoric acid[1,2]. In 1854 Clermont prepared tetraethyl pyrophosphate (TEPP) by heating the silver salt of pyrophosphoric acid with ethyl chloride although the powerful insecticidal properties of this compound were not discovered until some 80 years later.

Serious investigations into the synthesis of toxic organophosphorus compounds as potential nerve gases began during the second World War. At Cambridge Saunders and his colleagues[3] studied alkyl fluorophosphates such as tetramethylphosphoro-diamidic fluoride or dimefox (1), while in Germany Schrader made the highly active nerve gases tabun (2) and sarin (3).

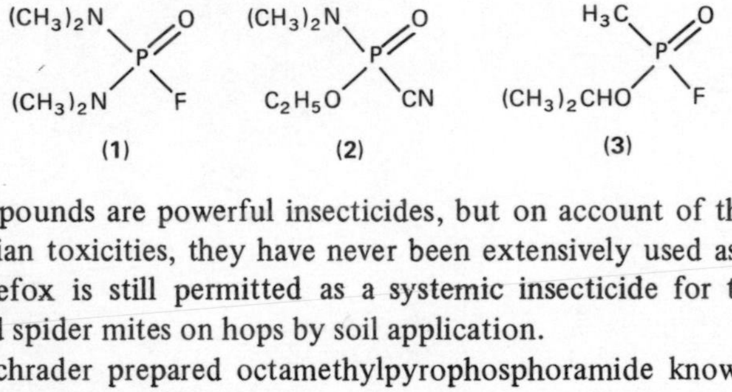

(1)   (2)   (3)

All these compounds are powerful insecticides, but on account of their extremely high mammalian toxicities, they have never been extensively used as insecticides[1]. However dimefox is still permitted as a systemic insecticide for the control of aphids and red spider mites on hops by soil application.

In 1941 Schrader prepared octamethylpyrophosphoramide known as schradan or 'Pestox' (4) by controlled hydrolysis of the phosphorochloridate (5).

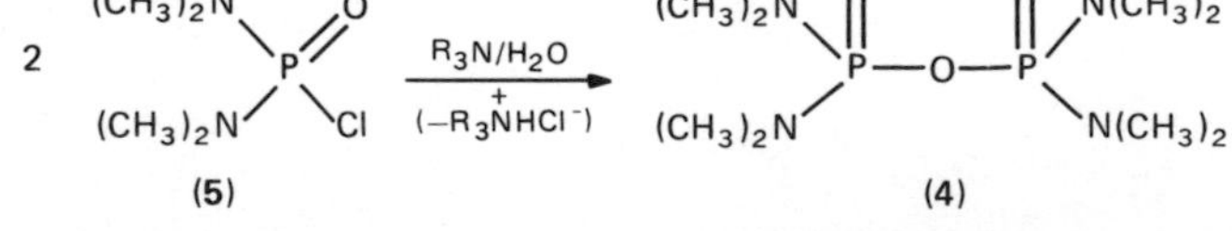

(5)   (4)

Schradan can be manufactured by a one-stage process from phosphorus oxychloride and dimethylamine without isolating the intermediate chloridate. Historically schradan was the first organophosphorus compound recognized to be a potent systemic insecticide, though dimefox (1) is also systemically active. However schradan has a high mammalian toxicity: LD_{50} (oral) to rats $\simeq 8$ mg/kg, and it has been replaced by the less toxic systox series.

Schrader[2,4] synthesized O,O-diethyl O-p-nitrophenyl phosphate or paraoxon (6; X = O) by reaction of O,O-diethyl phosphorochloridate with sodium p-nitrophenate:

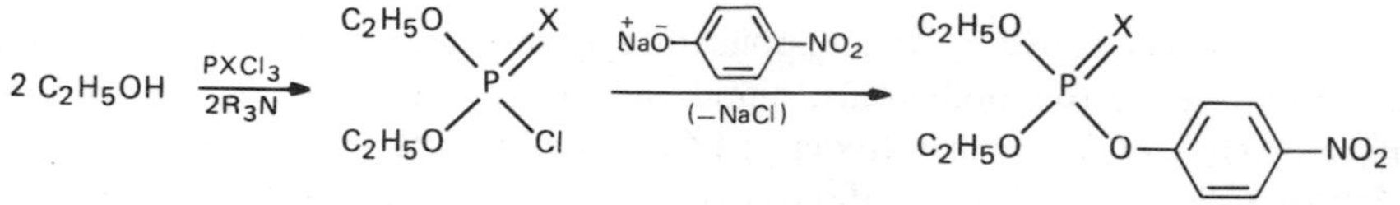

(6)

Paraoxon is a powerful insecticide with some systemic activity, moderate persistency, and very high mammalian toxicity: LD_{50} (oral) to rats $\simeq 3$ mg/kg. Another important organophosphorus insecticide was parathion (6; X = S), which can be obtained by a similar preparative route using thiophosphoryl chloride. It has a very broad spectrum of insecticidal activity, but again a high mammalian toxicity: LD_{50} (oral) to rats $\simeq 6.4$ mg/kg, and it has been largely superseded by less toxic materials. It has been discovered[1] that the introduction of a nuclear chlorine atom substantially reduced the mammalian toxicity; thus Chlorthion (7) has LD_{50} (oral) to rats of 400 mg/kg[4].

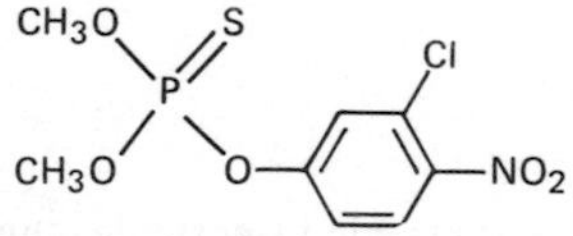

(7) 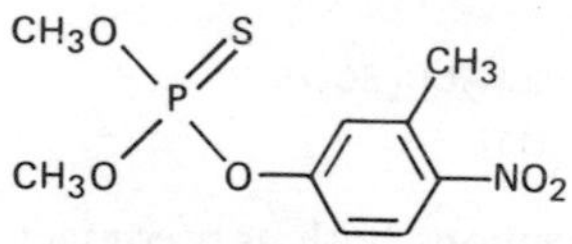   (8)

Replacement of the chlorine atom of (7) by a methyl group gives Sumithion (8) which has an even lower toxicity: LD_{50} (oral) to rats $\simeq 500$ mg/kg and a very similar spectrum of activity as a contact and stomach insecticide to that of parathion (6; X = S). Sumithion is valuable for controlling flies and mosquitoes that have developed resistance to organochlorine insecticides. A large number of phosphorothioates have found use as insecticides; one interesting example is bromophos (9), very effective against ectoparasites in livestock and as a general agricultural insecticide and acaricide:

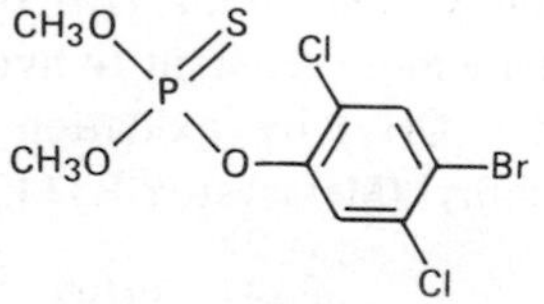

(9)

Bromophos has a remarkably low mammalian toxicity: LD_{50} (oral) to rats $\simeq 4000$ mg/kg. Profenofos was introduced by Ciba–Geigy Ltd. (1975) as a new insecticide for control of important cotton and vegetable pests:

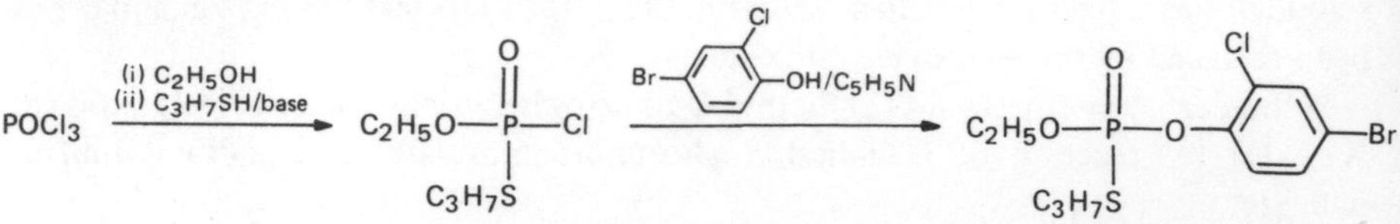

Profenofos is very active by spray application against chewing and sucking insects and mites, e.g. cotton bollworms, aphids, cabbage looper, and thrips. It has a comparatively low mammalian toxicity: LD_{50} (oral) to rats 360 mg/kg.

Systox (10; $R = C_2H_5$) and Metasystox or demeton-methyl (10; $R = CH_3$) were first synthesized by Schrader in 1950. The commercial products are mixtures; thus Metasystox contains 70% of the thionate (demeton-O-methyl, 10; $R = CH_3$) and 30% of the thiolate (dimeton-S-methyl, 11; $R = CH_3$). Metasystox is manufactured by condensation of O,O-dimethylphosphorochloridothionate and 2-ethylthio-ethanol [1,2]:

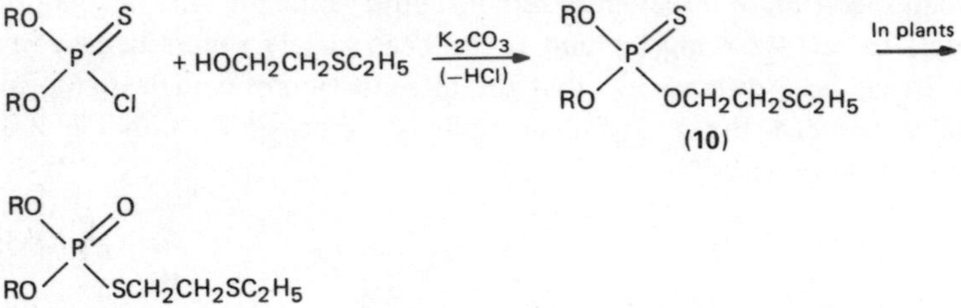

Metasystox functions both as a contact and a systemic insecticide, the LD_{50} value (oral) to rats $\simeq$ 180 mg/kg. In plants Metasystox rearranges to the thiolate form (11; $R = CH_3$) which is largely responsible for the systemic activity. The pure thiolate (demeton-S-methyl 11; $R = CH_3$) may be prepared by alkylation of the corresponding phosphate:

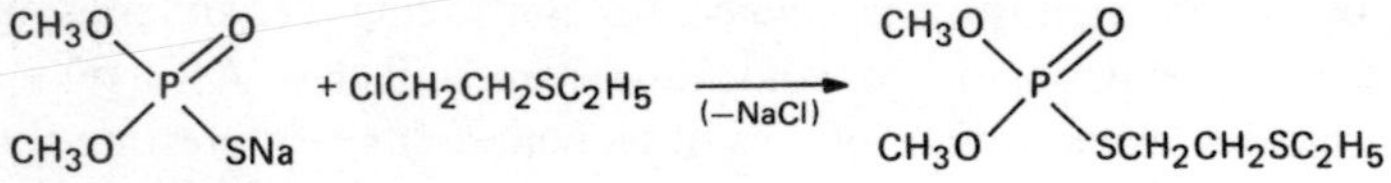

The oral toxicity (LD_{50}) of demeton-S-methyl to rats is 50 mg/kg and it has a fairly persistent action probably due to in vivo oxidation in plants to other insecticidal compounds which are more resistant to hydrolysis.

Demeton-S-methyl (11; $R = CH_3$) by oxidation with hydrogen peroxide, is converted into oxydemeton-methyl (Metasystox R) (12; $R = CH_3$):

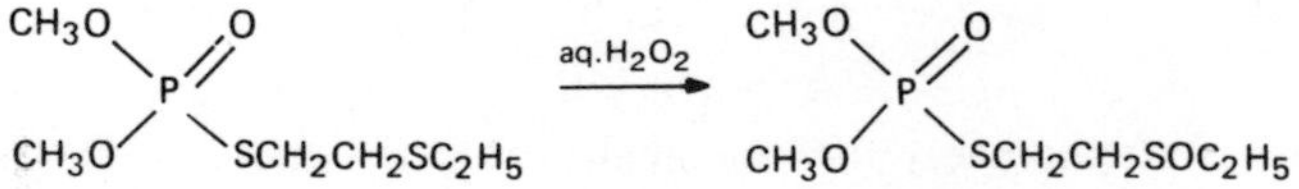

Mestaystox R has specific systemic activity against aphids, red spider mites, and leaf hoppers on most agricultural and horticultural crops. In general the systox group of organophosphorus insecticides is quite rapidly translocated in plants chiefly via the transpiration stream in the xylem.

Phorate or Thimet (O,O-diethyl-S-2-ethylthiomethylphosphorodithioate) (13) is prepared[1,2] by condensation of chloromethylethyl sulphide with sodium O,O-diethylphosphorodithioate (14):

$$4 C_2H_5OH + P_2S_5 \longrightarrow 2(C_2H_5O)_2PSSH + H_2S$$

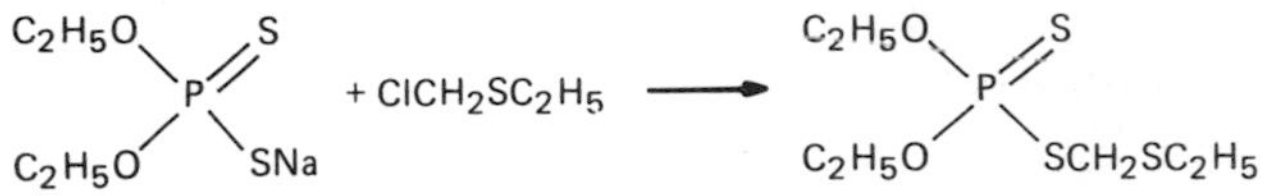

(14)   (13)

An alternative preparative route involves reaction of O,O-diethylphosphorodithioic acid with formaldehyde, followed by condensation with ethyl mercaptan:

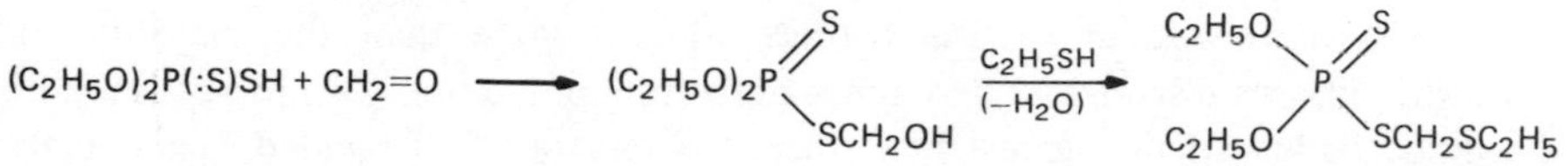

(13)

Phorate has both systemic and contact insecticidal action and is a very toxic compound: LD_{50} (oral) to rats 2 mg/kg. When phorate is absorbed and translocated in plants, it is oxidatively metabolized (see p. 88). Phorate is employed for the control of aphids, carrot fly, fruit fly, and wireworm in potato. It protects plants for a fairly long period due to the greater persistency of the sulphoxide metabolite in plants. A minimum interval of six weeks must be observed between the last application and harvesting of edible crops.

Malathion (15), introduced in 1950 by the American Cyanamid Company, is synthesized by addition of O,O-dimethylphosphorodithioic acid to diethyl maleate:

$$4 CH_3OH + P_2S_5 \longrightarrow 2(CH_3O)_2P(=S)SH + H_2S$$

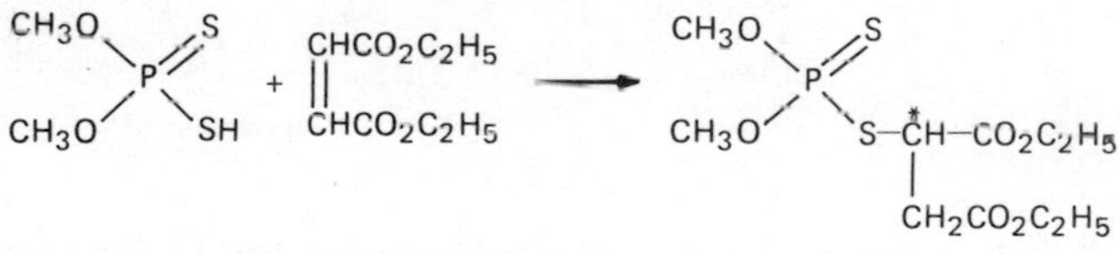

(C* = asymmetric carbon atom)   (15)

Commercially the preparation is carried out as a one-stage process in which the dithoic acid (obtained from the reaction of methanol and phosphorus pentasulphide) is added to diethyl maleate in the presence of catalytic quantities of base and hydroquinone to prevent polymerization of the maleate. Malathion (15) is an important and widely used contact insecticide and acaricide for the control of aphids, red spider mites, leaf hoppers, and thrips on a wide range of vegetable and other crops. It was important in the history of the development of organophosphorus insecticides since it was the first member with a good spectrum of insecticidal activity combined with a remarkably low mammalian toxicity: LD_{50} (oral) to rats $\simeq 1300$ mg/kg. The selectivity arises from metabolic activation in insects to the phosphoryl analogue, malaoxon, which is more toxic to both insects and mammals (LD_{50} 88 mg/kg) (see p. 85). Malaoxon is also formed by chemical oxidation of malathion (15) with nitric acid.

Dichlorvos or Vapona (2,2-dichlorovinyl dimethyl phosphate) (16) is prepared by the Perkow reaction from trimethyl phosphite and chloral:

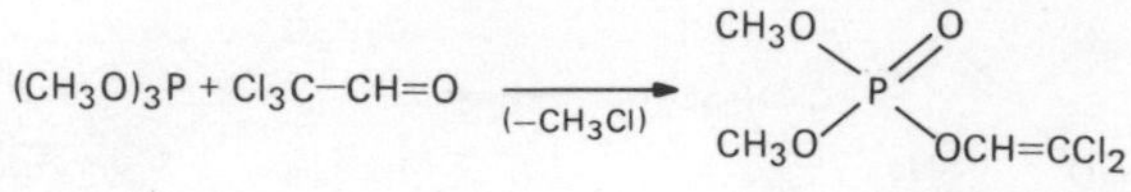

$$(CH_3O)_3P + Cl_3C-CH=O \xrightarrow[(-CH_3Cl)]{} \begin{array}{c} CH_3O \\ \diagdown \\ CH_3O \diagup \end{array} P \begin{array}{c} \diagup O \\ \diagdown OCH=CCl_2 \end{array}$$

(16)

The Perkow reaction (see p. 79) is very useful for the synthesis of vinyl phosphates, e.g. chlorfenvinphos, tetrachlorvinphos, and mevinphos.

Dichlorvos is a volatile contact and stomach insecticide and acaricide of short persistence, and is some thousand times more volatile than the majority of organophosphorus insecticides and hence has a fumigant action. Dichlorvos is used[5] as a domestic insecticide against flies since it is very rapidly degraded in mammals to non-toxic products (see p. 88). It is also useful against mosquitoes, aphids, caterpillars, thrips, and red spider mites in glasshouses. Care must be taken in its application as it has a fairly high mammalian toxicity LD_{50} (oral) to rats 80 mg/kg.

Chlorfenvinphos or Birlane may be regarded as a phenyl derivative of dichlorvos (16) and is synthesized from triethyl phosphite and 2,4-α,α-tetrachloro-acetophenone[2,5]:

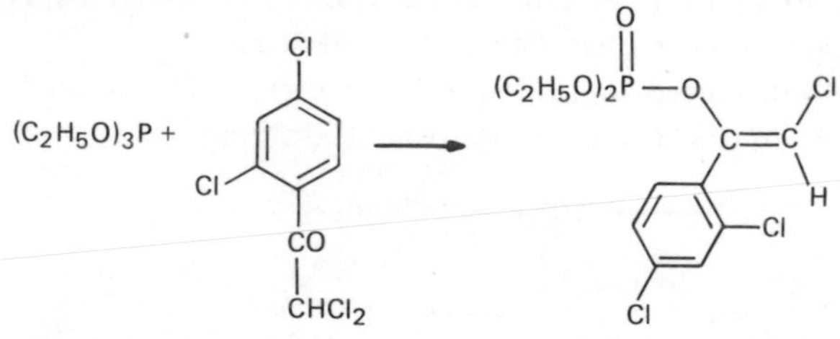

Birlane (trans isomer)

Commercial Birlane is a mixture of geometrical isomers, mainly the trans isomer as shown above containing the chlorine atoms on the opposite sides of the double bond. It is relatively stable towards hydrolysis and is useful as a soil insecticide against, for instance, cabbage root fly, carrot fly, mushroom fly, and wheat bulb fly. Birlane is specially toxic to rats: LD_{50} (oral) 15 mg/kg, but is less toxic to other mammals (LD_{50} (oral) to mice 117–200 mg/kg).

A closely related compound is tetrachlorvinphos or Gardona which was introduced by Shell Research Ltd. (1966) and is synthesized by the Perkow reaction from trimethyl phosphite and 2,4,5-α,α-pentachloroacetophenone[2,5]:

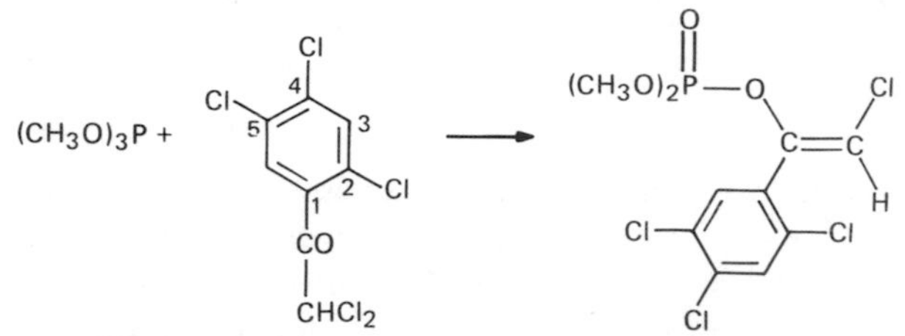

Gardona (*trans* isomer)

The crude product is a mixture of the geometric isomers; the *trans* isomer (with the chlorine atoms on opposite sides of the double bond) is obtained (98%) by recrystallization[2]. Gardona is an extremely safe insecticide: LD_{50} (oral) to rats $\simeq$ 4000 mg/kg and finds similar uses to malathion for the control of domestic and garden pests, for instance flies in dairies and livestock barns. Generally it is effective against codling and winter moths, caterpillars, flea beetles, and weevils[6,10] (Plate 6).

Another organophosphorus insecticide containing a polychloroaromatic residue is fenchlorphos or Korlan, introduced by the Dow Chemical Company (1954) as an animal systemic insecticide. Fenchlorphos can be prepared from 2,4,5-trichlorophenol and thiophosphoryl chloride as follows:

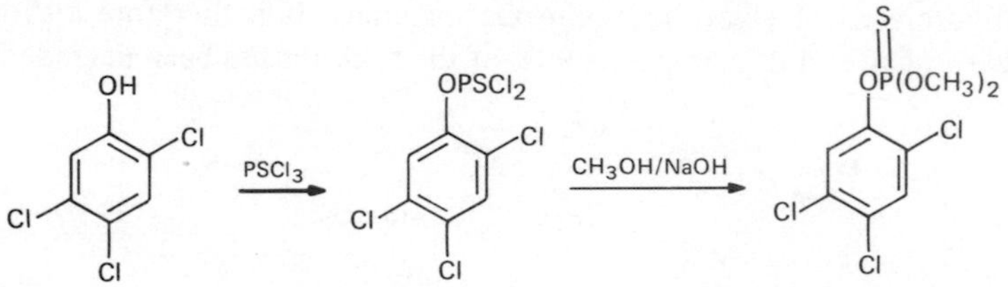

Korlan

Fenchlorphos has a very low mammalian toxicity (LD_{50} (oral) to rats 1250–1750 mg/kg) and it can be used for treatment of dairy cattle against internal and external pests, e.g. grubs, lice, and ticks, by oral application. It also has good contact insecticidal activity against flies and cockroaches. Korlan is rapidly degraded and excreted in cattle and is metabolized by P-O-phenyl and P-O-methyl bond cleavage[2]:

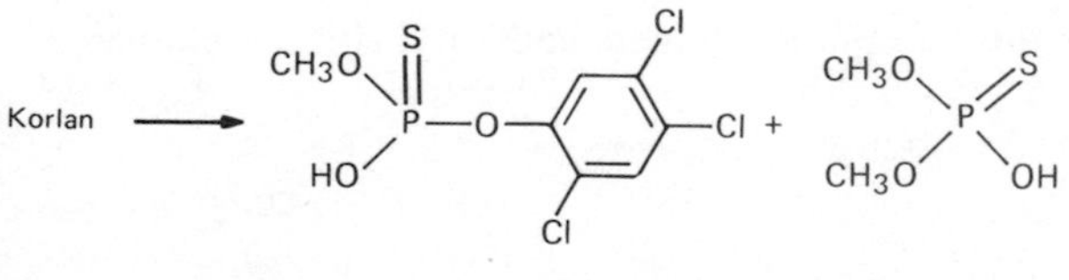

Another important member of this series is mevinphos or Phosdrin (**17**) developed by Shell Research in 1953[5], and is prepared as the *trans* isomer by condensation of *O,O*-dimethylphosphorochloridate with the sodium enolate of methyl acetoacetate:

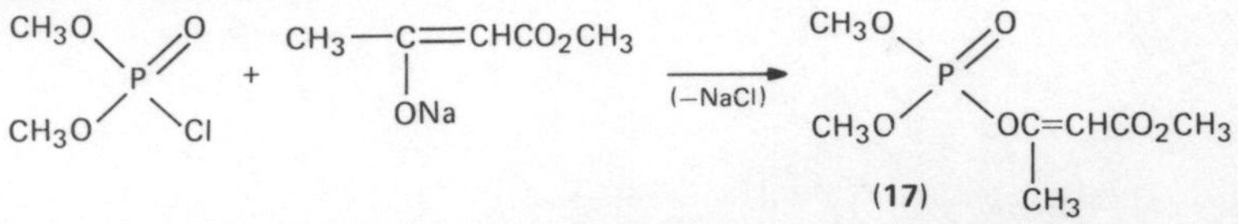

$$CH_3O \underset{CH_3O}{\overset{}{>}}P\overset{O}{\underset{Cl}{<}} \; + \; CH_3-\underset{ONa}{\overset{|}{C}}{=}CHCO_2CH_3 \xrightarrow[(-NaCl)]{} CH_3O\underset{CH_3O}{\overset{}{>}}P\overset{O}{\underset{OC=CHCO_2CH_3}{<}}$$

(17) CH_3

or by the Perkow reaction of trimethyl phosphite and methyl α-chloroaceto-acetate; the latter route has the advantage of giving mainly the *cis* isomer (see p. 79). The commercial product is a mixture of the *cis* and *trans* isomers containing some 60% of the *cis* form which is about 100 times more toxic to both insects and mammals than the *trans* form:

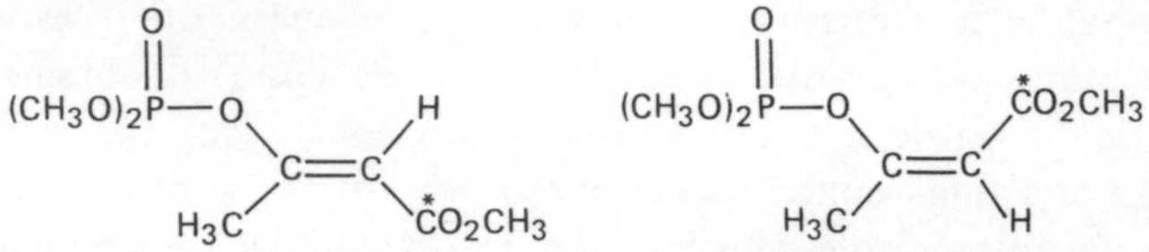

$$(CH_3O)_2\overset{O}{\overset{\|}{P}}-O \qquad\qquad (CH_3O)_2\overset{O}{\overset{\|}{P}}-O$$

cis isomer (methyl groups on same side of double bond)   *trans* isomer (methyl groups on opposite sides of the double bond)

Mevinphos (17) which superseded tetraethyl pyrophosphate (p. 68), is used as a contact and systemic insecticide and acaricide for control of aphids, caterpillars, beet leaf miners, and mites where a rapid kill of the pest is required just before harvest. This compound, although highly toxic: LD_{50} (oral) to rats 5 mg/kg, is rapidly hydrolysed in plants to non-toxic materials. It is therefore a short-acting insecticide and after 4 days in plants 90% of the toxicant has been degraded.

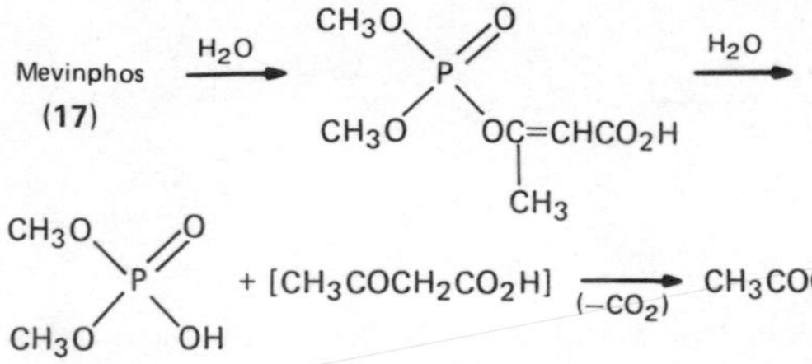

$$\text{Mevinphos} \xrightarrow{H_2O} CH_3O\underset{CH_3O}{\overset{}{>}}P\overset{O}{\underset{OC=CHCO_2H}{<}} \xrightarrow{H_2O}$$

(17) CH_3

$$CH_3O\underset{CH_3O}{\overset{}{>}}P\overset{O}{\underset{OH}{<}} + [CH_3COCH_2CO_2H] \xrightarrow[(-CO_2)]{} CH_3COCH_3$$

The Ministry of Agriculture recommends[6] that a minimum period of three days should be observed between the last application of mevinphos and the harvesting of edible crops.

Another important insecticide is *O,O*-dimethyl-*S*-methylcarbamoylmethyl phosphorodithioate known as dimethoate or Rogor (18). This can be prepared from sodium *O'O*-diethylphosphorodithioate and ethyl chloroacetate:

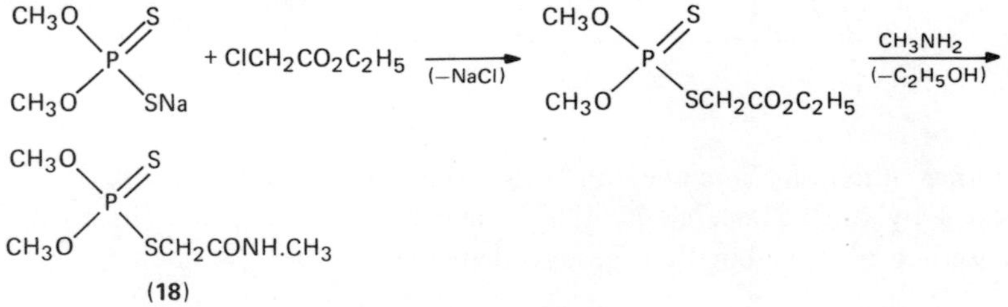

$$CH_3O\underset{CH_3O}{\overset{}{>}}P\overset{S}{\underset{SNa}{<}} + ClCH_2CO_2C_2H_5 \xrightarrow[(-NaCl)]{} CH_3O\underset{CH_3O}{\overset{}{>}}P\overset{S}{\underset{SCH_2CO_2C_2H_5}{<}} \xrightarrow[(-C_2H_5OH)]{CH_3NH_2}$$

$$CH_3O\underset{CH_3O}{\overset{}{>}}P\overset{S}{\underset{SCH_2CONH.CH_3}{<}}$$

(18)

Dimethoate (18) functions as a systemic insecticide and acaricide, effective against aphids and red spider mites and thrips on most agricultural and horticultural crops, also plum and apple sawflies, olive and wheat bulb flies[6]. Rogor has a moderate mammalian toxicity: LD_{50} (oral) to rats $\cong$ 230 mg/kg, and unlike the majority of organophosphorus insecticides dimethoate is not absorbed by the lipid phase and hence it has good residual properties. The minimum interval to be observed between the last spraying and harvesting of edible crops is one week[6]. In plants and animals dimethoate (18) is metabolized to the rather more toxic phosphoryl analogue known as O-methoate which is used for the control of aphids on hops.

An important and versatile group of organophosphorus insecticides is obtained by condensation of the alkali or ammonium salt of O,O-dimethylphosphorodithioic acid with a heterocyclic chloromethyl compound[7]. One example is afforded by menazon (19) developed by Imperial Chemical Industries Ltd in 1961[4,7,8] and prepared by reaction of sodium O,O-dimethylphosphorodithioate with 2-chloro-methyl-4,6-diamino-1,3,5-triazine:

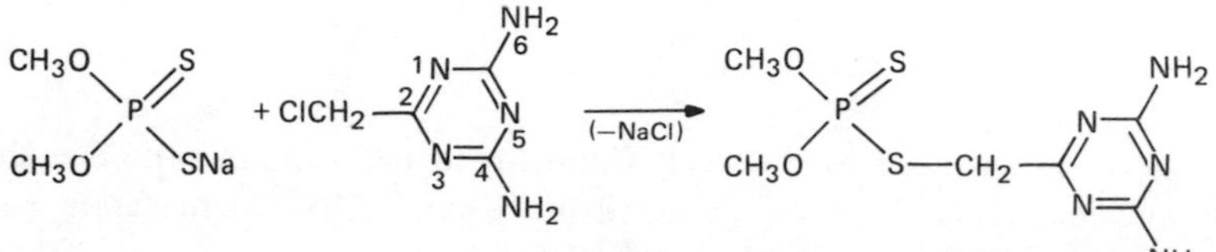

Menazon (19) is a selective systemic aphicide and acaricide effective by spray or root application for control of aphids on a wide range of crops. Menazon is a remarkably safe insecticide: LD_{50} (oral) to rats $\cong$ 1900 mg/kg, and is a poor cholinesterase inhibitor *in vitro* but is selectively activated *in vivo* in aphids. Other examples include phosphorylated derivatives of the heterocycles benzotriazine, pyrimidine, pyridine, coumarin, and quinazoline. One of the most active members is azinphos-methyl or Gusathion (20), obtained from N-chloromethylbenzazimide (21):

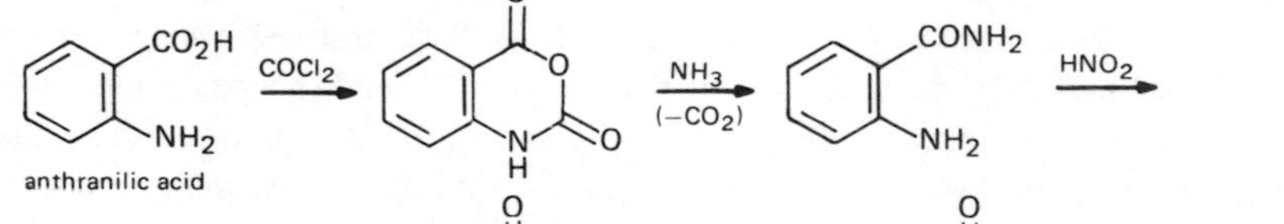

N-Chloromethylbenzazimide (21) is obtained from anthranilic acid:

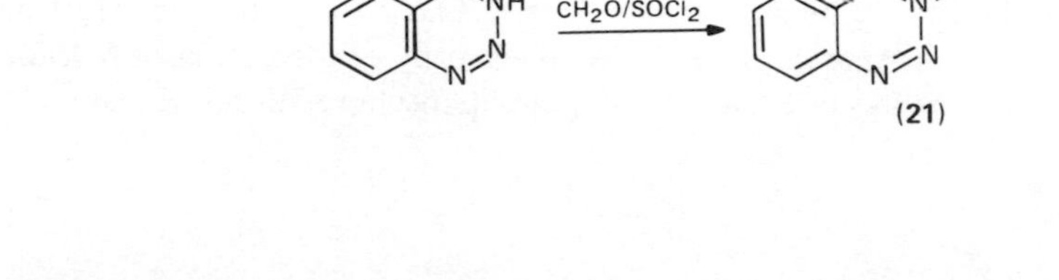

Gusathion (20), discovered by Baeyer A-G (1953), is a contact and stomach insecticide of high mammalian toxicity: $LD_{50} \cong 15$ mg/kg, with greater residual activity than the majority of organophosphorus insecticides. It is valuable for control of codling, tortrix and winter moths, fruit flies, weevils, caterpillars, and aphids[6]. In insects gusathion is metabolized to the more toxic phosphoryl analogue. Azinphos-ethyl is also used as an insecticide and is more effective against red spider mites.

Miral, introduced by Ciba–Geigy (1974), is also a triazine derivative and acts as a broad-spectrum soil insecticide which is harmless to earthworms. Miral should be a valuable substitute for the persistent organochlorine soil insecticides, like aldrin, which are being phased out of use. It is also active against sucking and chewing pests, and is a valuable systemic nematicide (see Chapter 11, p. 189).

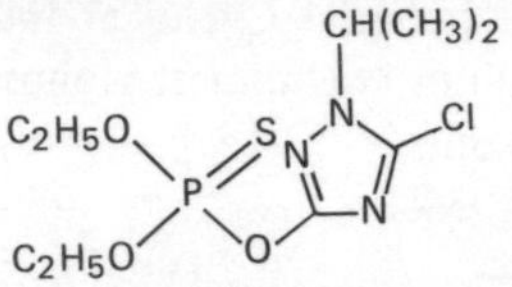

Miral

Diazinon (22) was introduced by the Geigy Company in 1952 and incorporates the pyrimidine nucleus; it is obtained by condensation of ethyl acetoacetate and isobutyramidine[2,9]. Isobutyramidine is obtained as follows:

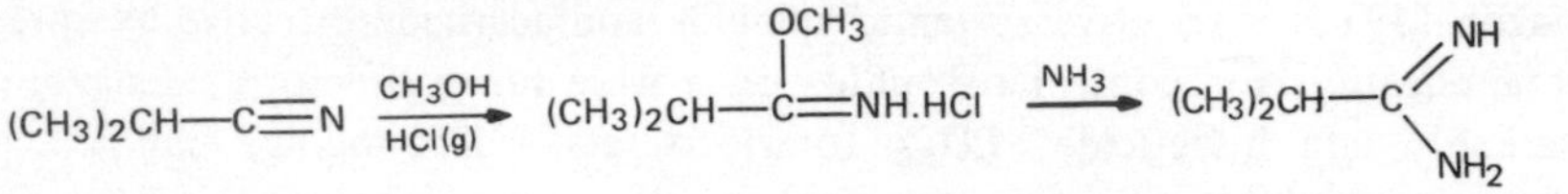

isobutyramidine

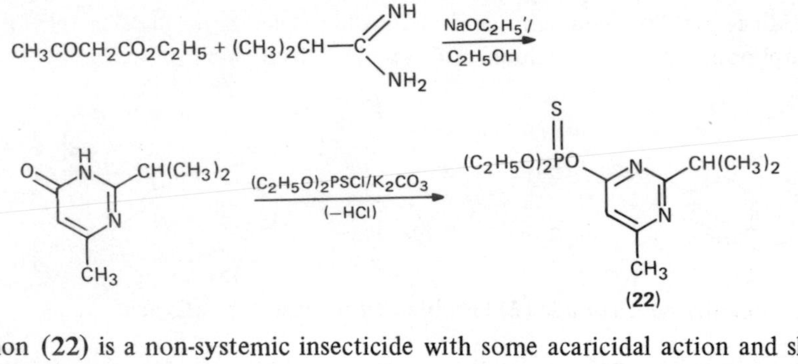

(22)

Diazinon (22) is a non-systemic insecticide with some acaricidal action and shows fairly good residual activity; also it is sufficiently volatile to be active against flies[7]. It is effective against a number of soil, fruit, vegetable, and rice pests, e.g. cabbage root, carrot and mushroom flies, aphids, spider mites, thrips and scale insects together with domestic and livestock pests[2,5,6,10]. A minimum period of two weeks is needed between the last treatment and harvesting of edible crops[6]. Diazinon has fairly low mammalian toxicity[2]: LD_{50} (oral) to rats $\cong 150$ mg/kg. Pirimphos-methyl or Actellic is similarly prepared by condensation of N,N-diethyl-guanidine, ethyl acetoacetate, and O,O-dimethylphosphorochloridothioate:

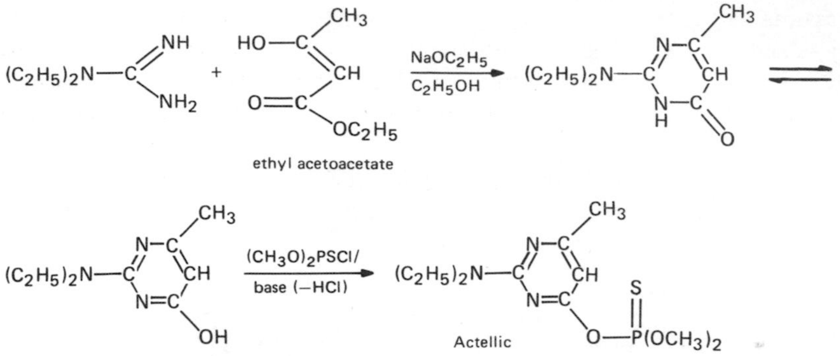

Actellic was introduced by Imperial Chemical Industries in 1970 as a broad-spectrum insecticide effective against pests in stored products, e.g. beetles, weevils, mites, and moths, and against insects affecting public health, e.g. flies, cockroaches mosquitoes, lice, bed bugs and fleas. It has very low mammalian toxicity: LD_{50} (oral) to rats 2000 mg/kg. Actellic will control insects which have become resistant to organochlorine insecticides and malathion; it is fast acting and has both contact and fumigant action.

Chloropyrifos or Dursban (23) was discovered by Dow Chemical Company in 1965[2,5] and contains the pyridine nucleus; it is prepared by reaction of *O,O*-diethylphosphorochloridothioate with 3,5,6-trichloro-2-hydroxypyridine[5]:

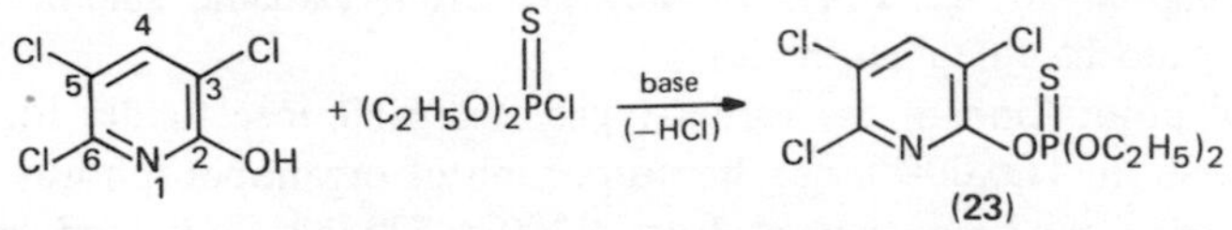

Dursban (23) is a non-systemic insecticide with a wide spectrum of activity, by contact, ingestion, and vapour action[5,6]. It is moderately persistent and retains its activity in soil for 2–4 months and is valuable against mosquito and fly larvae, cabbage root fly, aphids, and codling and winter moths on fruit trees. Chloropyrifos has become one of the most widely applied insecticides in homes and restaurants against cockroaches and other domestic pests[10]. It is a comparatively safe insecticide, the mammalian toxicity LD_{50} (oral) to rats $\simeq$160 mg/kg and the chemical is rapidly detoxified in animals.

An example of an insecticide containing the coumarin nucleus was Potasan, synthesized by Schrader (1947) from ethyl acetoacetate and resorcinol:

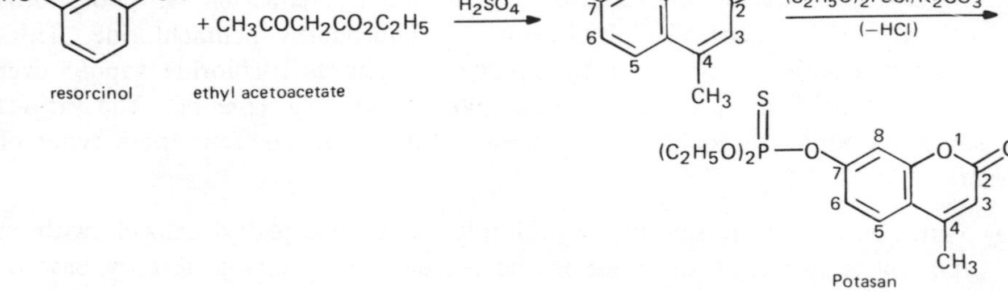

Potasan is particularly effective as a contact insecticide against Colorado beetles on potato and has fairly high mammalian toxicity: LD_{50} (oral) to rats $\simeq 30$ mg/kg. However the 6-coumarinyl isomer is almost non-toxic to both insects and mammals, while the 3-chloro derivative has low mammalian toxicity and is useful for control of animal parasites[1,2].

Quinalphos or Bayrusil, discovered by Bayer A-G (1965) is obtained by reaction of *o*-phenylenediamine, chloroacetic acid, and *O,O*-diethylphosphoro-chloridothioate:

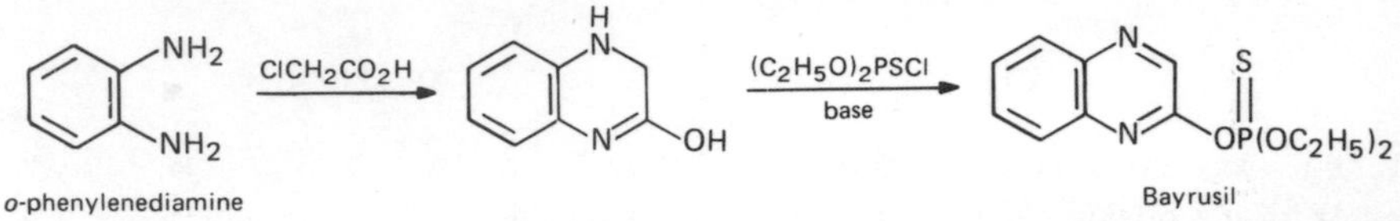

It is a contact, non-persistent, insecticide and acaricide effective against caterpillars and other vegetable pests and for control of mosquitoes and mites. Bayrusil is degraded within plants in a few days. The mammalian toxicity LD_{50} (oral) to rats is 66 mg/kg. More than one hundred organophosphorus compounds have been used as pesticides[2]. They are generally considerably more toxic to mammals than organochlorine insecticides, but they have the important advantage of being relatively non-persistent owing to the ease of their biodegradation (cf. organo-chlorines Chapter 14, p. 214), so they are often valuable substitutes for the persistent organochlorine insecticides.

The total production of the earlier types of organic insecticides for the period 1951–61 averaged 160,000 tons; the proportion of organophosphorus insecticides increased during the same period from 3200 to 22,000 tons, and in 1966 was 54,500 tons of which about half the total was parathion-methyl and parathion[11].

Elementary phosphorus is obtained by reduction of phosphate minerals by fusion with carbon and silica in an electric furnace:

$$Ca_3(PO_4)_2 + 3\,SiO_2 + 5\,C \xrightarrow{1400°C} 3\,CaSiO_3 + 5\,CO + 2\,P$$

Key phosphorus compounds for synthesis of organophosphates are obtained as follows. Phosphorus pentasulphide by heating a mixture of phosphorus and sulphur at 350°C; phosphorus trichloride by passing chlorine over molten phosphorus and subsequent distillation from the residual solid phosphorus pentachloride; phosphoryl chloride or phosphorus oxychloride by oxidation of phosphorus trichloride or by controlled hydrolysis of phosphorus pentachloride. Thio-phosphoryl chloride is prepared by passing phosphorus trichloride vapour over molten sulphur in the presence of a suitable catalyst, e.g. charcoal. The various organophosphorus insecticides are generally synthesized *via* four main types of reaction[12]:

(a) condensation of phosphorus oxychloride, or thiophosphoryl chloride with an alcohol, thioalcohol, or amine in the presence of a suitable tertiary base or other acid-binding agent such as sodium carbonate (e.g. synthesis of parathion, p. 69):

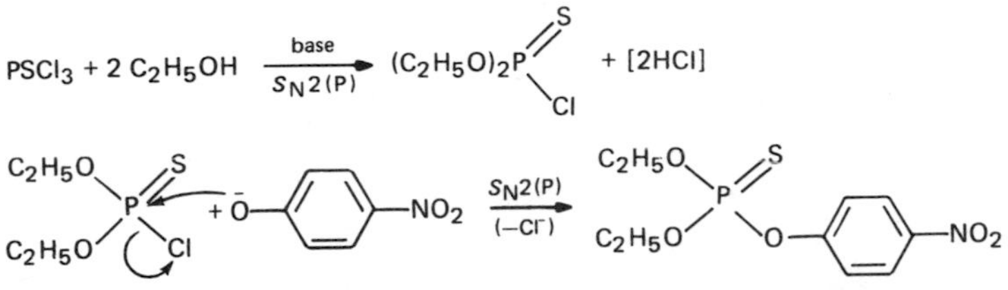

$$PSCl_3 + 2 C_2H_5OH \xrightarrow[S_N2(P)]{base} (C_2H_5O)_2P\overset{S}{\underset{Cl}{\diagup}} + [2HCl]$$

parathion

Such reactions are all essentially bimolecular nucleophilic substitutions at an electrophilic phosphorus centre and are designated $S_N2(P)$;

(b) reaction of the appropriate alcohol with phosphorus pentasulphide gives the corresponding O,O-dialkylphosphorodithoic acid:

$$P_2S_5 + 4 ROH \xrightarrow{warm} 2 (RO)_2P(=S)SH + H_2S$$

Examples are the synthesis of phorate (p. 71) and malathon (p. 71);

(c) pyrophosphates are prepared by partial hydrolysis of the appropriate phosphorochloridate in the presence of a small amount of water (half a molar equivalent) and a tertiary amine:

$$2(RO)_2P(=O)Cl \xrightarrow[(-HCl)]{H_2O/R_3N} (RO)_2\overset{O}{\overset{\|}{P}}-O-\overset{O}{\overset{\|}{P}}(OR)_2$$

An example is the synthesis of schradan (p. 68). Pyrophosphates may also be obtained by reaction of the phosphorochloridate and the sodium salt of a phosphate:

$$(RO)_2P(=O)Cl + (RO)_2P(=O)ONa \xrightarrow{(-NaCl)} (RO)_2\overset{O}{\overset{\|}{P}}-O-\overset{O}{\overset{\|}{P}}(OR)_2$$

(d) phosphites can be obtained by reaction of phosphorus trichloride with alcohols in presence of a tertiary amine:

$$3 ROH + PCl_3 \xrightarrow{3R_3N} (RO)_3P + [3 HCl]$$

An example of the use of phosphites in pesticide synthesis is provided by dichlorvos (p. 72). This is obtained from trimethyl phosphite and chloral by the Perkow reaction believed to involve attack of the nucleophilic $P^{(III)}$ atom on the electrophilic carbonyl carbon atom followed by rearrangement – the final stage being an Arbusov reaction[4]:

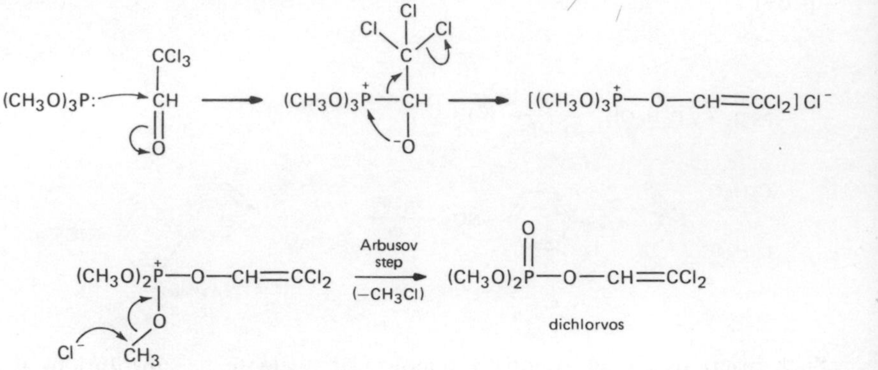

dichlorvos

A wide variety of $P^{(III)}$ compounds will react with α-halogenocarbonyl compounds to give vinyl phosphates.

Mode of Action of Organophosphorus Insecticides

The insecticidal organophosphorus compounds apparently inhibit the action of several enzymes, but the major action *in vivo* is against the enzyme acetylcholinesterase[2,13-15]. This controls the hydrolysis of the acetylcholine (24), generated at nerve junctions, into choline (25). In the absence of effective acetylcholinesterase, the liberated acetylcholine accumulates and prevents the smooth transmission of nervous impulses across the synaptic gap at nerve junctions. This causes loss of muscular coordination, convulsions, and ultimately death (see Chapter 3, p. 34). Acetylcholinesterase is an essential component of the nervous systems of both insects and mammals so the basic mechanism of toxic action of the organo-

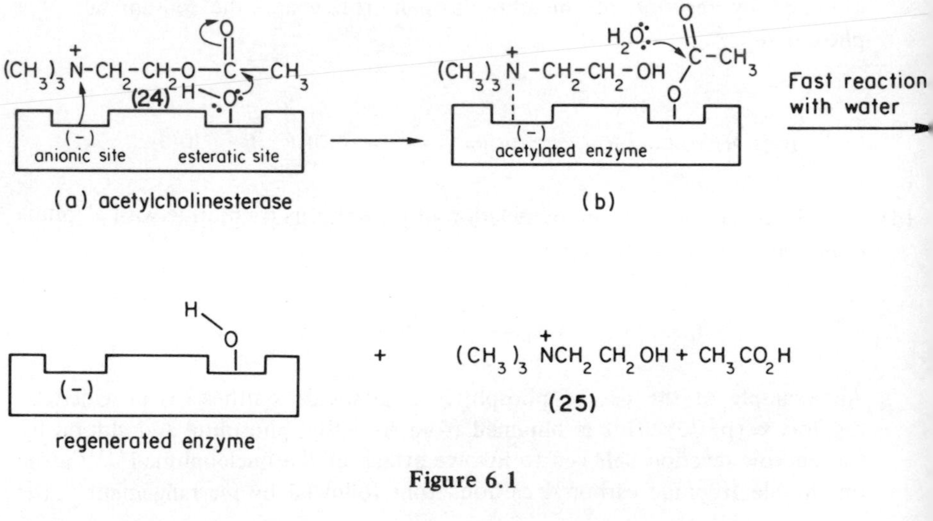

Figure 6.1

phosphorus compounds is considered to be essentially the same in insects and mammals. The active centre of the enzyme acetylcholinesterase contains two main reactive sites: an 'anionic site' which is negatively charged and binds onto the cationic part of the substrate (acetylcholine, **24**), and the 'esteratic site' containing the primary alcoholic group of the amino acid serine which attacks the electrophilic carbonyl carbon atom of the substrate. The normal enzymic hydrolysis of acetylcholine (**24**) to choline (**25**) may therefore be illustrated as shown (Figure 6.1).

Figure 6.1(a) depicts the formation of the initial enzyme—substrate complex by the orientation of the active centres of acetylcholinesterase to the substrate (acetylcholine). Figure 6.1(b) shows formation of the acetylated enzyme, which is subsequently rapidly hydrolysed to choline (**25**) and acetic acid leaving the enzyme with both its active sites intact, so permitting it to repeat the enzymic hydrolytic process on further substrate molecules releasing several thousand choline molecules per second[7].

The majority of active organophosphorus compounds have the general structure (**26**):

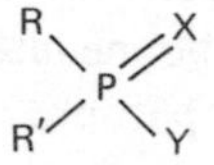

$$R,R' \diagdown \underset{R'}{\overset{R}{\diagup}} P \overset{X}{\underset{Y}{\diagdown}}$$

(26)

where R,R′ are generally lower alkyl, alkoxy, alkylthio, or substituted amino groups; X is oxygen or sulphur; and Y is good leaving group or one capable of being metabolized into such a group.

The organophosphorus compound (**26**) mimics the natural substrate acetylcholine (**24**) by binding itself to the esteratic site of acetylcholinesterase (Figure 6.1). The subsequent reaction between the enzyme (ECH_2OH) and the organophosphorus compound is of the bimolecular S_N2 type and mirrors the normal three-stage reaction between the enzyme and acetylcholine:

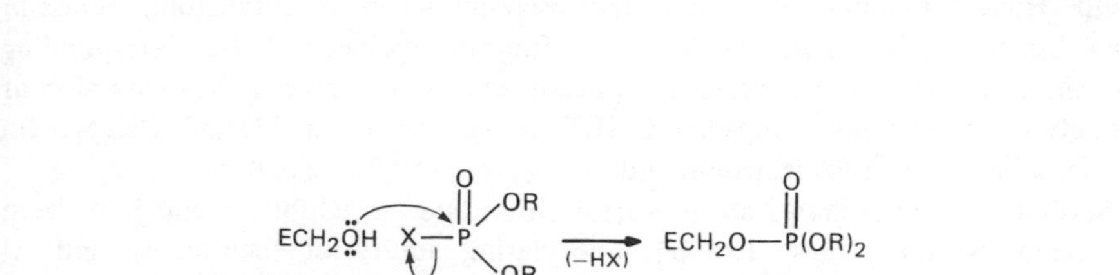

$$\underset{\text{enzyme}}{(RO)_2\overset{O}{\overset{\|}{P}}\!-\!X + ECH_2OH} \underset{}{\overset{(1)}{\rightleftharpoons}} (RO)_2\overset{O}{\overset{\|}{P}}X\cdot ECH_2OH \xrightarrow[(-HX)]{(2)} \underset{(27)}{(RO)_2\overset{O}{\overset{\|}{P}}OCH_2E} \xrightarrow[\text{slow}]{(3)\ H_2O}$$

$$(RO)_2\overset{O}{\overset{\|}{P}}OH + ECH_2OH$$

$$ECH_2\overset{..}{\overset{.}{O}}H \quad X\!-\!\overset{O}{\overset{\|}{P}}\overset{OR}{\underset{OR}{\diagdown}} \xrightarrow[(-HX)]{} ECH_2O\!-\!\overset{O}{\overset{\|}{P}}(OR)_2$$

(27)

Initially a complex is formed between the enzyme and the phosphate (step 1) which subsequently gives the phosphorylated enzyme (27) (step 2) and the latter is slowly hydrolysed to the free enzyme (step 3). In contrast, the normal reaction between acetylcholinesterase and acetylcholine (24) is shown below:

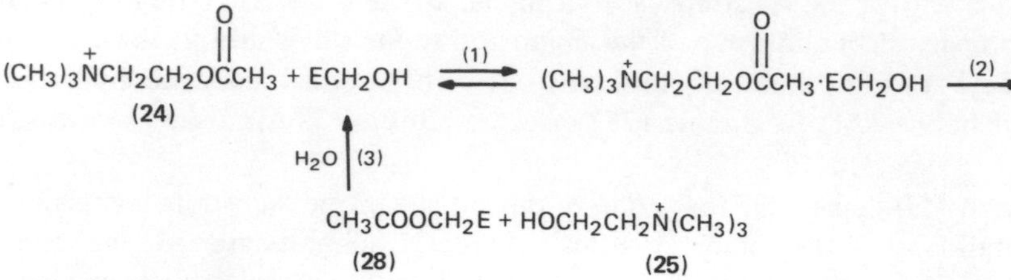

$$(CH_3)_3\overset{+}{N}CH_2CH_2O\overset{O}{\overset{\|}{C}}CH_3 + ECH_2OH \underset{}{\overset{(1)}{\rightleftharpoons}} (CH_3)_3\overset{+}{N}CH_2CH_2O\overset{O}{\overset{\|}{C}}CH_3 \cdot ECH_2OH \overset{(2)}{\longrightarrow}$$

(24)

$$H_2O \uparrow (3)$$

$$CH_3COOCH_2E + HOCH_2CH_2\overset{+}{N}(CH_3)_3$$

(28) (25)

The acetylated enzyme (28) is very rapidly hydrolysed by water (step 3) so that the active enzyme is quickly regenerated enabling the hydrolysis of acetylcholine to choline to be effectively catalysed by the enzyme. On the other hand, when an organophosphate is present the inactive phosphorylated enzyme (27) is only very slowly hydrolysed to the active enzyme because the phosphorus—oxygen bond is much stronger than the carbon—oxygen bond of the acetylated enzyme (28). So the organophosphate effectively poisons the enzyme by phosphorylation and thus blocks efficient hydrolysis of acetylcholine into choline[4,7,15].

Although the majority of organophosphorus compounds active as phosphorylating agents show some insecticidal activity, it is not possible to predict how effective they will be in practice, so it is still necessary to synthesize a wide range of analogous compounds for biological screening in order to identify the optimum insecticide by trial and error. Sometimes closely related compounds display markedly different types and degrees of activity; thus specific 2,4-dichlorophenyl- and 2,4,5-trichlorophenylphosphorothioates function as nematicides, acaricides, or soil insecticides[1]. The introduction of a chlorine atom into the 3-position of potasan (see p. 77) increases the toxicity towards mosquito larvae 10,000 times[1]. Similarly the 3-methyl and chloro derivatives of parathion (p. 69) have substantially reduced mammalian toxicities. It would not be possible to predict these effects since the substituents do not greatly alter the physicochemical properties[7].

Generally there is evidence[2] that compounds with m-substituents are much more effective as insecticides than would be predicted from consideration of electronic effects, and this may be a reflection of the steric properties of m-substituents. Many commercial organophosphorus insecticides contain the thiophosphoryl (P=S) group. These compounds are usually very weak inhibitors of acetylcholinesterase *in vitro*, but are activated *in vivo* by mixed-function oxidases to the corresponding phosphates (P=O); another type of important *in vivo* oxidation is the conversion of strongly electron donor groups like C_2H_5S (in e.g. phorate, p. 71) and $(CH_3)_2N$ (in e.g. schradan, p. 68) into electron-withdrawing groups[2,16] (see p. 84).

Several enol phosphates are powerful cholinesterase inhibitors and have been developed as insecticides. The phosphorylating ability of such compounds is enhanced by protonation and may be due to interaction of the carboxylic group with the esteratic site of the enzyme (see Figure 6.1, p. 80). Such interaction would probably be sensitive to steric factors — thus *cis*-mevinphos (p. 74) is substantially more active as an anticholinesterase agent than the *trans* isomer, probably because interaction between the carboxylic group and the enzyme would be sterically hindered in the *trans* isomer by the presence of the *cis*-diethyl-phosphoryl moiety[2]. The high activity of the *cis* isomer is therefore explicable in terms of obtaining a good fit of the toxicant molecule to the anionic and esteratic sites of the enzyme (see p. 80). The distance between the active sites of insect acetylcholinesterase is 4.5–5.9 Å and may be related to the distance between the phosphorus atom and the carbonyl carbon atom in the geometric isomers of mevinphos as follows: *cis* isomer P–C* distance = 4.3–5.2 Å (good fit) whereas for the *trans* isomer the corresponding distance is 2.2–4.4 Å (bad fit) (see p. 74).

Generally it is found[1,2] that phosphorylating ability is reduced by the presence of bulky groups attached to phosphorus and for this reason most insecticides contain lower alkoxy groups. Another example of the importance of obtaining a good fit of the toxicant molecule to the anionic and esteratic sites of acetylcholinesterase is provided by the insecticide amiton (29) prepared from O,O-diethylphosphorochloridothioate and sodium 2-diethylaminoethoxide by a route involving the thiono—thiolo rearrangement[2]:

$$(C_2H_5O)_2PSCl + NaOCH_2CH_2N(C_2H_5)_2 \xrightarrow{(-NaCl)}$$

$$\underset{\displaystyle (C_2H_5O)_2\overset{\displaystyle S}{\overset{\|}{P}}-OCH_2CH_2N(C_2H_5)_2}{} \xrightarrow{70-80^\circ} \underset{\displaystyle (C_2H_5O)_2\overset{\displaystyle O}{\overset{\|}{P}}-SCH_2CH_2N(C_2H_5)_2}{}$$

$$(29)$$

Amiton (29) is a water-soluble, persistent systemic insecticide against aphids, scale insects, and mites[2], but suffers from a very high mammalian toxicity: LD_{50} (oral) to rats $\simeq$ 3 mg/kg. At physiological pH values amiton is ionized and such ionic compounds tend to be selectively toxic to mammals because their ionic character appears to restrict their entry through the sheath protecting the insect nervous system, whereas this ion-impermeable barrier is absent in mammals[2,7]. Amiton (29) is a phosphoryl analogue of acetylcholine and the high anticholinesterase activity of the compound arises from the excellent affinity of the quaternary nitrogen atom for the anionic site of the enzyme (see p. 80).

Asymmetry in an insecticidal molecule may also influence its activity and this is to be expected since enzymes are themselves asymmetric molecules. The valuable properties of malathion (15) (p. 71) stimulated extensive investigation of analogous compounds, and it was discovered that the asymmetry in the succinate group due to the presence of the chiral carbon atom (C*) was associated with differences in insecticidal potency: *dextro*-malathion was appreciably more toxic to both houseflies and mice as compared with the *laevo* isomer. It was also a more active inhibitor of acetylcholinesterase and liver carboxyesterase[2,7].

Studies of the metabolism of selected types of organophosphorus insecticides in plants, animals, and insects have proved invaluable in elucidating the mode of action, predicting probable metabolites for related organophosphates, and accounting for selective toxicity. Many organophosphorus insecticides show little *in vitro* anticholinesterase activity and the *in vivo* activity is the result of net metabolic activation. Examples of activation processes are enzymic oxidations of phosphoro-thioates, like parathion diazinon, and malathion, involving conversion of P=S to P=O, of sulphides such as phorate, involving the transformation: $\geq\!S \rightarrow \geq\!SO \rightarrow \geq\!SO_2$, and of amides, such as schradan, to the N-oxide or methylol:

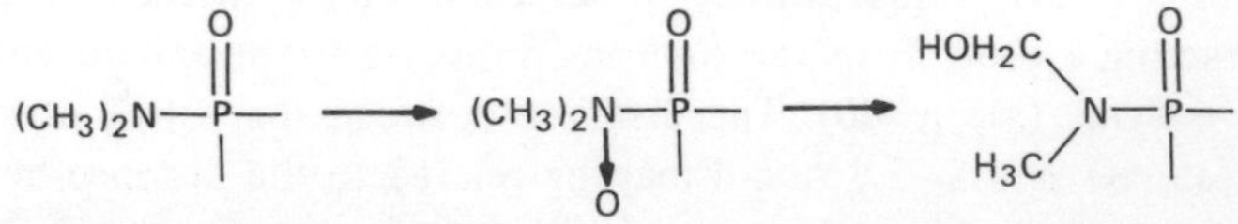

Such *in vivo* activation processes occur principally in the insect gut and fat body tissues and in the mammalian liver[16,17]. The effectiveness of a given insecticide to a specific insect will depend on the balance of biochemical activation and detoxification processes occurring in the insect species. It is important to realize that the metabolic pathways will often differ in different organisms, e.g. insects, mites, soil microorganisms, plants, or vertebrates.

Enzymes known as mixed-function oxidases (MFO) occur in animal, fish, and insect cell microsomes and, in the presence of molecular oxygen and reduced nicotinamide adenine dinucleotide phosphate (NADPH) or reduced nicotinamide adenine dinucleotide (NADH), they will oxidize a variety of lipophilic substrates such as steroids, lipids, and foreign organic compounds[1,2]. One of the oxygen atoms is incorporated into the substrate (RH) while the other is reduced to water:

$$RH + NADPH + H^+ + O_2 \xrightarrow{\text{MFO}} ROH + NADP^+ + H_2O$$

These enzymes occur in the vertebrate liver and fat body. They are not specific and only require a substrate of high lipophilicity. However, the microsomal hydroxy-lation of foreign compounds requires the presence of a special microsomal pigment, cytochrome *P*-450, while liver microsomes have a different electron transport system requiring cytochrome b_5[2]. Examples of processes effected by MFO are the following:

(a) hydroxylation: $RH \longrightarrow ROH$

(b) *O*- and *N*-dealkylation:

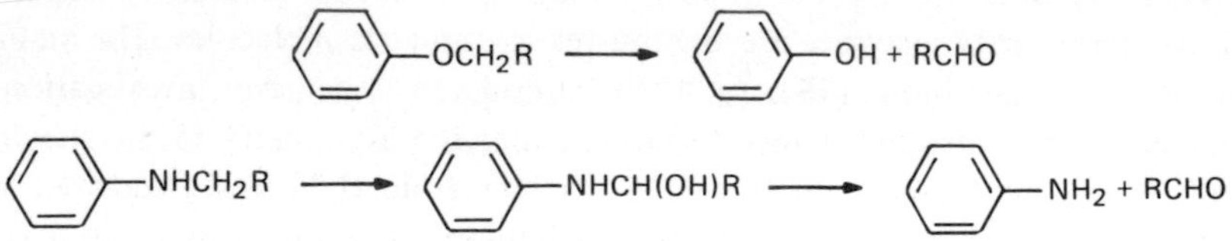

(c) oxidative desulphuration: $\geq\!P\!=\!S \longrightarrow \geq\!P\!=\!O; \quad \geq\!C\!=\!S \longrightarrow \geq\!C\!=\!O$

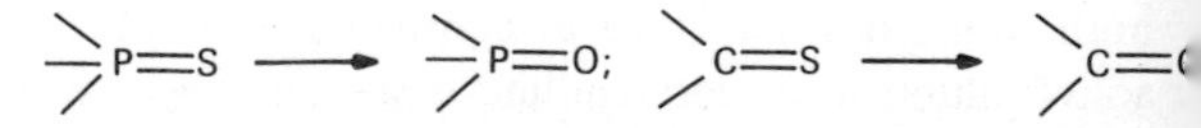

(d) oxidation of sulphides: $RR'S \longrightarrow RR'SO \longrightarrow RR'SO_2$

(e) deesterification:

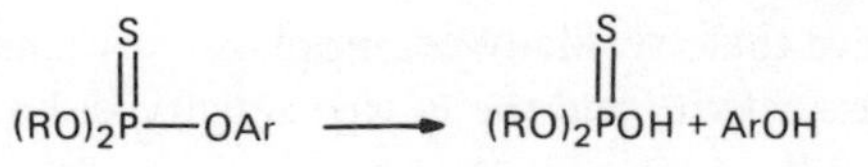

(f) epoxidation:

(g) oxidation of tertiary amines:

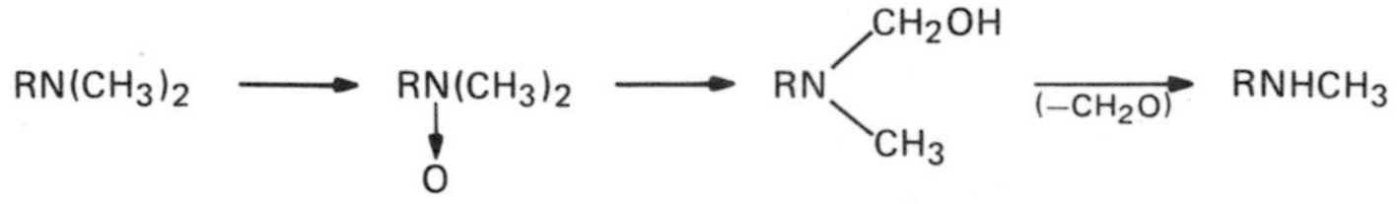

The following are illustrative examples of the metabolism of some important organophosphorus insecticides. The metabolism of parathion (see p. 69) is shown in Scheme 6.1.

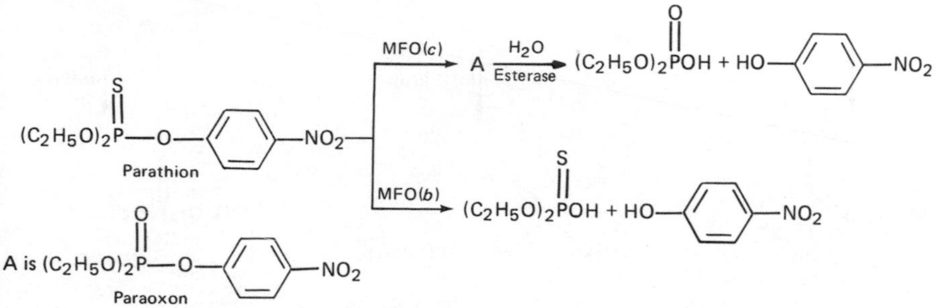

Scheme 6.1

Phosphorothioates like parathion are poor inhibitors of acetylcholinesterase, whereas their oxo analogues are very active anticholinesterase agents. The toxicity of such compounds as parathion is therefore due to their *in vivo* oxidative desulphuration (Scheme 6.1, path *c*) by microsomal MFO in animals and insects; in plants the oxidation is probably achieved by peroxidases[2,16,18]. Paraoxon is subsequently deactivated by esterase-catalysed hydrolysis, and microsomal MFO can also effect some detoxification of parathion by dearylation (Scheme 6.1, path *b*). In some cases organophosphorus compounds may show differential toxicity as a result of different metabolism in mammals and insects. One of the first examples was malathion which, although an effective contact insecticide, showed remarkably low mammalian toxicity (p. 71). Malathion is rapidly activated to malaoxon by oxidative desulphuration in insects and mammals[17,18]. Malaoxon is a highly potent anticholinesterase agent which is very toxic to insects and mammals. Malathion and its toxic metabolite, malaoxon, are detoxified by carboxyesterases which hydrolyse the carboethoxy moiety leading to polar, water-soluble, compounds that are excreted (Scheme 6.2). Vertebrates show a greater carboxyesterase activity, as compared with insects, so that the toxic agent malaoxon builds up more in insects than in mammals accounting for the selective toxicity of malathion (15) towards insects. The major metabolic pathways for malathion (15) are shown in Scheme 6.2.

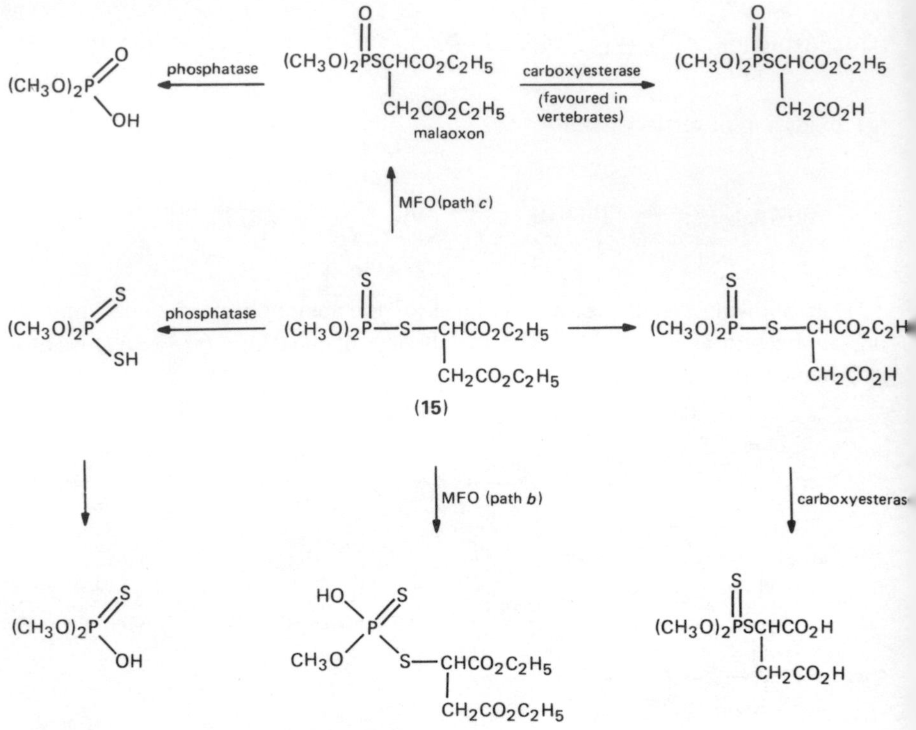

$(CH_3O)_2P$ (=O, OH) ←— phosphatase — malaoxon: $(CH_3O)_2PSCHCO_2C_2H_5$ (=O) / $CH_2CO_2C_2H_5$ —carboxyesterase (favoured in vertebrates)→ $(CH_3O)_2PSCHCO_2C_2H_5$ (=O) / CH_2CO_2H

↑ MFO (path c)

$(CH_3O)_2P$ (=S, SH) ←— phosphatase — $(CH_3O)_2P-S-CHCO_2C_2H_5$ (=S) / $CH_2CO_2C_2H_5$ **(15)** —→ $(CH_3O)_2P-S-CHCO_2C_2H$ (=S) / CH_2CO_2H

↓   ↓ MFO (path b)   ↓ carboxyesterase

$(CH_3O)_2P$ (=S, OH)   HO / CH_3O — P(=S) — S—$CHCO_2C_2H_5$ / $CH_2CO_2C_2H_5$ $(CH_3O)_2PSCHCO_2H$ (=S) / CH_2CO_2H

Scheme 6.2

The metabolities shown (Scheme 6.2) agree with the main compounds isolated from treated mammals and insects. The relative proportion of metabolites obtained indicate a greater PS → PO conversion (path c) in insects than in mammals followed by hydrolysis of the P–S–C linkage, especially in houseflies. On the other hand, in mammals the major degradation is via hydrolysis of the carboethoxy linkage. The selectivity of malathion (15) towards insects is thus explained by the differences in the rates and routes by which it is metabolized in insects and mammals.

Rogor or dimethoate (18, p. 74) has been extensively applied for insect control owing to its good insecticidal activity combined with comparatively low mammalian toxicity. The metabolism of dimethoate has been studied in plants, insects, and vertebrates[1,2,17]. The major features of the metabolism are similar in all organisms; the metabolic reactions occurring include O- and N-dealkylation by MFO (path b) (p. 84); hydrolysis of P–O and P–S bonds by phosphatases; activation by oxidative desulphuration by oxidases (path c, p. 84) resulting in conversion of P=S → P=O; and deamination by amidases. The metabolism is summarized in Scheme 6.3. Dimethoate is generally more rapidly degraded in mammals and eliminated in the urine as water-soluble hydrolysis products, for instance, in sheep amidases give the metabolite (30), but in rats and

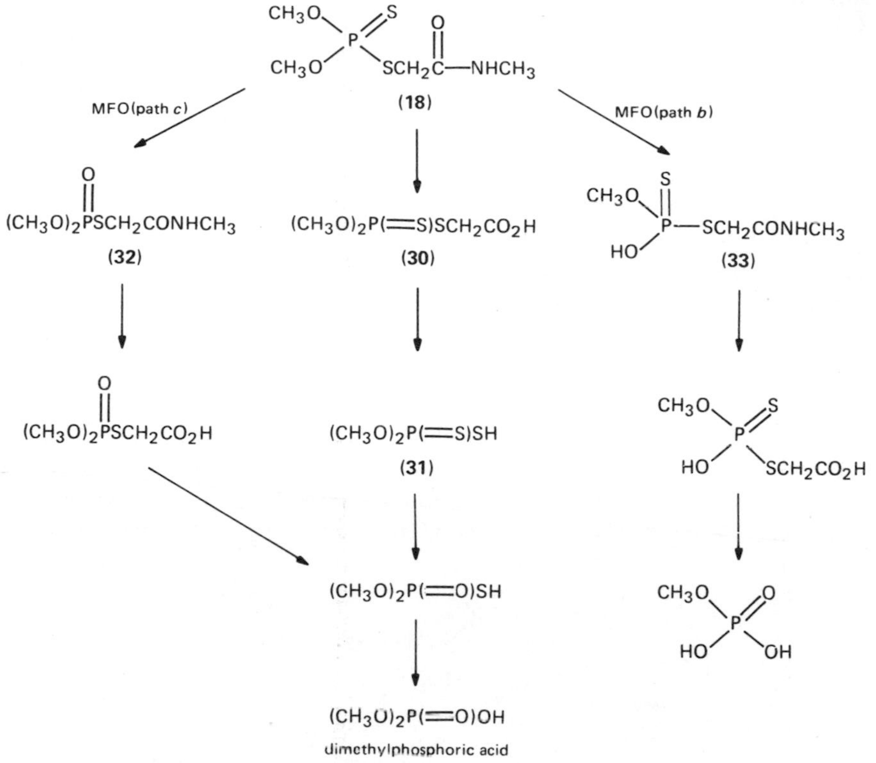

Scheme 6.3

mice both (30) and (31) were isolated while in guinea pigs the major metabolite was the dithioic acid (31). In mammals the metabolism of rogor takes place mainly in the liver[17]. In insects, both houseflies and cockroaches metabolized rogor (18) slowly accounting for the selective toxicity to these insects. In several insect species attack by amidases is of less importance than that by phosphatases, but this does not apply to the boll weevil where, as in mammalian metabolism, the first degradation is amidase attack. The special toxicity of rogor (18) towards houseflies depends upon its rapid penetration into the fly and the substantial conversion to the more toxic oxo analogue (32) by oxidase desulphuration coupled with the sensitivity of fly acetylcholinesterase to phosphorylation. In the olive fruit fly dimethoate (18) is also converted to the oxo compound (32) which is ultimately degraded to dimethylphosphoric acid. The metabolism of dimethoate (18) has also been studied[17] in various plants — cotton, corn, pea, potato, and olive trees. In plants, it is probable that total hydrolysis is due to phosphatases rather than carboxyesterases or amidases and a major metabolic pathway appears to involve O-demethylation by oxidases (p. 84, path b). Thus in cotton plants which have been treated with dimethoate by root application the major metabolite was the dithioate (30), whereas by foliar application it was the O-monomethyl compound (33).

However, in general the same metabolites were found in plants as were isolated from treated mammals and insects, and all the plant species examined showed appreciable amounts of the oxo metabolite (32) showing that dimethoate is also activated *in vivo* in plants.

Dichlorvos or Vapona (16, p. 72) is a volatile organophosphorus insecticide with a rapid 'knock down' action on flies, mosquitoes, and moths, which is widely used as a domestic insecticide. Although quite toxic to mammals, it has little persistency because of rapid hydrolysis to inactive compounds — thus it is metabolized by hydrolytic splitting of either the vinyl or methyl ester linkages to give polar metabolites such as methyl phosphate, dimethyl phosphate (34), and phosphoric acid (35) which are eliminated in urine (Scheme 6.4). Dichloroethanol (36) is also excreted as the glucuronic acid conjugate (37) Such conjugates (glucuronides) can be formed from alcohols, phenols, carboxylic acids, amines, and thiols by the action of glucuronyl transferases and they have been observed in the *in vivo* metabolism of several organophosphorus insecticides, e.g. dichlorvos and parathion[2]. The metabolism of dichlorvos (16) is summarized in Scheme 6.4.

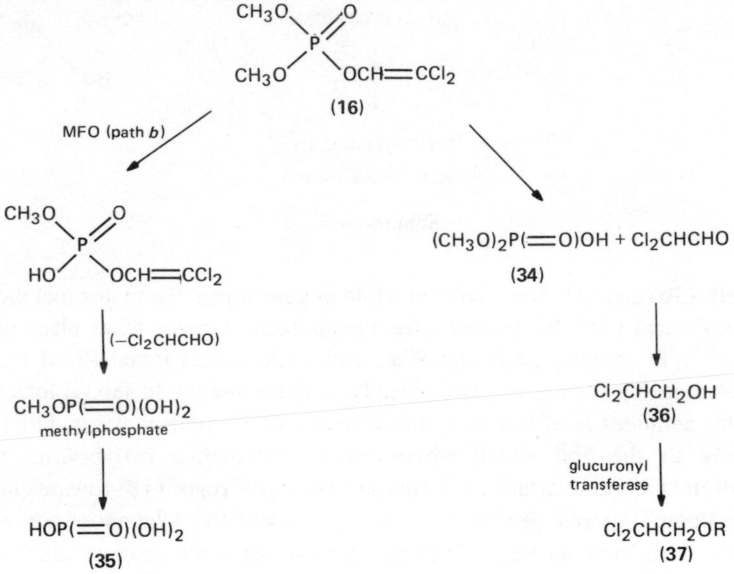

Scheme 6.4

The systemic insecticide phorate (13, p. 71) provides an example of the metabolism of a sulphide group. The various metabolites are shown in Scheme 6.5[2,17]; in plants, animals, and insects the sulphide group is oxidized to the corresponding sulphoxide (38) and sulphone (39) (path *d*, p. 84). In plants (e.g.

cotton and pea) there was also oxidative desulphuration in which P=S was converted to P=O (path *c*, p. 84) by oxidases. On the other hand, in insects (e.g. cockroaches and bugs) oxidative desulphuration does not occur.

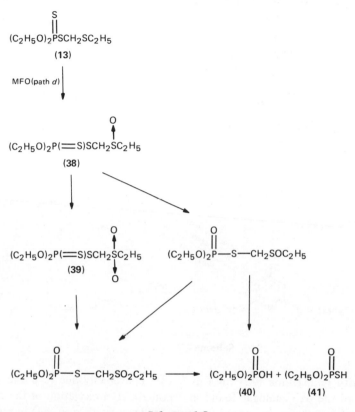

Scheme 6.5

Hydrolysis by phosphatase cleavage of the P—S bond was especially significant in boll worms and weevils where the principal metabolites were diethylphosphoric acid (40) and diethylphosphorothioic acid (41)[17]. The initial oxidation of the sulphide to the sulphoxide occurs rapidly, whereas the subsequent oxidation to the sulphone is slow in plants, although it is rapid in mammals. This oxidation somewhat increases the activity of the original compound and enables plants to be protected rather longer due to the toxicity of the sulphoxide and sulphone metabolites. However the activation is not as great as that effected by the process of oxidative desulphuration, i.e. the conversion of P=S → P=O.

The metabolism of diazinon (22, p. 76) provides an example of an organophosphorus insecticide containing a heterocyclic moiety, and has been studied in mammals, insects, and plants – the major metabolites are shown in Scheme 6.6; those marked * are insecticidal.

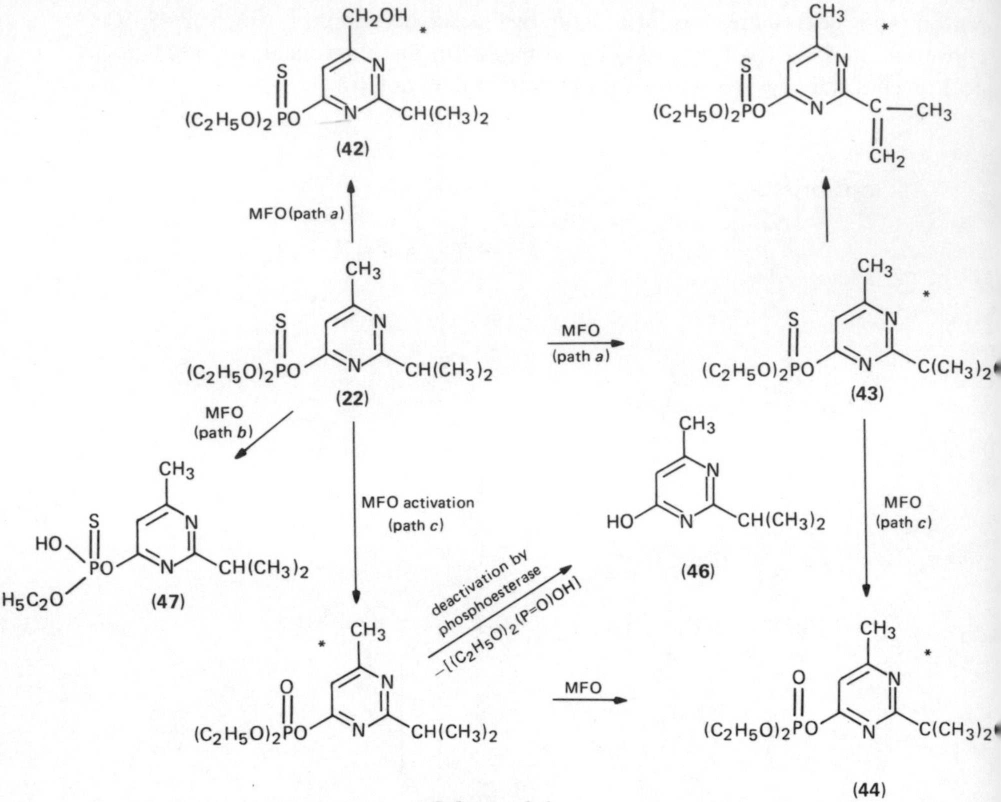

$$\text{(42)} \qquad \text{(43)}$$

MFO (path a)

MFO (path a)

(22)

MFO (path b)

MFO (path c)

(43)

(47)

MFO activation (path c)

deactivation by phosphoesterase
$-[(C_2H_5O)_2(P=O)OH]$

(46)

MFO (path c)

MFO

MFO

(44)

Scheme 6.6

The selectivity of diazinon (22) towards insects is probably due to the relatively lower level of the oxo analogue found in mammals. Hydroxylation of the methyl and isopropyl side-chains by mixed-function oxidases (path a, p. 84) in rat liver, sheep, and cockroaches gives a series of active metabolites (42–44, Scheme 6.6).

Diazinon (22) was detoxified in rat liver and cockroaches by P–O–aryl cleavage promoted by glutathione (GSH) to give diethylphosphorothioic acid (41) and the pyrimidinyl glutathione (45).

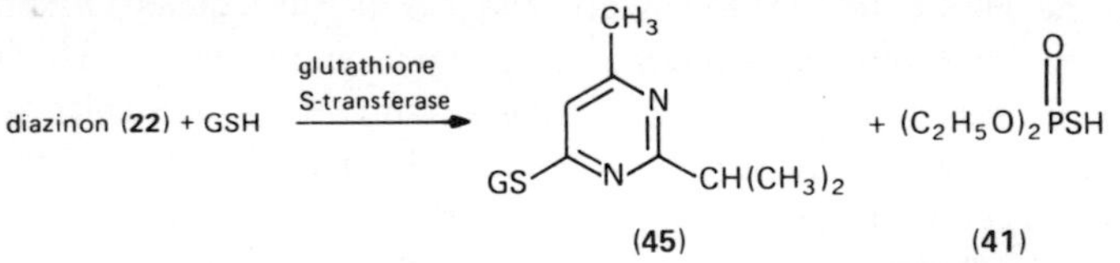

$$\text{diazinon (22)} + \text{GSH} \xrightarrow[\text{S-transferase}]{\text{glutathione}} \quad \text{(45)} \quad + \quad (C_2H_5O)_2\overset{\text{O}}{\underset{}{\text{PSH}}} \quad \text{(41)}$$

Insects generally have poor transferase activity, although a diazinon-resistant strain of houseflies owed their tolerance to the presence of glutathione S-transferase[2]. Rat liver, but not insects, contains a phosphotriesterase with high specificity towards diazinon (22) which it hydrolyses to the inactive compounds diethylphosphoric acid and the 6-pyrimidinol (46)[1,2]. The presence of this enzyme in mammals probably accounts for the selective toxicity of diazinon towards insects. The phosphoramidate, schradan (**4**, p. 68), was one of the first systemic insecticides known and is still applied to a limited extent for control of aphids and mites on citrus, apple, and hops, but it suffers from the disadvantage of very high mammalian toxicity.

Schradan itself is not an anticholinesterase agent and owes its high mammalian toxicity and insecticidal activity to *in vivo* activation by oxidation of the tertiary amino group. Oxidative activation can also be demonstrated by chemical treatment of schradan with oxidants like permanganate, dichromate, hydrogen peroxide, or bromine water[17]. The first step in the metabolism of schradan by insects and mammals is oxidation of the tertiary amino group by mixed-function oxidases (path *g*, p. 85) to give the *N*-oxide (**48**) which probably rearranges to the *N*-methylol (**49**), the latter subsequently giving formaldehyde and the hepta-methylpyrophosphoramide (**50**), so that the bioxidation results in overall demethylation:

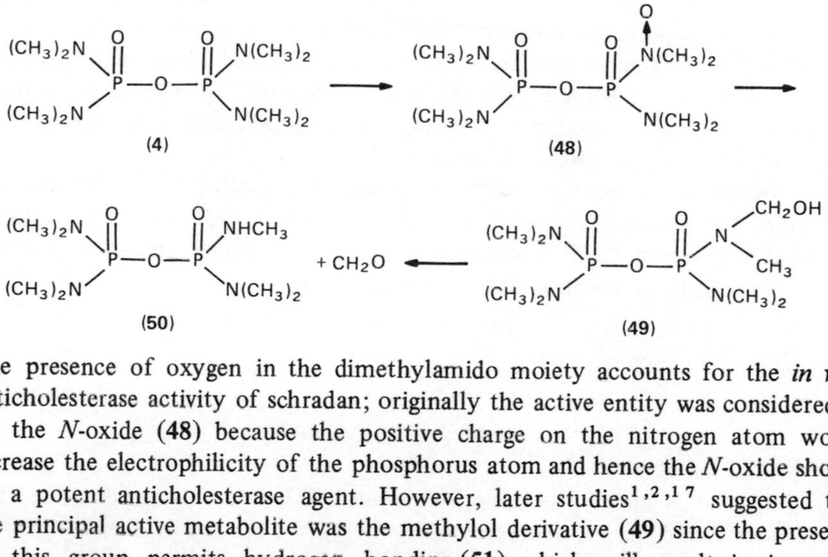

The presence of oxygen in the dimethylamido moiety accounts for the *in vivo* anticholesterase activity of schradan; originally the active entity was considered to be the *N*-oxide (**48**) because the positive charge on the nitrogen atom would increase the electrophilicity of the phosphorus atom and hence the *N*-oxide should be a potent anticholesterase agent. However, later studies[1,2,17] suggested that the principal active metabolite was the methylol derivative (**49**) since the presence of this group permits hydrogen bonding (**51**) which will result in increased electron drift from the phosphorus atom and should therefore enhance its phosphorylating ability:

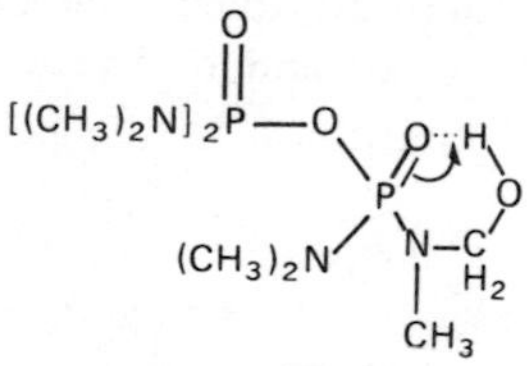

(**51**)

The different metabolic pathways and the resultant balance of activation and detoxification processes, in many instances, account for the selective toxicity exhibited by several organophosphorus compounds. The most generally important metabolic activation process is microsomal oxidative desulphuration (path c, p. 84) effected by oxidases whereby phosphorothioates are converted into the much more active phosphates (i.e. P=S → P=O). This activation has been noted in the previous discussion of the metabolism of such insecticides as parathion (p. 85), malathion (p. 85), dimethoate (p. 86), and diazinon (p. 89).

Selectivity can sometimes be attributed to the differing activities of the oxidases from various insect species towards a given organophosphorus substrate. For instance, isopropyl parathion easily undergoes oxidative desulphuration by housefly oxidase but not by the bee enzyme, and this together with the greater sensitivity of housefly cholinesterase to isopropyl paraoxon accounts for the selective toxicity of isopropyl parathion to houseflies as cf. bees. Steric effects may also be significant in some cases; for instance in the thiono analogues of mevinphos (p. 74), the *cis* isomer was selectively oxidized (activated) by mouse liver oxidase as cf. the *trans* isomer.

Metabolic oxidation of sulphides by MFO leading to formation of the corresponding sulphoxides and sulphones also leads to some activation of the original insecticide since oxidation enhances electron removal from the phosphorus atom. Examples are phorate (p. 88) and metasystox (p. 70); such sulphide oxidation is valuable because its occurrence in plants and soils increases the persistency of action of systemic insecticides.

The metabolic oxidation of amide groups by MFO (path g, p. 85) also results in activation. The classic example of this effect is schradan where overall oxidative demethylation occurs (p. 91). Dimefox (1, p. 68) is similarly activated. Another interesting example is dicrotophos or Bidrin (52):

$$(CH_3O)_2\overset{\overset{\displaystyle O}{\|}}{P}O \underset{H_3C}{\diagdown}C=C\overset{H}{\underset{CON(CH_3)_2}{\diagup}}$$

(52)

This compound is mainly the *cis*-crotonamide (52), which is more active than the *trans* form, and was introduced by Ciba–Geigy in 1963 and by Shell in 1965. It acts as a systemic insecticide and acaricide of moderate persistence which is very effective against sap-feeding insect pests[5]. The metabolism of bidrin (52) involves oxidative demethylation to another insecticide monocrotophos or Azodrin (53) developed particularly for control of caterpillars and boll worms on cotton[5]. This is finally degraded to the unsubstituted amide (54) with the production of at least four active metabolites which contribute to the persistency of action of Bidrin (52):

$$RN(CH_3)_2 \xrightarrow[\text{(path }a)]{MFO} RN\overset{CH_2OH}{\underset{CH_3}{\diagdown}} \xrightarrow{(-CH_2O)} R-N\overset{H}{\underset{CH_3}{\diagdown}} \longrightarrow$$

(52)   (53)

$$(R = (CH_3O)_2P(=O)OC(CH_3)=CHCO—)$$

$$R\text{—}N\overset{H}{\underset{CH_2OH}{\diagup}} \quad \longrightarrow \quad RNH_2 + CH_2O$$

(54)

The side-chains of arylalkanes may be oxidized by MFO (path *a*, p. 84) to give alcohols which may be further metabolized by elimination or conjugation, e.g. the metabolism of diazinon (p. 89).

Hydroxylation may sometimes lead to activation; thus tri-*o*-tolyl phosphate (TOCP) is metabolized *in vivo* to the active neurotoxic *o*-tolyl saligenin cyclic phosphate (55; R = *o*-tolyl); generally *o*-tolyl *O*-arylphosphates yield saligenin cyclic phosphates (55):

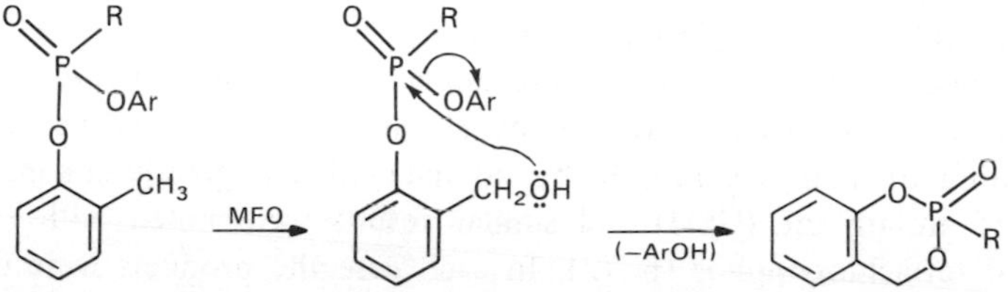

(R = alkoxy, aryloxy, or alkylamido) (55)

This discovery led to the development in 1968 of the insecticide Salithion (56) by the Sumitomo Chemical Company in Japan. Salithion is obtained by phosphorylation of *o*-hydroxybenzyl alcohol with methyldichlorido-phosphorothioate:

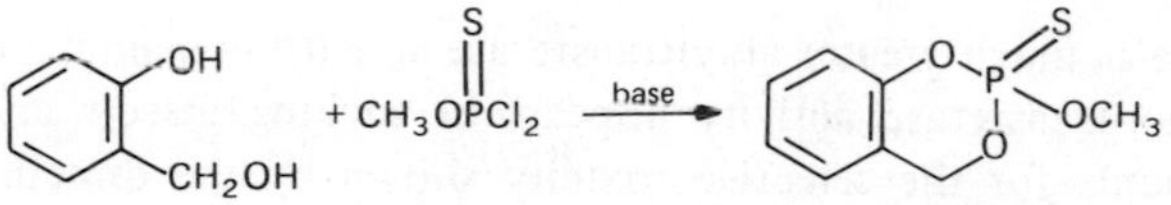

(56)

It is a short-lived insecticide used for protection of fruits and vegetables.

Various metabolic processes also result in detoxification of organophosphorus insecticides. Such processes generally involve phosphorus ester bond cleavage which introduces a negative charge into the molecule so destroying activity as a phosphorylating agent. The products are also much more water soluble and so are readily excreted in the urine. Phosphorus ester bonds are cleaved by hydrolysis catalysed by phosphoroesterases which occur widely in mammalian tissues, insects, and microorganisms. These esterases also hydrolyse some carbamic and carboxylic esters. Phosphatases only hydrolyse partial esters of orthophosphoric acid but other esterases hydrolyse a variety of neutral phosphorus compounds cleaving the bond between the phosphorus atom and the most acidic group or the so-called leaving group. As well as phosphorus ester bonds, bonds like

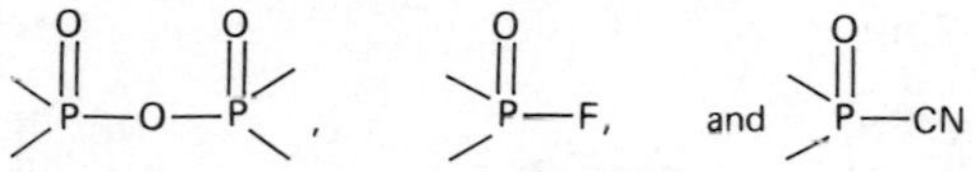

are cleaved. Examples are seen in the metabolism of parathion (p. 85) and diazinon (p. 89); the latter also illustrates the enzymic cleavage of a P—O—aryl bond in the presence of glutathione. Liver MFO often detoxify insecticides by oxidative dealkylation (path *b*, p. 84); for instance chlorfenvinphos (p. 72) is de-ethylated:

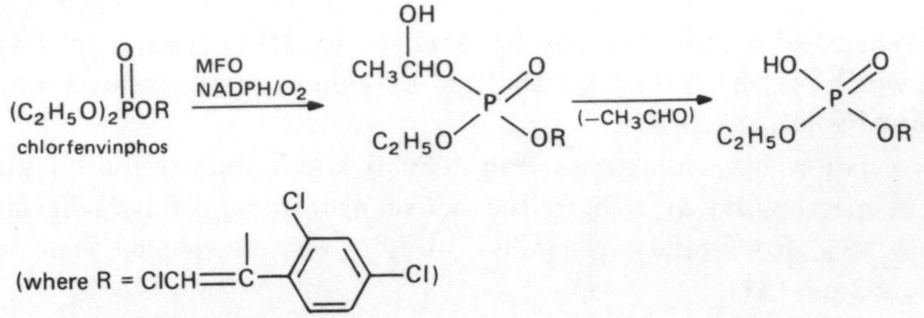

The enzyme catalyses dealkylation of dimethyl, diethyl, diisopropyl, and di-n-butyl esters, although generally ethyl esters appear to be the favoured substrates. Demethylation more often is effected by glutathione *S*-alkyltransferase, thus the demethylation of methyl parathion by rat liver MFO is greatly enhanced by the presence of glutathione (GSH) and similar results were noted with mevinphos (p. 73) and tetrachlorvinphos (p. 72). In each case the products were the mono-demethylated pesticide and S-methylglutathione:

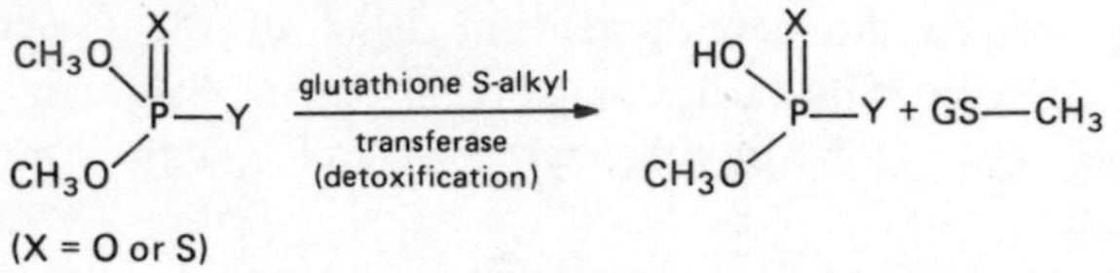

Generally there is much greater alkyltransferase activity in mammalian liver than in insects, and no transferase activity appears in sucking insects and mites which probably accounts for the selective toxicity shown by the dimethyl phosphorus esters towards this species.

The metabolic cleavage of *S*-alkylphosphorothiolate bonds occurs by an oxidative mechanism for thiolothionates or by a hydrolytic mechanism for thiolate esters, although the detoxification of malathion (15) is chiefly due to hydrolysis by carboxyesterase (p. 85).

Menazon (19, p. 75) in rats is mainly degraded at the P—S bond and the main metabolite excreted is the 2-methylsulphinylmethyl compound (57):

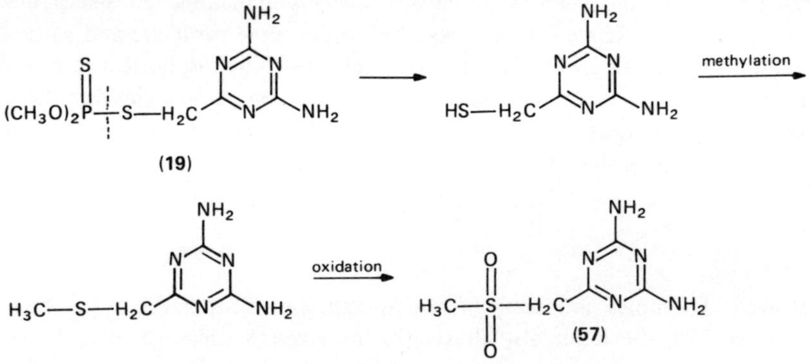

The early organophosphorus insecticides like parathion, schradan, and tetraethyl pyrophosphate (TEPP) were highly active compounds but were also extremely toxic to mammals, and were the most dangerous chemicals ever to be used in agriculture. In applying such compounds, operators must wear full protective clothing and a respirator[6]; they have caused several human fatalities and any birds or small mammals covered by the spray are killed. However, organophosphorus compounds are comparatively rapidly biodegraded to non-toxic, water-soluble compounds which are quickly excreted by animals. Consequently, unlike the organochlorine insecticides, they do not accumulate in the environment (see Chapter 14, p. 215)[18].

Many organophosphorus compounds function as systemic insecticides which enables smaller amounts of the active ingredient to be used more effectively and reduces the harmful effects on natural predators. To this end efforts are currently directed towards the discovery of new organophosphorus compounds showing selective toxicity like menazon (1961) which is a specific systemic aphicide (p. 75). To combat insect resistance (see p. 101) a whole range of systemic organophosphorus insecticides are needed showing low mammalian toxicities, varying degrees of persistency, and specific activity against the target insects[7].

The selective toxicity of a given insecticide depends on a number of factors[15], for instance, the balance of metabolic activation and deactivation processes occurring in a given insect species is often critical. Various insect species may have different enzyme systems exhibiting different levels of activity towards the organophosphorus substrate so metabolism may be responsible for selectivity, e.g. malathion (p. 85). In some cases, the organophosphorus compound may penetrate the insect cuticle more readily than the mammalian skin. Certainly, organophosphorus insecticides are rapidly biodegraded in mammals to water-soluble metabolites which are excreted in the urine; in several instances the toxicant is appreciably more persistent in insects. Salithion (p. 93) is so rapidly degraded in mice that only 2.4% of the original dose remained 3 hours after treatment, whereas with treated houseflies 10% of the insecticide was present 24 hours after treatment[2]. Such differential degradation and excretion will lead to selective toxicity to houseflies as compared with mice.

Another consideration is the transport of the toxicant to the site of action. Study of a series of 2-chlorovinyl phosphate insecticides showed that although chlorfenvinphos (p. 72) has a fairly high mammalian toxicity the introduction of a chlorine atom at the 5-position of the phenyl ring resulted in a substantial reduction in mammalian toxicity but the insecticidal activity was not appreciably affected. The low mammalian toxicity of tetrachlorvinphos is attributed to its low solubility in water and organic solvents leading to slow penetration and transport from the point of application to the active site.

Acetylcholine and ionized anticholinesterase agents such as amiton (p. 83) are extremely toxic to mammals and to certain insects, for instance aphids, mites, and scale insects but are not effective against other species. Schradan (p. 68) shows a similar spectrum of activity, and the selectivity arises from differences in the composition of the nervous systems of insects and vertebrates. In insects, the nerve junctions are protected from ionic materials by a lipid nerve sheath and the thickness of this lipid barrier appears to be the decisive factor determining the resistance of insects towards schradan. Tolerant insects like American cockroaches and houseflies have a thick sheath whereas susceptible insects, such as rice bugs and green leaf hoppers have only a thin membrane.

The actual receptor site may also vary in different organisms so that, in spite of the specifically similar mechanism of toxicity of organophosphorus compounds (see p. 80) in animals and insects, there are still a number of factors which can be exploited to obtain compounds showing selective toxicity towards a given target insect pest.

Carbamates

The successful development of organophosphorus insecticides stimulated examination of other compounds known to possess anticholinesterase activity. One such compound is the alkaloid physostigmine (58), the active ingredient in calaban beans which has been used for trial by ordeal in West Africa[8,19]. The physiological properties of this alkaloid were supposed to be based on the phenylmethyl-carbamate part of the structure and led to the discovery of a number of parasympathomimetic drugs like neostigmine (59).

The compounds being quite strong bases are ionized in aqueous solution and therefore have very low lipid solubility. Consequently they are unable to penetrate the ion-impermeable sheath surrounding the insect nervous system (cf. p. 95). Therefore, efforts were made to synthesize compounds in which the N-substituted carbamate part of the molecule was attached to a less basic, more lipophilic moiety, since such compounds should show greater insecticidal activity. In 1951 the Geigy Company introduced Isolan (1-isopropyl-3-methylpyrazolyl-5-dimethylcarbamate, 60)[12]. This water-soluble compound was a most effective systemic aphicide and was also active against houseflies, but showed a very high mammalian toxicity, (LD_{50} (oral) to rats $\simeq 12$ mg/kg) so the compound was not extensively developed[12]:

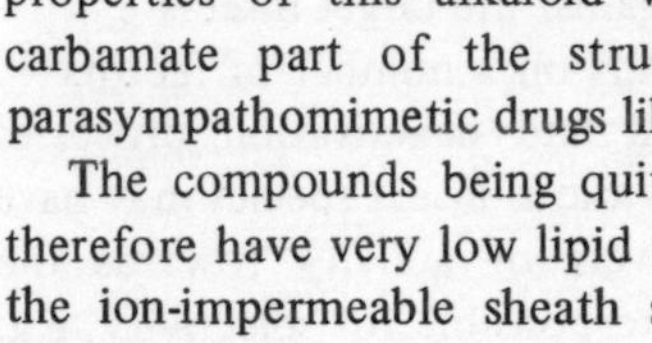

(58) 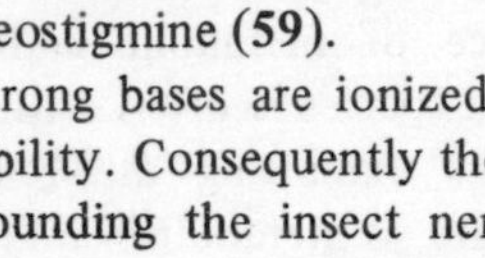   (59) 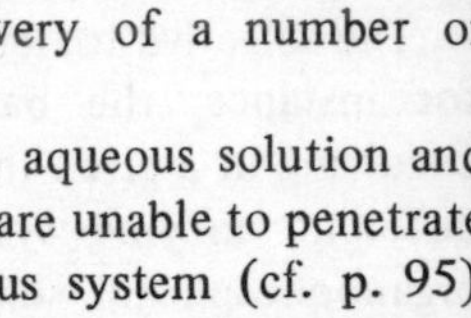   (60)

Later work showed[16] that in the dimethylcarbamate series heteroaromatic derivatives, especially those containing the pyrimidine nucleus, had much lower mammalian toxicities and in 1968 Imperial Chemical Industries Ltd. introduced pirimicarb (61) synthesized from ethyl α-methyl acetoacetate and dimethylamino-acetamidine:

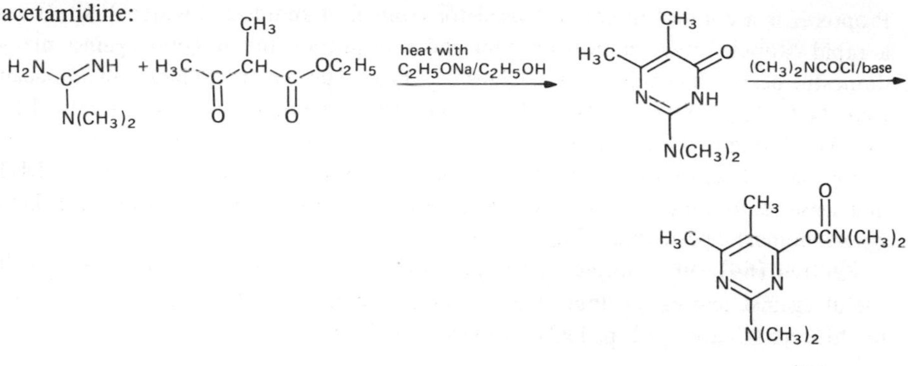

(61)

Pirimicarb is a fast-acting specific systemic aphicide and is also effective against aphids which have developed resistance to organophosphorus insecticides[5]. The chemical is generally applied as a foliar spray, but is also taken up by plant roots and translocated in the xylem. It has only moderate mammalian toxicity (LD_{50} (oral) to rats $\simeq$ 147 mg/kg) (Plate 9).

Many monomethylcarbamates also show valuable insecticidal properties and are much more easily hydrolysed than the dimethylcarbamates. Phenol carbamates are especially useful in insecticides; the first member of this group was the α-naphthyl compound, carbaryl or Sevin **(62)** introduced by the American Union Carbide Company in 1956[9,12,16,19], and was the first successful commercial carbamate insecticide. Carbaryl is prepared from α-naphthol as shown below:

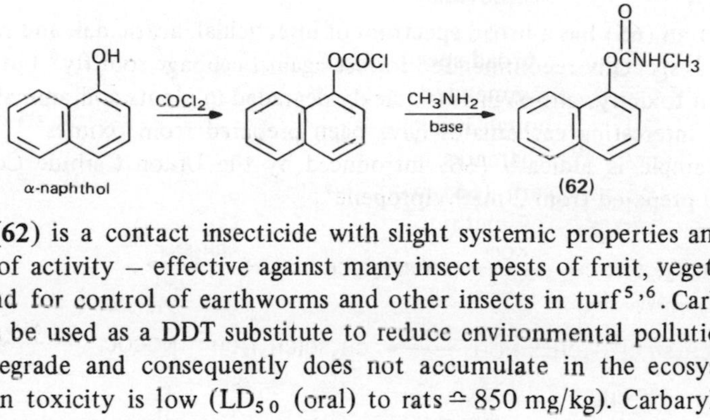

(62)

Carbaryl **(62)** is a contact insecticide with slight systemic properties and a broad spectrum of activity — effective against many insect pests of fruit, vegetables, and cotton, and for control of earthworms and other insects in turf[5,6]. Carbaryl may sometimes be used as a DDT substitute to reduce environmental pollution since it does biodegrade and consequently does not accumulate in the ecosystem. The mammalian toxicity is low (LD_{50} (oral) to rats $\simeq$ 850 mg/kg). Carbaryl has been used all over the world to a greater extent than all the other carbamate insecticides combined[10]. Other interesting N-methylcarbamates include propoxur or Baygon **(63)** introduced by Bayer Ltd. (1959) and prepared from catechol[5]:

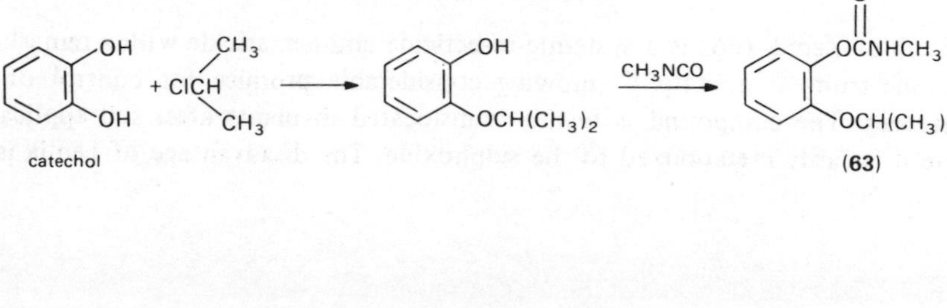

(63)

Propoxur is a contact insecticide used for control of aphids and white flies[6]. It has a rapid 'knock down' action on flies and mosquitoes and is good against many domestic pests especially cockroaches[8,10]. Propoxur has a moderate mammalian toxicity (LD_{50} (oral) to rats $\simeq$ 150 mg/kg), and is being extensively examined by the World Health Organization for control of mosquitoes. It shows good residual action, as well as rapid 'knock down', and appears a reasonable substitute for DDT in disease vector control. Ficam W was developed by Fisons in 1971 and is excellent against domestic pests (Plate 7).

Zectran (64), an example of several amino derivatives, is a contact insecticide useful against leaf-eating insects and also as a moluscicide[8,10]; for the latter use methiocarb (Chapter 12, p. 192) is especially effective.

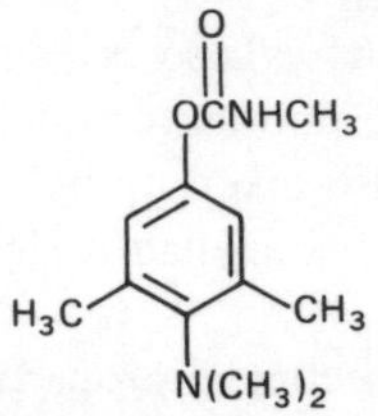

(64) 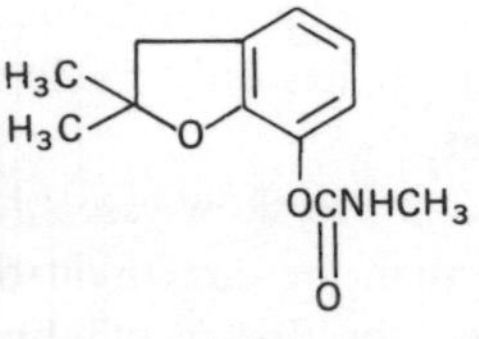   (65) 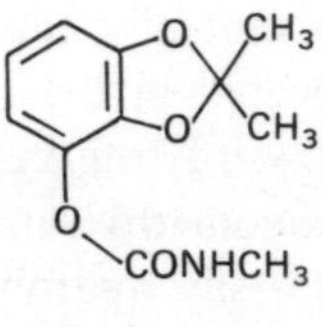   Ficam W

Several N-methylcarbamates are systemic in plants and are translocated from roots and leaves; examples are carbofuran (65), aldicarb or Temik (66), and methomyl or lannate (67)[10,16].

Carbofuran (65) has a broad spectrum of insecticidal, acaricidal, and nematicidal activity it is specially recommended for use against cabbage root fly[6] but has a high mammalian toxicity, although it is quickly degraded in plants and animals[8].

Several interesting carbamates have been prepared from oximes[8,12]; the best known example is aldicarb (66) introduced by the Union Carbide Company in 1965[5] and prepared from 2-methylpropene[9]:

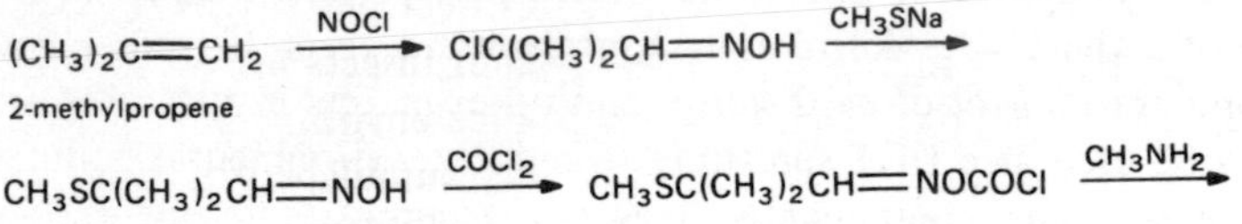

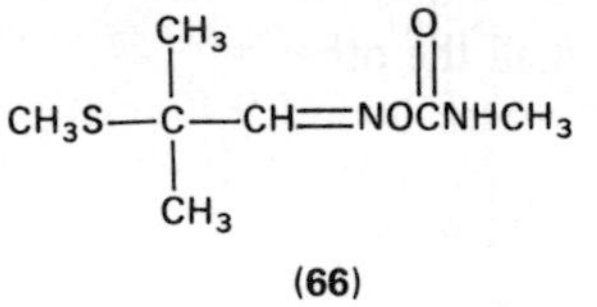

(66) 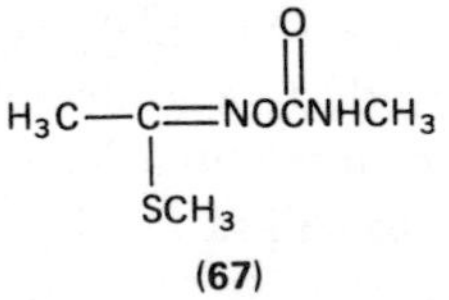   (67)

Aldicarb or Temik (66) is a systemic insecticide and nematicide with a remarkably wide spectrum of activity[5,6] showing considerable promise for control of soil nematodes. The compound is readily translocated in plants after soil application where it is easily metabolized to the sulphoxide. The disadvantage of Temik is the very high mammalian toxicity (LD_{50} (oral) to rats $\simeq$ 1 mg/kg). Consequently Temik is only marketed in granular formulations[8].

Methomyl or Lannate (67) has similar biological activity and is especially useful for soil and seed treatment[5,8]. It is used for control of many insect pests[6] generally by foliar spray, but unfortunately methomyl also shows high mammalian toxicity (LD_{50} (oral) to rats $\simeq$ 20 mg/kg).

Carbamates are metabolized by two basic mechanisms, both involving breakage of the carbamate ester linkage, namely by direct esterase attack (path *a*), or by initial oxidation by mixed-function oxidases (MFO) followed by hydrolytic breakdown of an unstable intermediate[17] (path *b*):

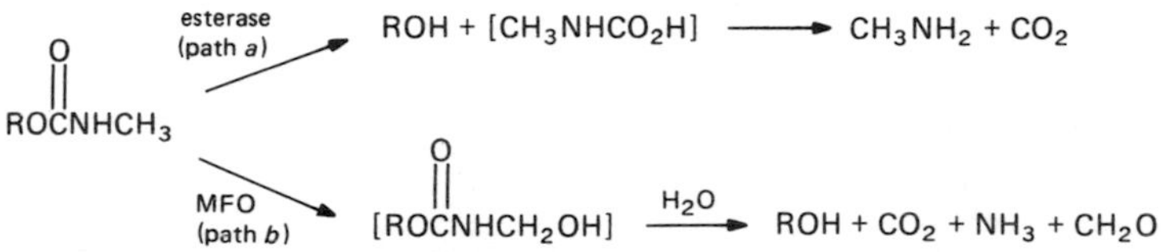

The metabolism of carbaryl (62) has been extensively examined in insects and mammals[20]. Scheme 6.7 shows the metabolic pathways determined by experiments *in vivo* in rabbits, houseflies, and cockroaches and *in vitro* by rat, mouse, and rabbit liver microsomes:

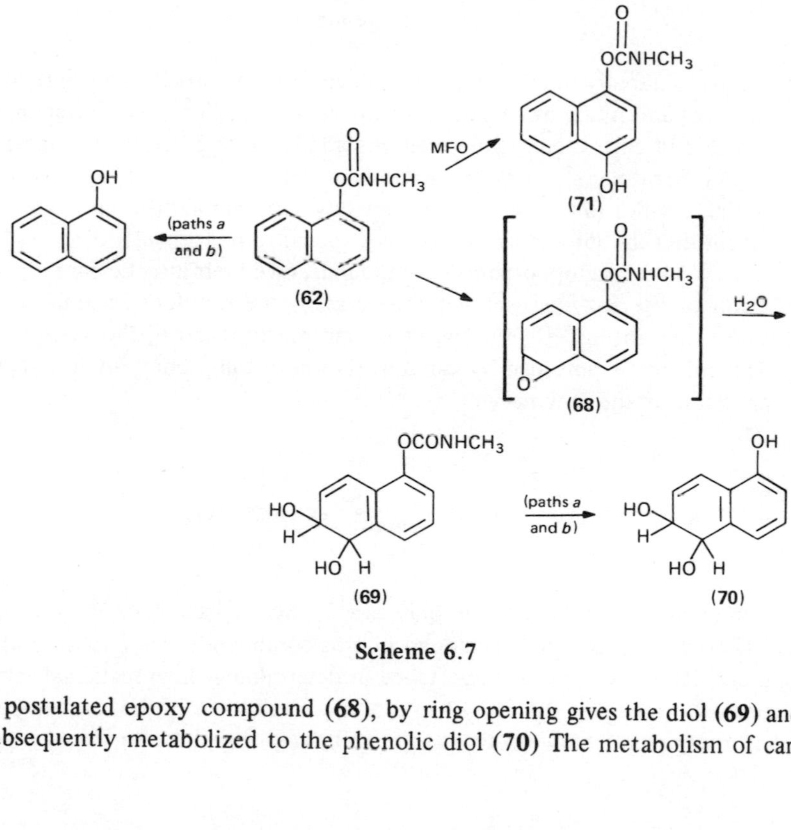

Scheme 6.7

The postulated epoxy compound (68), by ring opening gives the diol (69) and this is subsequently metabolized to the phenolic diol (70) The metabolism of carbaryl is complex and thin-layer chromatography of the urine from treated rabbits indicated the presence of some five additional metabolites which may be conjugates. Carbaryl is certainly rapidly metabolized and excreted in mammals[17].

The oxime carbamate aldicarb (66) is metabolized in cotton plants and houseflies in a similar manner to organophosphorus thioether derivatives such as phorate (see p. 88) (Scheme 6.8):

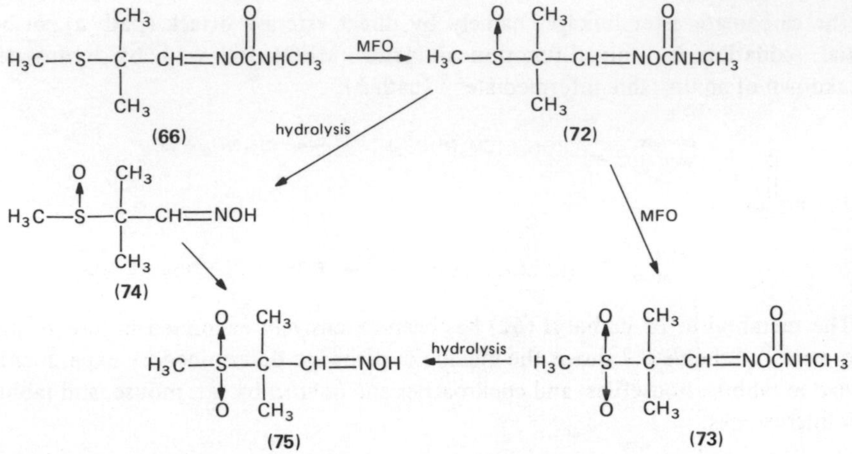

Scheme 6.8

Aldicarb (66) is rapidly metabolized by mixed-function oxidases (MFO) to the sulphoxide (72) and much more slowly to the sulphone (73)[20]. The latter appears relatively stable in cotton and is the main metabolite isolated from the plants two months after treatment. Enzymic hydrolysis of (72) and (73) yields the corresponding oximes (74) and (75) respectively. The metabolism in mammals is similar except that the sulphone (73) does not appear to be formed[17].

Carbamates, like organophosphorus compounds, owe their insecticidal properties to inhibition of the enzyme acetylcholinesterase – the resultant accumulation of acetylcholine preventing effective nervous transmission across the synapse (see p. 34). The enzyme is poisoned by carbamoylation of the primary hydroxyl group of a serine residue of the enzyme:

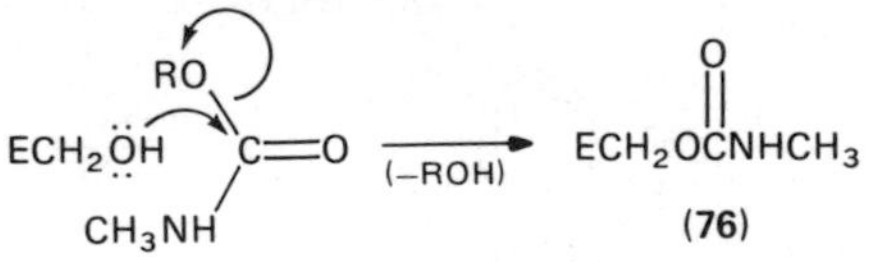

The carbamoylated enzyme (76) is only slowly hydroylsed back to the active enzyme. However, unlike the organophosphorus compounds, the structure of the leaving group RO⁻ is of critical importance in determining the insecticidal activity of carbamates[9]. Generally the rate of hydrolytic breakdown of the carbamoylated enzyme is intermediate between that of the acetylated and phosphorylated enzyme so that acetylcholinesterase is inactivated for a significant time. However, as with the organophosphates, the *in vitro* anticholinesterase properties of carbamates often bear little relation to their *in vivo* insecticidal activity because of the importance of such additional factors as ease of penetration and metabolism.

For insecticidal activity, carbamates appear to require a degree of structural resemblance to the natural enzyme substrate acetylcholine, so that the carbamate competes strongly with acetylcholine for the reactive sites on acetylcholinesterase. Furthermore, activity appears to be assisted by the presence of a bulky side-chain group situated some 5 Å away from the carbonyl group. These features are illustrated by the structural formulae of aldicarb (**66**) and Baygon (**63**):

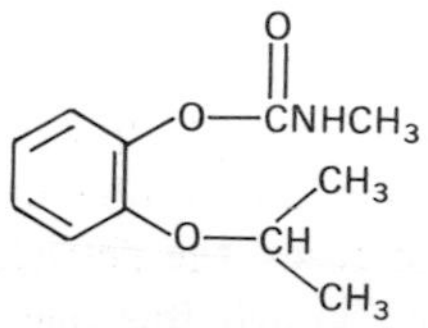

(63) 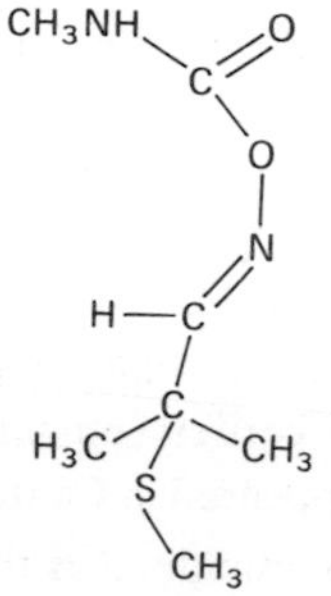   (66) 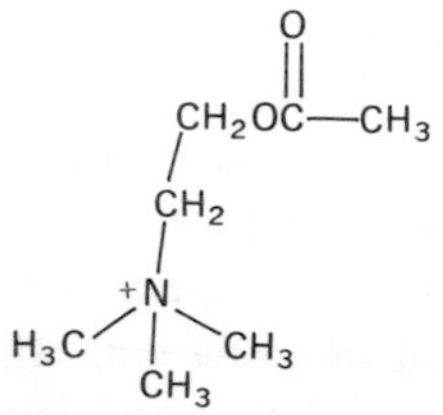   acetylcholine

The major biochemical mechanism of insect resistance to carbamate insecticides appears to be detoxification *via* enzymic hydrolysis; thus carbaryl-resistant houseflies show an abnormally high concentration of the enzyme carbamate esterase which converts carbaryl (**62**) to the inactive α-naphthol[20]:

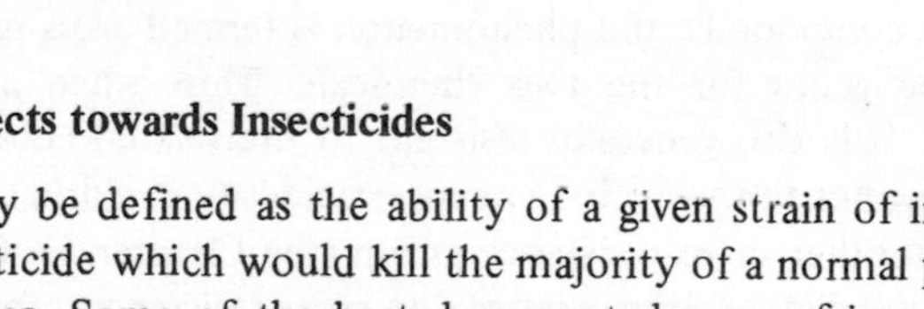

(62)   α-naphthol

Resistance of Insects towards Insecticides

Resistance may be defined as the ability of a given strain of insects to tolerate doses of an insecticide which would kill the majority of a normal population of the same insect species. Some of the best documented cases of insect resistance have been observed with DDT and other persistent organochlorine insecticides (see Chapter 5, p. 55), though serious resistance to organophosphorus and other insecticides has also been noted and has caused serious control problems[21,22].

By 1946 some strains of DDT-resistant houseflies had been discovered and in 1950, 5 to 11 species had acquired tolerance to one or more insecticides. In 1969 there were 102 resistant insect species: 55 to DDT, 84 to dieldrin, and 17 to organophosphorus compounds. Further some insects were resistant to all three types of insecticide. In addition 20 species of mites and ticks had developed tolerance to acaricides. By 1974, it has been estimated[23] that over 250 species had become resistant to one or more insecticides. One of the early examples of an insect acquiring tolerance to an insecticide was recorded in California in the 1920s when scale insects infesting citrus orchards become resistant to hydrogen cyanide.

It was however not expected that the introduction of the new synthetic insecticides in the later 1940s would induce such rapid insect resistance. The reason was probably that these chemicals had extremely high initial toxicity and so they quickly killed all the susceptible individuals in the pest population leaving the small number of naturally resistant pests available to reproduce explosively with little competition because these non-selective insecticides often eliminated many of the natural predators.

Pesticides do not produce resistance, they merely select resistant individuals already present in the natural pest population. The tolerant individuals confer resistance to their progeny in the genes so succeeding generations of insects will also be resistant to the pesticide. In the majority of cases, the pesticide probably does not induce mutations which confer resistance, though this may be true for warfarin-resistant rats which have appeared in Central Wales[2] (see Chapter 10, p. 180).

In screening a new potential insecticide, it is therefore important to see whether it is effective against strains of the target pest which are already tolerant to established insecticides, and also how quickly a strain resistant to the new chemical develops[12].

The inheritance of specific resistance is generally comparatively simple and often monofactorial, although the influence of the principal gene may sometimes be modified by secondary genes. Brown and Pal[22] obtained data on the genetic characteristics of the resistance of different insect species towards DDT, dieldrin, and organophosphorus compounds. In a total of 17 species the inherited resistance to DDT was monofactorial in 13 species; to dieldrin in 16 species; and to organophosphorus compounds in 5 species. The relatively simple mode of inheritance is probably a further reason for the rapid growth of resistance in field populations of pest insects often enabling the insect to acquire resistance to several insecticides simultaneously. When a specific detoxification mechanism confers resistance to two compounds, the phenomenon is termed cross-resistance since this involves the same genes for the two chemicals. Thus when an insect becomes resistant to DDT, it is also generally resistant to the related compounds DDD and methoxychlor, but not to the cyclodiene insecticides (e.g. aldrin), or lindane (HCH) which fall into another cross-resistance group (see Chapter 5). Organophosphorus insecticides can be divided into two major cross-resistance groups illustrated by parathion and malathion, and the insects are often also resistant to carbamate insecticides (see p. 96). A fifth cross-resistance group is afforded by the pyrethrins although resistance to these is less extensive.

When different resistance mechanisms exist in a given insect, it is said to show multiple resistance. This can be induced when the insect population has been exposed to different insecticides, and also in some cases when the pressure seems to have come from only one insecticide and the multiple resistance is morphological or behavioural in origin, when it can generally be overcome by quite small increases in dosage of the toxicant.

In the native insect population only a few individuals are pre-adapted against the insecticide and there was no reason to favour this mutation in the absence of the insecticide. On the other hand, when the insecticide was introduced into the environment, the tolerant strains survived, reproduced, and simultaneously there was a general selection of a genotype better adapted to the prevailing insecticidal environment.

The various physiological resistance mechanisms are also very important; in DDT-resistant insect strains, tolerance is often due to an abnormally high concentration of the enzyme DDT-dehydrochlorinase which converts DDT to the non-insecticidal DDE (see Chapter 5, p. 55). Mechanisms associated with resistance to organophosphorus compounds are more complex. Many organophosphates, like schradan, have to be metabolized *in vivo* to the active insecticide. They may also be deactivated by enzymes — phosphatases, and carboxyesterases. So the toxicity of the given compound depends on the balance of activating and deactivating enzymes within the insect. With malathion, for instance, the low mammalian toxicity is ascribed to the higher carboxyesterase activity in mammals in comparison with the low activity of this enzyme in susceptible insects (Chapter 6, p. 85). Insects exhibiting resistance to malathion generally show no cross-resistance to other insecticides, suggesting that tolerance depends on high carboxyesterase activity, which is supported by the discovery that carboxyesterase inhibitors function as malathion synergists and almost eliminate resistance[2].

The resistance to organophosphates shown by several strains of houseflies and blowflies is associated with exceptionally low levels of aliesterase activity which is controlled by a single gene, whereas normally houseflies have large quantities of an aliesterase[2].

Microsomal mixed-function oxidases (MFO) play a vital role in both the activation and degradation of organophosphates. It has been shown that the remarkable selectivity of chlorfenvinphos (Chapter 6, p. 72) from one organism to another depends on the differing ability of their liver enzymes to de-ethylate the chemical. The greater the activity of the liver enzymes, the lower will be the effectiveness of chlorfenvinphos LD_{50} (oral) dogs $>$ 12,000; mouse 170–200; rat 15 mg/kg[1,2]. A microsomal oxidation involving NADP and oxygen may be one cause of DDT and organophosphorus resistance in some insects. The level of microsomal oxidases varies considerably from one strain of insects to another. The responsible gene determines the level of NADP-dependent microsomal oxidase activity, which detoxifies certain chlorinated hydrocarbons, pyrethroids, carbamates, and organophosphorus insecticides. The oxidase inhibitor piperonyl butoxide prevented the metabolic degradation of these insecticides.

The resistance shown by houseflies to certain organophosphorus insecticides, such as dichlorvos and bromophos, appears to be related to the decreased penetration of the toxicant to the thoracic ganglionic complex of the resistant flies. Resistance may also be due to behavioural patterns; certain strains of anopheline mosquitoes will not settle on surfaces which have been treated with DDT and others will not enter buildings which have been sprayed with DDT. Similarly codling moth larvae have envolved the habit of discarding the initial bite of the apple when they bore into it to avoid the stomach poisons which which the apples have been treated[21].

The birth of generations of DDT-resistant insects resulted in the substitution first of lindane or γ-HCH, followed by organophosphates (e.g. malathion) for their control. However after about 6 years the pests, in many cases, developed strains showing multiple resistance. Carbamate insecticides like carbaryl were often useful, but unfortunately pests which had acquired tolerance to organophosphates often showed cross-resistance to carbamates. Many important insect pests and disease vectors have multiple resistance and are consequently difficult to control. Red spider mites in many parts of the world are an outstanding example of this problem.

The growth in the population of resistant anopheline mosquitoes has resulted in outbreaks of malaria in areas in which the disease was previously thought to have been practically eradicated. Malaria control campaigns have been also hindered by the emergence of strains of bed bugs which are resistant to DDT.

When a particular insecticide is no longer applied, the resistant strain of insect often reverts to the natural susceptible strain. However, when the original insecticide is reintroduced resistance very quickly reappears. There is some advantage to be gained by changing from one insecticide to another every 5–6 generations.

The addition of synergists is also often helpful in overcoming resistance; for instance, DDT-dehydrochlorinase inhibitors such as WARF Antiresistant (77):

$$Cl—\langle\!\!\bigcirc\!\!\rangle\!—\!\!\underset{\underset{O}{\|}}{\overset{\overset{O}{\|}}{S}}\!—N(C_4H_9)_2$$

(77)

have restored the toxicity of DDT to populations of resistant houseflies. Piperonyl butoxide inhibits microsomal enzymes and has been useful against insects that have developed tolerance to some organophosphorus and carbamate insecticides while with malathion-resistant insects, the most effective synergists were triphenyl phosphate, tributyl phosphorotrithioate (DEF), and several dibutylcarbamates[2]. The major biochemical detoxification mechanisms operating in dimethoate-resistant insects can be inhibited by methylene dioxyphenyl synergists. However recent evidence[23] indicates that insects can become resistant to the synergists themselves which can substantially reduce the effectiveness of synergist–insecticide mixtures.

References

1. Fest, C. and Schmidt, K. J., *The Chemistry of Organophosphorus Pesticides*, Springer-Verlag, Berlin, 1973.

2. Eto, M., *Organophosphorus Pesticides: Organic and Biological Chemistry*, CRC Press, Cleveland, Ohio, U.S.A., 1974.
3. Saunders, B. C., *Some Aspects of the Chemistry and Toxic Action of Organic Compounds Containing Phosphorus and Fluorine*, Cambridge University Press, 1957, p. 91.
4. Emsley, J. and Hall, D., *The Chemistry of Phosphorus*, Harper and Row, London, 1976, p. 494.
5. *Pesticide Manual* (Eds. Martin, H. and Worthing, C. R.), 4th edn., British Crop Protection Council, 1974.
6. Approved Products for Farmers and Growers, Ministry of Agriculture, Fisheries, and Food, 1978.
7. Cremlyn, R. J. W. *Internat. Pest Control*, **16** (6), 5 (1974).
8. Martin, H., *The Scientific Principles of Crop Protection*, 6th edn., Arnold, London, 1973, p. 242.
9. Kay, I. T., Snell, B. K., and Tomlin, C. D. S., 'Chemicals for agriculture' in *Basic Organic Chemistry* (Eds. Tedder, J. M., Nechvatal, A., and Jubb, A. H.), Wiley, London, 1975, p. 426.
10. Ware, G. W., *Pesticides*, W. H. Freeman and Co., San Francisco, 1975, p. 42.
11. Frear, D. E. H., *Pesticides Handbook Entoma*, 20th edn., College Science Publisher, State College, Pennsylvania, 1968.
12. Hartley, G. S. and West, T. F. *Chemicals for Pest Control*, Pergamon Press, Oxford, 1969, p. 64.
13. O'Brian, R. D., *Toxic Phosphorus Esters*, Academic Press, New York, 1960.
14. Triggle, D. J. 'Chemical aspects of the autonomic nervous system' in *Theoretical and Experimental Biology*, Vol. 4, Academic Press, New York, 1965.
15. Corbett, J. R., *The Biochemical Mode of Action of Pesticides*, Academic Press, London and New York, 1974, p. 107.
16. Metcalf, R. L., 'Chemistry and biology of pesticides' in *Pesticides in the Environment* (Ed. White-Stevens, R.), Dekker, New York 1971, p. 1.
17. Fukuto, T. R. and Sims, J. J., 'Metabolism of insecticides and fungicides' in *Pesticides in the Environment* (Ed. White-Stevens, R.), Dekker, New York 1971, p. 145.
18. Walker, C. H., 'Variations in the intake and elimination of pollutants in *Organic Chlorine Insecticides: Persistent Organic Pollutants* (Ed. Moriarty, F.), Academic Press, London, 1965, p. 73.
19. Hassall, K. A., *World Crop Protection: Pesticides*, Vol. 2, Iliffe Books Ltd., London, 1969, p. 138.
20. Beynon, K. I., 'The relevance of pesticide metabolism', *Proc. 8th Brit. Insectic. and Fungic. Conf.*, Brighton, 3, 771 (1975).
21. Woods, A., *Pest Control*, McGraw-Hill, London, 1974, p. 124.
22. Brown, A. W. A. and Pal, R., *Insect Resistance in Arthropods*, World Health Organization, Geneva, 1941.
23. Sawicki, R. M., *Chem. and Ind.*, **1974**, 980.

Chapter 7
Fungicides

The majority of commercial fungicides used at present belong to the class known as protectant or surface fungicides. They are usually applied to plant foliage as dusts or sprays[1-4]. Such materials do not appreciably penetrate the plant cuticle and are not translocated within the plant, whereas the much more recent systemic fungicides or plant chemotherapeutants as they are sometimes called are absorbed by the plant via the roots, leaves, or seeds and are translocated within the plant.

Most pathogenic fungi penetrate the cuticle and ramify through the plant tissues, so if the fungicide is to combat the fungal infection a protectant fungicide must be applied *before* the fungal spores reach the plant. However a few fungi such as powdery mildews are restricted to the surface of the leaf and in these cases surface fungicides may also possess eradicant action.

If a given candidate chemical is to be an effective protectant fungicide, the following conditions must be satisfied[5]:

(a) it must have very low phytotoxicity otherwise too much damage will be caused to the host plant during application;
(b) it must be fungitoxic *per se* or be capable of conversion into an active fungitoxicant within the fungal spore and must act quickly before the fungal infection penetrates the plant cuticle;
(c) generally the fungicide must be able to penetrate the fungal spore and reach the ultimate site of action in the fungus;
(d) most agricultural protectant fungicides are applied as foliar sprays, and they must be capable of forming tenacious deposits which are resistant to the effects of weathering over long periods.

Considering these criteria in rather more detail, the most difficult problem is the attainment of the desired selective toxicity to the fungus because of the close relationship between plants and fungi. In fact none of the protectant fungicides currently on the market is completely non-phytotoxic. A possible route to a non-phytotoxic fungicide would be the discovery of compounds that interfere with the biosynthesis of chitin which is found in the cell walls of most parasitic fungi but is absent in higher plants[6]. The cell membrane is semipermeable often consisting of two monolayers of lipids surrounded on either side by a layer of protein so a degree of lipid solubility is an important factor aiding penetration. For instance, in a series of 2-alkylimidazolines the fungitoxicity reached a maximum with increasing length of the alkyl chain up to C_{17} and then decreased[6], and there are many similar cases in which addition of inert lipophilic substituents increases the fungicidal activity[5]. Once the compound has gained entry into the fungal cell and reached the critical reactor site, it must then exert its toxic action on the fungus by either a chemical or physical mechanism. With chemical toxicants there will be a chemical reaction with perhaps a vital enzyme which ultimately kills the fungus. On the other hand, with physical toxicants the deleterious effect on the fungus is caused by the compound possessing the correct hydrophilic – lipophilic balance so that it dilutes the biophase and thus physically inhibits vital cellular processes. In both mechanisms, the success of the material does require the optimum oil–water solubility balance enabling the fungicide to reach the critical site in the fungal cell.

The majority of protectant fungicides are directly toxic to fungi and so will show up as active against spore germination *in vitro* tests. Fungicides may be applied to fruits, foliage, or seeds as dusts or sprays: in dusting a uniform coverage is most important and this requires a small particle size (approximately 5 μm)[4,5]. Spraying is a much more widely used method of application; the spray may be a solution or a fine suspension of the material, and in the latter case reduction of particle size leads to increased effectiveness in disease control which has been illustrated in the control of tomato early blight (*Alternaria solani*) by dichlone. A more uniform coverage is also aided by small spray droplets which reduces 'run off', though too fine a spray results in substantial loss through evaporation. Most fungicides are formulated with wetting agents or 'spreaders' which are particularly beneficial when the chemical is being applied to a waxy leaf surface (see Chapter 2, p. 14). The spreading property of the spray on foliage is an important factor in determining the tenacity of the dried deposit on which the persistency of protectant fungicides largely depends. The fungicide, as a dried deposit on the leaf surface, must generally be stable towards photochemical oxidation, hydrolysis, and carbonation, and instability in sunlight probably accounts for the ineffectiveness of chloranil as a foliage fungicide as compared with dichlone which is more stable[5]. Of course in some cases, the products of decomposition may be more active than the original compound; Thus the fungistatic nabam is activated by oxidation on the leaf surface to the active fungicide (see p. 114).

The earliest fungicides were inorganic materials like sulphur, lime-sulphur, copper, and mercury compounds. Elemental sulphur has been recognized as fungicidal for at least 170 years. Hence, in 1803 the Royal Gardener Forsyth recommended the use of a sulphur spray against mildew on fruit trees. In the nineteenth century sulphur was increasingly employed against powdery mildew on fruit and later for control of powdery mildew on grapes[1]. It may be applied as dusts or sprays, and the more finely divided the sulphur, the more effective is the formulation against the disease, so colloidal sulphur is formulated by grinding up flowers of sulphur with a mineral diluent (e.g. kaolin) so that there is a sulphur content of 40%, with most particles of less than 6 μm in diameter. A much more widely used liquid sulphur product is 'lime-sulphur' obtained by boiling sulphur with an aqueous suspension of slaked lime. This is a clear orange-coloured liquid consisting mainly of calcium polysulphides which break down on exposure to air to release elemental sulphur. The polysulphide content of a lime-sulphur mixture is the best indication of its effectiveness and the official specification contains at least 24% weight/volume of the polysulphide[2,4].

In 1958 the weight of sulphur used against fungi was four times that of all other fungicides combined, although since that time the amount has fallen off due to the development of more modern organic fungicides. However sulphur and lime-sulphur are still extensively applied against powdery mildews and apple and pear scab[7].

There has been much speculation regarding the mode of fungitoxic action of sulphur. At first it was assumed that sulphur could not be the toxic agent, although Sempio (1932) took the contrary view, and the fungicidal action was ascribed to the production of various sulphur derivatives. The earliest idea was that the active fungitoxicant was hydrogen sulphide and experiments showed that fungal spores can readily reduce sulphur to hydrogen sulphide which was shown to be toxic to the spores[5]. But in 1953 this theory was disproved when it was shown conclusively that colloidal sulphur was appreciably more fungitoxic than the equivalent quantity of hydrogen sulphide. Another hypothesis ascribed the fungitoxicity of sulphur to various oxidation products such as sulphur dioxide, sulphuric acid, thiosulphuric acid, or pentathionic acid[5]. The importance of these products has however been largely discounted because their toxicity can be accounted for on the basis of their hydrogen ion concentrations. Since no sulphur derivative appears responsible for the activity we have now come back full circle to the idea of Sempio that sulphur itself as the S_8 molecule may be toxic to fungi[4,6]. Certainly work on plants and fungi indicates that sulphur is not biologically inert and probably becomes involved in biological redox systems. There is considerable evidence that sulphur can penetrate fungal spores and the improved activity observed in the presence of urea, hydrocarbons, soft soap, or lime is probably due to enhanced penetration[5]. Fungal spores can take up large amounts of sulphur, practically all of which is subsequently evolved as hydrogen sulphide, and the toxicity of sulphur is probably due to it functioning as hydrogen acceptor thereby interfering with vital hydrogenation and dehydrogenation reactions within the fungal cell. It remains difficult to understand why sulphur has a specific toxicity to fungi; unlike copper and other heavy metal fungicides, sulphur is practically non-toxic to mammals. This may possibly be due to differential availability at the site of action within the fungus and lipid-rich fungi such as powdery mildews may retain more sulphur than other types of organisms[4].

Among the heavy metals, only the compounds of copper and mercury have been used widely as fungicides, although silver is the most toxic metal cation to fungi. The relative toxicity of the various metal cations to fungi is as follows: $Ag > Hg > Cu > Cd > Cr > Ni > Pb > Co > Zn > Fe > Ca$. The toxicity of the metals has been related to their position in the Periodic Table — toxicity within a given group generally increases with the atomic mass. It has also been related to the relative chelating powers of the metals, the stability of the metal sulphide, and the electronegativity of the cations. Recent studies[5,6] indicate that the latter criterion provides the best correlation because the degree of electronegativity is a measure of the stability of metal bonds with cellular constituents which will in turn affect the stability of metal chelates and sulphides. The degree of fungitoxicity is probably determined by the strength of covalent or coordinate bonding in unionized complexes at the cell surface.

Copper sulphate has been used since the eighteenth century as a seed treatment against cereal bunt, although now replaced by the much more effective organo-mercurial compounds. Copper ions in solution are toxic to all plant life; selective fungicidal action can therefore only be achieved by application of an insoluble copper compound to the foliage. Examples of compounds used include copper oxychloride, copper carbonate, cuprous oxide, and Bordeaux mixture. The latter is the most important of the copper fungicides and was discovered by Millardet in 1882 (Chapter 1, p. 4). Bordeaux mixture was effective against vine downy mildew (*Plasmopara viticola*), a disease which had been introduced into France from the United States of America on vine rootstocks imported because they were resistant to *phylloxera*. The mildew threatened the French vineyards and so the situation was extremely serious. The botanist Millardet, who recommended the use of American vines, discovered that he had merely exchanged one disease for another, so his accidental discovery of Bordeaux mixture came at just the right time to control the mildew. Bordeaux mixture, named from the locality of its origin, consists of $CuSO_4$ (4.5 kg), $Ca(OH)_2$ (5.5 kg) in 454 litres of water. Once the mixture has been prepared it should be sprayed on to the crop as soon as possible since the fungitoxicity of the mixture decreases on standing. It is rather difficult to apply because the precipitate tends to block the spray nozzles.

The chemistry of Bordeaux mixture and the precise mode of fungicidal action are complex. The proportions of the ingredients used and the method of preparation have considerable influence on the fungicidal effectiveness of the product. In particular, the fineness and the composition control the physical properties of the dried deposit which in turn greatly affects the tenacity of the spray deposit on the leaves.

The active ingredient is probably not cupric hydroxide, but rather a basic copper sulphate approximating to the formula $[CuSO_4 \cdot 3Cu(OH)_2]$. Bordeaux mixture is almost insoluble in water, so how is the copper mobilized to kill the fungus? Wain considered[8] that exudates both from the surface of the leaf and from the fungal spores can dissolve appreciable quantities of copper from the dried Bordeaux deposits due to the presence of certain compounds like amino and hydroxy acids which can form chelates with the copper. It has been demonstrated[5] that exudates from the spores of *Neurospora sitophila* react with Bordeaux mixture giving a soluble copper complex which dissociates in solution yielding toxic cupric ions. Similarly[8] when copper fungicides are employed as seed dressings the exudates from the seed dissolved copper from the dressing which accounts for the fungicidal protection provided. Soluble copper compounds are too phytotoxic to be useful as foliar sprays and different insoluble copper compounds ('fixed coppers') vary considerably in their fungitoxic properties. These have the advantage of easier application as compared with Bordeaux mixture and are now widely used. Two well-known examples are copper oxychloride, approximately $[CuCl_2 \cdot 3Cu(OH)_2]$, which is marketed both as a colloid and as a dispersable powder, the former being used with a wetting agent, and yellow cuprous oxide in finely divided form prepared by reduction of copper salts in alkaline solution. Such 'fixed coppers' are used for control of potato and tomato blight, vine and hop downy mildew, and many other common leaf and fruit diseases of horticultural crops[7]. In Europe their major use is in the protection of potatoes from late blight (*Phytophthora infestans*) about 15% of the potato acreage being sprayed annually.

The powerful bactericidal properties of mercuric chloride led to its examination as a fungicidal dressing for cereal seeds when it was shown to be effective against *Fusarium* disease of rye. For medicinal purposes mercuric chloride was quickly replaced by less poisonous organomercury derivatives and these were subsequently examined as cereal seed dressings[3].

Mercury forms stable organic derivatives containing carbon—mercury bonds; the majority of organomercurial seed dressings have the general formula RHgX (where R = alkyl/aryl radical and X = an anion). The fungicidal activity against wheat bunt (*Tilletia caries*) is almost entirely dependent on the nature of the hydrocarbon radical R and the active species is probably RHg^+. Those derivatives containing phenyl or other aryl groups were generally the most effective as seed dressings. Two well-known examples are phenylmercury chloride and the corresponding acetate. Such compounds are extensively applied as seed dressings against wheat bunt and smut on oats and barley[7]. The cereal seeds are either steeped in a suspension of the active ingredient or may be dusted with a powder (see Chapter 2, p. 16). Seed dressing is a very efficient method of applying the chemical; with organomercurial seed dressings only a few ounces of the active ingredient are needed per acre. A disadvantage is that the treated seed presents hazards especially to seed-eating birds (see Chapter 14, p. 213). In Britain where phenylmercury compounds are much more extensively used than alkyl derivatives, there have been comparatively few cases of birds poisoned by organomercurials[3]. On the other hand, in Sweden, where alkylmercury compounds are largely used as liquid seed dressings widespread mercury poisoning of birds and other mammals has occurred.

Phenylmercury compounds are also used as sprays against apple and pear scab and canker, as industrial fungicides, and to protect cotton and rice from fungal attack[7].

Arylmercury compounds may be obtained by direct electrophilic substitution (mercuration) of the aromatic substrate. Phenylmercury acetate, one of the most widely used of all organomercurials, is prepared by heating benzene and mercuric acetate in glacial acetic acid:

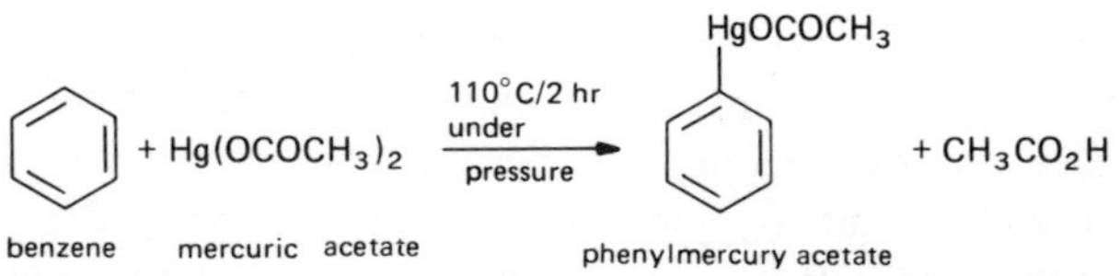

No other compounds can economically compete with organomercury derivatives as seed dressings for cereals. Their main disadvantage is that they are very poisonous and there are fears that their continued large-scale use may result in serious effects on the ecosystem. Efforts are therefore being made to find less toxic seed dressings.

Mercury, like copper and several other heavy metals, is known to affect respiration by poisoning essential sulphydryl respiratory enzymes in the fungal cell, and the toxicity can be reversed by the addition of sulphydryl compounds[5]. However it is not certain whether organomercurials are toxic *per se* or whether they just function as carriers of toxic mercury ions through lipid barriers to a reactive site.

Some alkyl-, but not arylmercury compounds produce severe brain lesions in mammals. They may destroy nervous tissue. Certain mercury compounds also undergo hydrolysis to inorganic mercury salts which are powerful cumulative poisons tending to accumulate in the kidneys[4]. These compounds should therefore be phased out as pesticides as soon as possible.

Inorganic tin compounds like stannic chloride are generally not fungicidal and in alkyltin derivatives of types $RSnX_3$, R_2SnX_2, and R_3SnX, the fungitoxicity increases as the number of alkyl groups goes up from one to three[4]. As with the organomercurials, the nature of the anion (e.g. Cl, OH, $OCOCH_3$) was not decisive and maximum activity always occurred with alkyl groups containing 3 or 4 carbon atoms, but the trialkytin compounds were too phytotoxic for use as foliage fungicides. Tributyltin oxide, $(C_4H_9)_3Sn-O-Sn(C_4H_9)_3$, is however used extensively for stabilization of transparent plastics against photochemical change, for rot proofing of fabrics, in antifouling paints to inhibit the growth of barnacles and algae on the bottom of ships, and for preservation of marine timbers.

Triphenyltin compounds were much less phytotoxic and still very strongly fungicidal[2-5]; thus triphenyltin acetate is the active ingredient of the foliage fungicide Brestan or fentin which has a wide spectrum of activity and is used specially for control of potato blight[7] and leaf spot diseases of sugar beet and celery; it also deters insects from feeding on the treated foliage. Although fentin has higher phytotoxicity than 'fixed' coppers, it is appreciably more fungicidal and can be successfully used at only about one-tenth of the dose required for the copper fungicides, and consequently the fungus is controlled with less damage to the host plant.

These organotin compounds can be prepared by a double decomposition reaction from phenylmagnesium bromide and stannic chloride:

$$3\ C_6H_5MgBr + SnCl_4 \xrightarrow{(C_2H_5)_2O} (C_6H_5)_3SnCl + 3\ MgBrCl \xrightarrow{CH_3CO_2Na}$$

$$(C_6H_5)_3SnOCOCH_3 + NaCl$$

fentin

The aryltin derivatives apparently do not build up in the environment since they undergo biodegradation to triphenyltin (IV) hydroxide. Consequently their application does not appear to damage the environment.

The development of purely organic fungicides really began with the discovery of the fungicidal activity of the dithiocarbamates which had been originally developed as vulcanization agents for the rubber industry. The dithiocarbamates and their derivatives are one of the most important groups of organic fungicides for controlling plant diseases. They were introduced by Tisdale and Williams in 1934: the most valuable members were derived from dimethyldithiocarbamic acid since other alkyl groups reduced the fungitoxicity[5,9]. Thiram or tetramethylthiuram disulphide (1) was the first compound to be applied as a fungicide and is still used, especially against grey mould on lettuce and strawberry and as a seed dressing against soil fungi causing damping off diseases[7]. Later work[9] resulted in the discovery of the fungicidal activity of the zinc and ferric salts of dimethyldithiocarbamic acid, known as ziram or ferbam (2; M = Zn, x = 2; or M = Fe, x = 3).

Disodium ethylenebisdithiocarbamate or nabam (3) was also fungicidal and is used for control of some root rots, but as a foliage fungicide it tends to be phytotoxic and has little persistence in rain and has been largely replaced by the insoluble zinc and manganese salts known as zineb and maneb (4; M = Zn or Mn respectively). They are some of the most widely used organic protectant fungicides and are applied for the control of a wide range of phytopathogenic fungi such as downy mildews, potato and tomato blight[7]. They have low mammalian toxicities with LD_{50} values to rats of approximately 7000 mg/kg and they now rank in tonnage use only below sulphur and fixed coppers and have partly replaced copper fungicides for control of potato blight. (Plate 11).

A more recent compound is sodium N-methyldithiocarbamate or metham-sodium (5) valuable as a soil sterilant for control of damping off diseases, potato cyst eelworm, and weed seedlings[7].

Thiram (1) is prepared by the interaction of carbon disulphide and dimethyl-lamine in the presence of sodium hydroxide solution to give sodium dimethyldi-thiocarbamate which is subsequently oxidized by air, hydrogen peroxide, chlorine, or iodine to thiram:

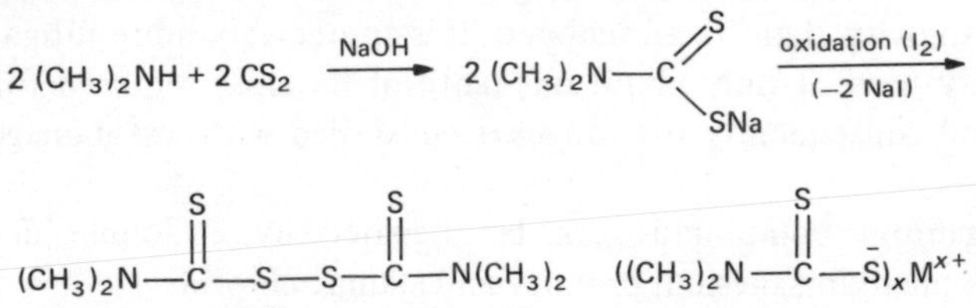

The ethylenebisdithiocarbamates, the most important dithiocarbamates as protectant fungicides[3], are obtained by reaction of ethylene diamine with carbon disulphide in the presence of sodium hydroxide:

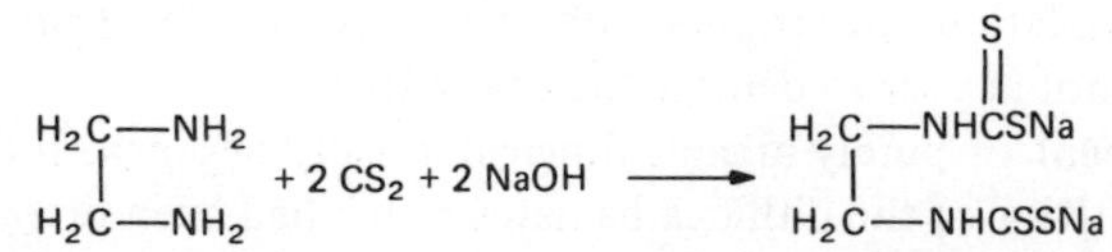

$$H_2C—NH—\overset{\displaystyle S}{\overset{\|}{C}}—\overset{-}{S}$$
$$H_2C—NH—\underset{\displaystyle S}{\underset{\|}{C}}—S \quad M^{2+}$$

$$CH_3NH—\overset{\displaystyle S}{\overset{\|}{C}}SNa$$

       (4)   (5)

Nabam (3) which is water soluble is converted into either zineb (4; M = Zn) or maneb (4; M = Mn), which are insoluble, by reaction with an aqueous solution of zinc or manganous sulphate.

The uses of dithiocarbamates and their derivatives as fungicides have been reviewed and their mode of fungicidal action has been the subject of considerable study[4,9]. There are some differences in their fungicidal properties which suggest that the N,N-dimethyldithiocarbamates may have a different mode of action from that of the ethylenebisdithiocarbamates, thus they possess a distinctive spectrum of activity against various species of fungi and histidine has been shown to antagonize the antifungal action of thiram (1) but not that of nabam.

The dithiocarbamates, such as thiram and ziram probably owe their fungitoxicity to their ability to chelate with certain metal ions, especially copper. Studies with the fungus *Aspergillus niger* have demonstrated that in the presence of cupric ions an increase in the concentration of sodium dimethyldithiocarbamate inhibits the growth of the fungus at two levels. The first inhibition level occurs when the copper:dithiocarbamate ratio is reasonably high, approximately 20:1 and is associated with formation of the 1:1 copper:dithiocarbamate complex (6). When the dithiocarbamate concentration is further increased, the fungitoxicity decreases due to formation of the 1:2 complex (7).

    (6)   (7)

The second inhibition level requires high concentrations of dimethyldithiocarbamate ($\simeq 50$ p.p.m.) and is due to the presence of free dimethyldithiocarbamate ions since no cupric ions are needed for this effect. On the other hand, the presence of copper is definitely required for high fungicidal activity with dimethyldithiocarbamates, and the 1:1 chelate (6) can penetrate lipid barriers in the fungal cell, whereas the 1:2 chelate (7) is apparently not fungicidal because it is too insoluble in water to permit penetration and its formation accounts for the bimodal curve observed for the graph of fungitoxicity against sodium dimethyldithiocarbamate concentration (Figure 7.1).

When the 1:1 complex (6) has entered the fungal cell it is probably converted to the free dimethyldithiocarbamate anion which can complex with certain vital trace metals. Alternatively the complex may itself be fungicidal, and has been shown to interfere with the uptake of oxygen by yeast cells and pyruvate accumulates in *Aspergillus niger* after treatment with sodium dimethyldithiocarbamate. The fungitoxicity of the complex (6) may thus arise from interference with the respiration of the fungus by inactivation of the pyruvate dehydrogenase system (see Chapter 3, p. 29).

The ethylenebisdithiocarbamates (e.g. nabam, maneb, and zineb) almost certainly act by a different mechanism involving oxidative decomposition of the chemical on the leaf surface to such products as thiuram disulphide, carbon disulphide, and possibly ethylene diisothiocyanate[6]:

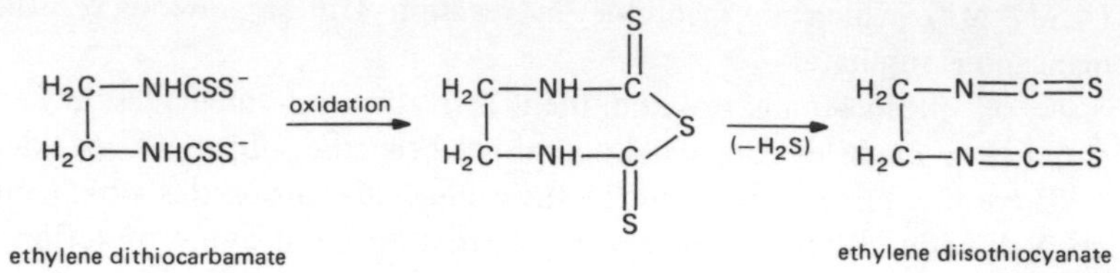

ethylene dithiocarbamate   ethylene diisothiocyanate

It has been shown[4] that pure ethylenebisdithiocarbamates are inactive *before* they have been exposed to air and also that isothiocyanates are fungicidal by virtue of their ability to react with vital thiol compounds within the fungal cell. The inhibitory action of nabam on the germination of fungal spores is strongly antagonized by the addition of thiols such as thiolglycollic acid and cysteine.

In 1951 Kittleson working for the Standard Oil Company of America discovered that certain compounds containing the N-trichloromethylthio group are powerful surface fungicides[10]. The best known example is captan or N-(trichloromethyl-thio)-4-cyclohexene-1,2-dicarboximide (8) which is easily synthesized from the readily available starting materials butadiene and maleic anhydride:

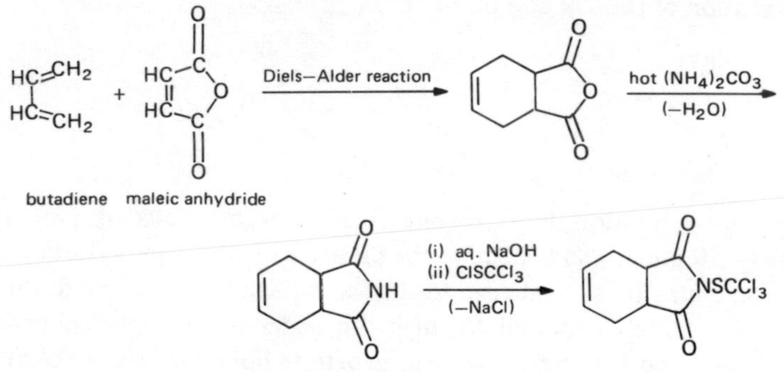

butadiene   maleic anhydride

tetrahydrophthalimide   (8)

Trichloromethylsulphenyl chloride is obtained from carbon disulphide:

$$CS_2 + 3\,Cl_2 \xrightarrow{I_2} Cl_3C\cdot SCl + SCl_2$$

Captan is a very effective and persistent foliage fungicide, especially for control of apple and pear scab, black spot on roses, and as seed dressings against many soil and seed-borne diseases. It reduces *Gloeosporium* rot of apples, stem rot of tomatoes, and *Botrytis* mould on soft fruit[1,7], but has little action against powdery mildews. Analogues which have been subsequently developed as foliar fungicides include

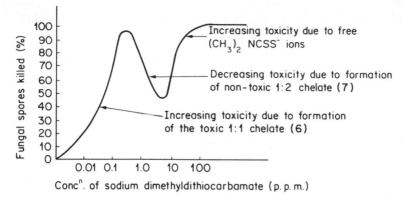

Figure 7.1   Relation between sodium dimethyldithiocarbamate concentration and fungitoxicity.

folpet (**9**) and Difolatan (**10**), which are more effective against potato blight. These are some of the safest of all fungicides with LD_{50} (oral) values to rats of about 10,000 mg/kg.

(9)   (10)

All the compounds originally listed by Kittleson contain the $>$N–S–CCl$_3$ group and he ascribed their fungitoxicity to the presence of this group. In contrast Rich[11] considered that the toxophore was probably the —CO—NR—CO— grouping and the —S—CCl$_3$ group was merely a convenient shaped charge to facilitate penetration of the compound into the fungal cell. The latter idea was however disproved by the discovery of a large number of fungicidal trichloromethylthio compounds which do not contain the imide moiety.

Lukens and Sisler (1958) showed that captan interacts with cellular thiols to produce thiophosgene which they regard as the ultimate toxicant[6]:

N—SCCl$_3$ + 2RSH $\longrightarrow$ NH + CSCl$_2$ + RSSR + HCl

thiophosgene

(8)

The evolved thiophosgene would finally poison the fungus by combination with vital sulphydryl-, amino- or hydroxyl-containing enzymes, and this hypothesis is supported by the fact that the fungitoxicity of captan and related compounds can be destroyed by addition of thiols. On the other hand, Owens and Novotny (1959) argued that the toxicity of captan to fungi arises from the reaction of the intact captan molecule with unprotected thiol groups in the fungal cell because they concluded that captan would react faster with cellular thiols than thiophosgene. The fungicidal activity of compounds of type R—S—CCl$_3$ probably depends on rupture of the R—S bond by reaction with cellular thiols which oxidizes thiol groups and simultaneously releases toxic products (e.g. thiophosgene) into the fungal cell. For maximum fungitoxicity, the toxic products must be released at the critical site of action in the fungal cell which in turn will be governed by such factors as the strength of the R—S bond and the oil/water solubility balance and both of these factors will be controlled by the nature of the R group[10].

Another theory proposed by Rich[11] suggested that captan owes its fungitoxicity to the formation of the transitory N-chlorotetrahydrophthalimide which subsequently reacts with vital enzymes in the fungal cell:

So perhaps the imide moiety does play a vital role in the activity of captan; certainly a large number of fungicidal compounds are known[12] which contain the imino group. One important mechanism of the fungitoxicity is probably the reaction with vital cellular thiols, but these trichloromethylthio compounds do not appear to act by a single biochemical mechanism. The idea of multisite action is supported by the fact that although N-trichloromethylthio fungicides have been extensively used for some 20 years they have not induced the emergence of resistant fungal strains (Chapter 7, p. 137).

Captan probably acts non-specifically because Neurospora crassa conidia accumulated the fungicide when they were exposed to a dose that killed 50% of the spores. The precise nature of the interactions between the toxicant and the fungal cell has not been definitely established, but thiols are the most likely reaction sites, since the intact captan molecule is unlikely to react with amino or hydroxy groups at cellular pH conditions but reacts easily with thiols.

The reactivity of captan (8) and the aromatic analogue folpet (9) with various biochemical systems has been clearly demonstrated. Thus interaction of a 20 μM solution of folpet with isolated yeast glyceraldehyde-3-phosphate dehydrogenase resulted in 82% reduction of enzymic activity due to combination with thiol groups in the enzyme. On the other hand, folpet also caused substantial inhibition of the activity of isolated α-chymotrypsin which does not contain thiol groups, suggesting that other reactions may be significant in the fungicidal action of these compounds. Experiments with rat liver mitochondria showed that both captan and folpet inhibited a number of mitochondrial reactions, including oxidative phosphorylation and the oxidation of the reduced form of nicotinamide adenine dinucleotide NADH$_2$.

Related surface fungicides of a similar type include the recently introduced dichloro- and tolyfluanids:

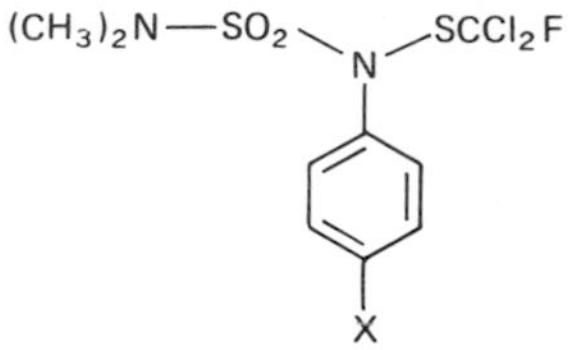

dichlorofluanid (X = H)
tolyfluanid (X = CH$_3$)

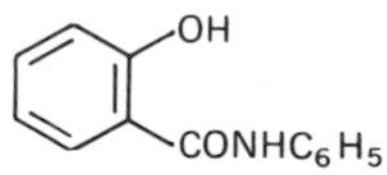

(11)

These are valuable against *Botrytis* on fruits. They do not contain the trichloro-methylthio group, but probably also owe their fungitoxicity to non-specific reaction with thiols.

Phenols

The majority of phenols, especially those containing chlorine are toxic to microorganisms; their bactericidal action has been known for a long time and many phenols are also fungicidal. However the majority of phenols are too phytotoxic to permit their use as agricultural fungicides. They are widely used as industrial fungicides: cresols contribute to the fungicidal action of creosote oil as a timber preservative (see Chapter 1, p. 4); chlorinated phenols such as pentachlorophenol and its esters are widely used as industrial biocides for the protection of such materials as wood, and textiles from fungal attack[13]. One of the earliest organic fungicides was salicylanilide or Shirlan (11) (1931) used to inhibit growth of moulds on cotton and against a number of leaf diseases, such as tomato leaf mould[2,5,8]. This has been superseded by other compounds, but is structurally closely related to one group of modern systemic fungicides (see p. 131).

Dinitrophenols are very versatile pesticides, thus 2,4-dinitro-*o*-cresol (DNOC) was first used as an insecticide in 1892 and in 1933 as a selective herbicide (see Chapter 1, p. 5). This is of course much too phytotoxic to be used as a foliage fungicide, but dinocap or Karathane 2,4-dinitro-6-(2-octyl)phenylcrotonate (12) introduced in 1946 is a non-systemic aphicide and contact fungicide which is effective for control of powdery mildew on many horticultural crops[7,14]. Unlike DNOC, which has a high mammalian toxicity (LD$_{50}$ (oral) to rats 30 mg/kg) and has done considerable damage to the environment (see Chapter 14, p. 210), dinocap (12) has low mammalian toxicity (LD$_{50}$ (oral) to rats 980 mg/kg). It is made by condensation of 2,4-dinitrophenol and 2-octyl alcohol followed by esterification with crotonyl chloride: Binapacryl (13) is closely related to dinocap and is used for control of red spider mites and powdery mildew on apple[4,7]. *o*-Phenylphenol or 2-hydroxybiphenyl is used on citrus fruit wrappings to inhibit rot, for the disinfection of seed boxes, and for control of apple canker.

The fungicidal action of the various phenols depends on their ability to uncouple oxidative phosphorylation and thus prevent the incorporation of inorganic phosphate into ATP without affecting electron transport. This action probably occurs at the mitochondrial cell wall and causes the cells to continue to respire but

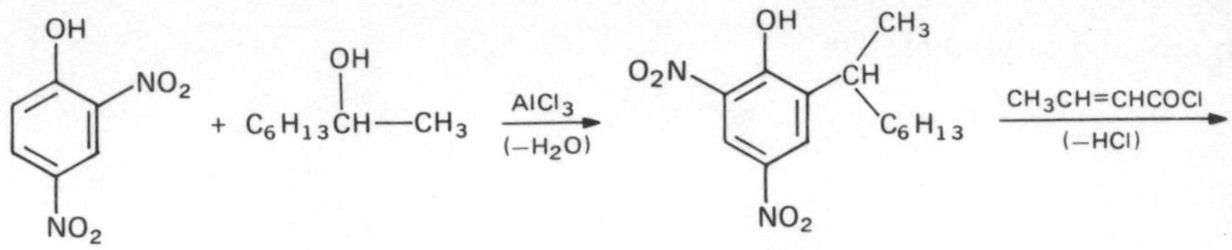

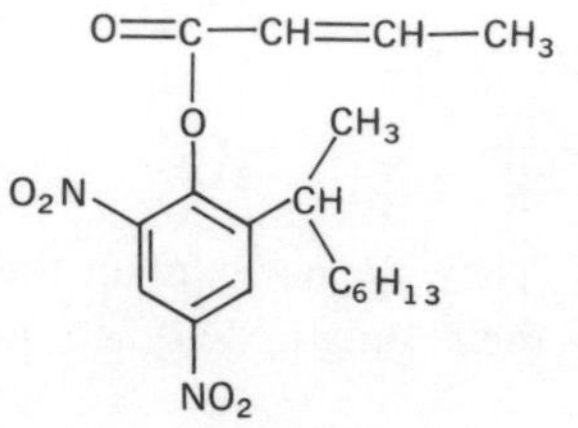

(12) 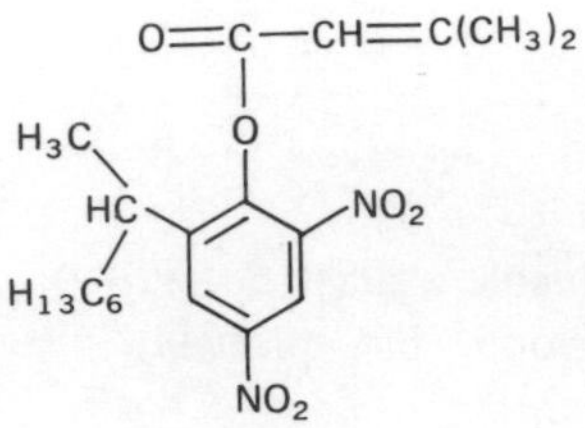   (13)

they are soon deprived of the ATP necessary for growth. The activity of phenolic esters like (12, 13) presumably arises from *in vivo* hydrolysis to the dinitrophenol within the fungal cell, the rest of the molecule merely functioning as a convenient shaped charge conferring the right degree of oil/water solubility to aid penetration of the fungal spore[4].

Oxine or 8-hydroxyquinoline is a protectant fungicide which, when suitably formulated, appears to possess limited systemic action. The sulphate has been used against *Rhizoctonia* and *Fusarium* on horticultural crops and the benzoate against the Dutch elm disease. Oxine has a striking capacity to form chelates with metals and the copper chelate (14) is a potent fungicide which is effective against a range of phytopathogenic fungi. The 1:2 chelate (14) has greater lipophilicity than oxine and the fungitoxicity of oxine is probably due to interaction with trace amounts of copper forming the chelate (14) which penetrates the fungal cell and then equilibrates with the 1:1 cupric ion:oxine complex which may be the actual toxicant. The proposed mode of action resembles that of the dithiocarbamates and oxine also exhibits a similar bimodal dosage–response graph (see p. 113)[2,4,14].

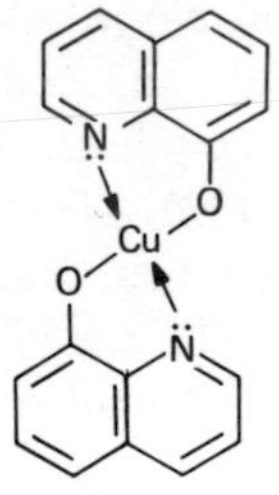

(14) 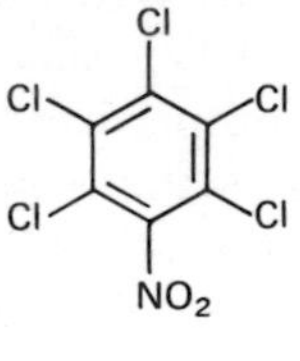   (15)

Chlorobenzenes and Related Compounds

2,6-Dichloro-4-nitroaniline or Allisan was marketed in 1959 by Boots Ltd. and is especially valuable for control of *Botrytis* on lettuce, tomatoes, and straw-berries[7,13], and against fungal organisms causing post-harvest decay of fruits[4]. It is

Plate 1. Application of the herbicide Asulox (asulam) by a motorized mist blower which effectively treats a wide swathe of bracken and the penetration achieved by the small droplets makes it possible to reduce the spray volume to 1 gal. of Asulox in 10 gal. of water per acre of bracken. (Photograph by courtesy of May and Baker Ltd) (See p. 150.)

Plate 2. Bracken spraying from the air with Asulox. (Photograph by courtesy of May and Baker Ltd) (See p. 150.)

Plate 3. A small crop sprayer. (Photograph by courtesy of May and Baker Ltd.) (See p. 17.)

Plate 4. Application of an experimental fungicide to apple trees. (Photo by courtesy of May and Baker Ltd) (See p. 17.)

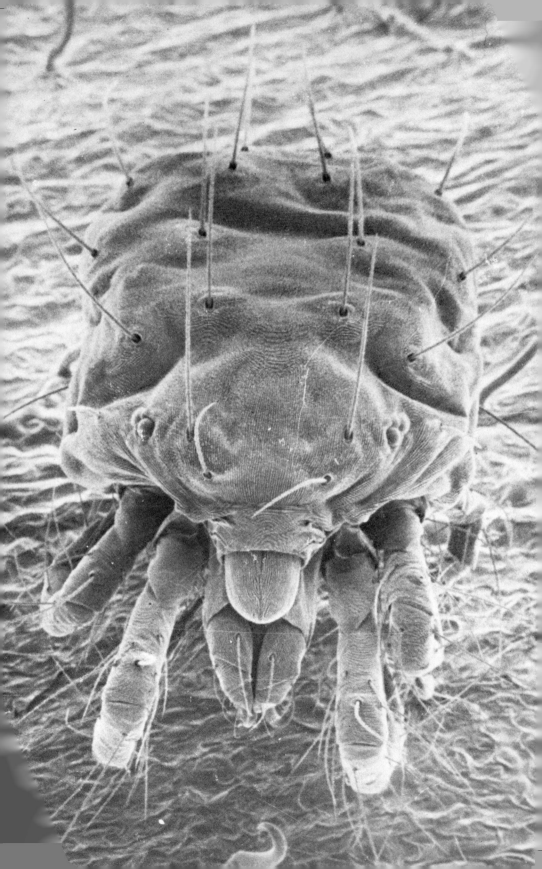

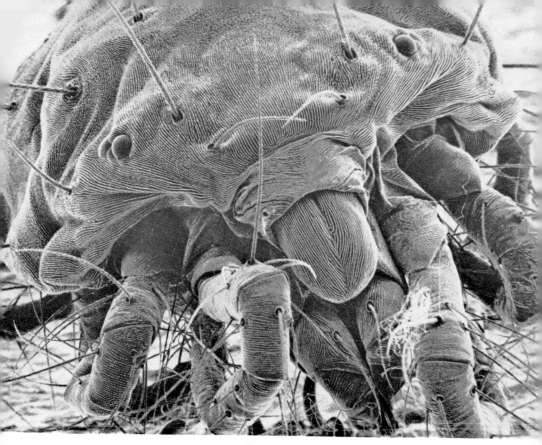

Plate 5. Scanning electron micrograph of a red spider mite. (Two views of the mite.) (This photograph is published by kind permission of the Agrochemicals Division of The Boots Company Ltd., Nottingham) (See p. 56.)

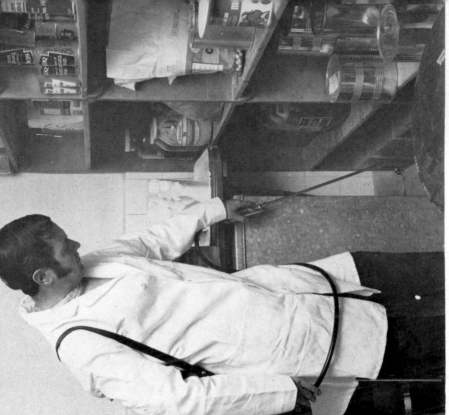

Plate 6. Safe spraying of vegetables with Gardona, a new insecticide from Shell; it can be applied safely without protective clothing and is effective against caterpillars and flea beetles. (A Shell photograph) (See p. 73.)

Plate 7. Spraying Ficam W, a methylcarbamate insecticide, in a food larder for control of cockroaches and other domestic pests. The non-odourous and non-staining properties of Ficam W ensure that there is no danger of food contamination, so long as food preparation surfaces are covered and food c

Plate 7. Rose after treatment with the carbamate insect... (Photograph by permission of ICI Ltd., Plant Pro... Division) (See p. 97.)

Plate 8. Aphids on rose (before treatment). (Photograph by permission of ICI Ltd., Plant Protection Division.)

Plate 10. Untreated peppers infected with *Phytophthora*. (Photograph by permission of ICI Ltd., Plant Protection Division.)

Plate 11. Peppers after treatment with a fungicide. (Photograph by permission of ICI Ltd., Plant Protection Division) (See p. 112.)

Plate 12. Mildew on untreated barley. (Photograph by permission of ICI Ltd., Plant Protection Division.)

Plate 13. After treatment of the barley with the systemic fungicide ethirimol. (Photograph by permission of ICI Ltd., Plant Protection Division) (See p. 132.)

Plate 14. Untreated potatoes. (Photograph by permission of ICI Ltd., Plant Protection Division.)

Plate 15. Potatoes after weed control by treatment with a chemical herbicide. (Photograph by permission of ICI Ltd., Plant Protection Division) (See p. 140.)

Plate 16. Left, broad bean plant infected with *Botrytis*; right, treated with the fungicide Rovral. (Photograph by courtesy of May and Baker Ltd) (See p. 120.)

Plate 17. Left, cereals treated with Carbyne (barban) for control of wild oats; right, untreated crop. (Photograph by courtesy of Fisons Ltd., Agrochemical Division) (See p. 149.)

Plate 18. Cattle graze land in North Wales after killing the bracken by aerial application of Asulox. Before treatment the land was virtually unusable due to bracken infestation. (Photograph by courtesy of

the urea herbicide Dicurane. Left, untreated; right, treated. (Photograph by courtesy of Ciba-Geigy Agrochemicals) (See p. 151.)

Plate 20. Blackgrass in wheat. (Photograph by permission of ICI Ltd., Plant Protection Division.)

(Photograph by permission of ICI Ltd., Plant Protection Division) (See p. 151.)

Plate 22. Control of couch grass in maize with Gesaprim (atrazine). Top, untreated; bottom, treated. (Photograph by courtesy of Ciba–Geigy Agrochemicals) (See p. 154.)

Plate 23. Mayweed control in barley with Mofix. Left, untreated; right, treated. Mofix is a mixture of the triazine terbuthylazine and bromofenoxine (3,5-dibromo-4-hydroxybenzaldehyde *O*-(2,4-dinitrophenyl)oxime). (Photograph by courtesy of Ciba–Geigy Agrochemicals) (See p. 155.)

Plate 24. Left, onions sprayed with the herbicide Totril (ioxynil octanoate); right, unsprayed. (Photograph by courtesy of May and Baker Ltd.) (See p. 162.)

Plate 25. Right-hand side, sugar beet sprayed pre-emergence with Venzar, a uracil herbicide. (Photograph by courtesy of Fisons Ltd., Agrochemical Division) (See p. 158.)

Plate 26. Rye grass in Suffolk after pre-emergence treatment with the herbicide Nortron against blackgrass. (Photograph by courtesy of Fisons Ltd., Agrochemical Division.)

Plate 27. The effect of a juvenile hormone mimic on *Pieris brassica* pupa; left, untreated; right, treated. (Photograph by permission of ICI Ltd., Plant Protection Division) (See p. 203.)

Plate 28. The effect of a juvenile hormone mimic on *Pieris brassica* pupa. (Photograph by permission of ICI Ltd. Plant Protection Division) (See p. 203.)

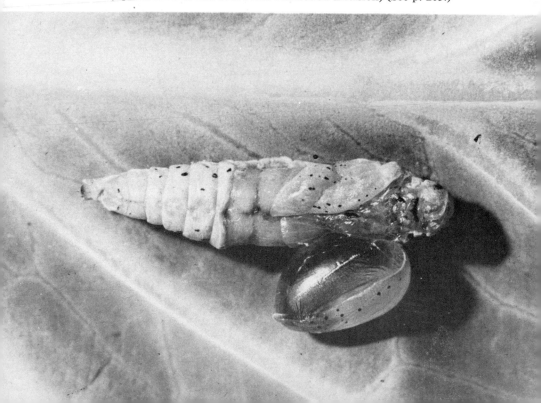

made by direct chlorination of *p*-nitroaniline. Other fungicidal chloronitrobenzenes include pentachloronitrobenzene (PCNB) called quintozene (15) which was introduced in the 1930s but has only recently become widely used as a soil fungicide against many pathogenic fungi such as damping off diseases[7], and is prepared by iodine-catalysed chlorination of nitrobenzene at 60°. 1,2,4,5-Tetrachloro-3-nitrobenzene (TCNB) or tecnazene[13] is specially useful for control of *Fusarium* rot of potatoes and inhibits sprouting during storage. Quintozene (15) shows selective toxicity to those fungi having chitinous cell walls and possibly owes its fungicidal properties to interference with chitin synthesis.

Quinones

A number of quinones occur in plants and are also products of fungal metabolism[2]. Certain members of this group have found use as agricultural fungicides such as tetrachlorobenzoquinone or chloranil, useful as a seed dressing, although as foliage fungicide it is too rapidly decomposed by sunlight. The related derivative of naphthaquinone known as dichlone (16) is more stable in light and has been used as a seed dressing, and a foliage spray especially against downy mildew and apple scab, and as an algicide. However these compounds tend to be rather too phytotoxic for extensive use as foliar fungicides. Dichlone (16) is manufactured by bubbling chlorine gas through a solution of 1,4-aminonaphthalenesulphonic acid in aqueous sulphuric acid:

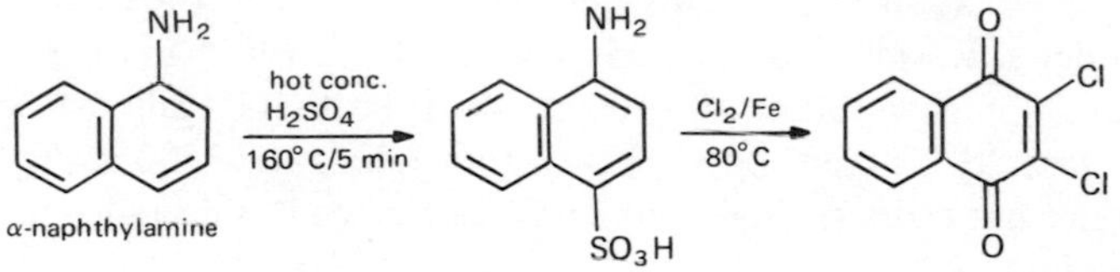

α-naphthylamine

hot conc. H$_2$SO$_4$ 160°C/5 min

Cl$_2$/Fe 80°C

(16)

Quinones are α,β-unsaturated ketones and probably owe their fungicidal properties to an addition reaction with vital sulphydryl-containing respiratory enzymes in the fungal cell, since their fungitoxicity can be antagonized by the addition of sulphydryl compounds like cysteine or glutathione[6]. The overall toxic mechanism therefore probably consists in binding the enzyme to the quinone nucleus by substitition or addition at the double bond[5,9]:

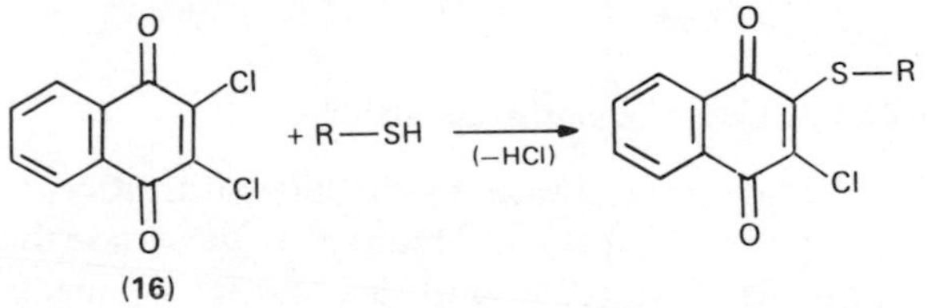

(16)

+ R—SH

(−HCl)

Dodine (n-dodecylguanidine acetate) (17) has been known as a bactericide since 1941[5] and more recently has been shown to have fungicidal activity, especially against apple and pear scab, cherry and blackcurrent leaf spot, and black spot on roses[7]. Dodine is a cationic surfactant and is generally formulated as a wettable powder containing the acetate. Dodine is prepared as follows[2,14]:

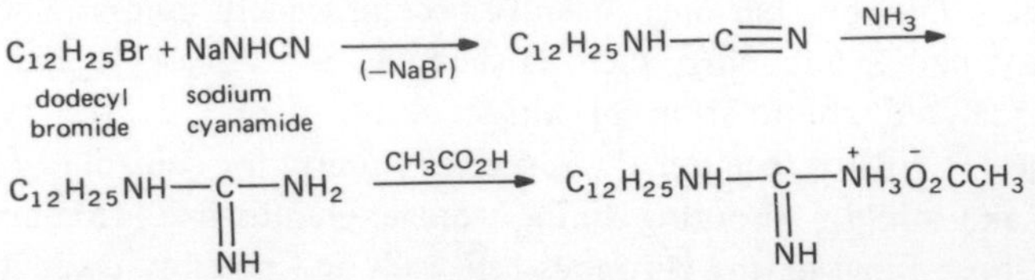

$$C_{12}H_{25}Br + NaNHCN \xrightarrow[(-NaBr)]{} C_{12}H_{25}NH—C\equiv N \xrightarrow{NH_3}$$

dodecyl   sodium
bromide   cyanamide

$$C_{12}H_{25}NH—\underset{\underset{NH}{\|}}{C}—NH_2 \xrightarrow{CH_3CO_2H} C_{12}H_{25}NH—\underset{\underset{NH}{\|}}{C}—\overset{+}{N}H_3\overset{-}{O}_2CCH_3$$

(17)

It is a protectant foliage fungicide of low mammalian toxicity, LD_{50} (oral to rats is about 1500 mg/kg. The fungitoxicity of a series of alkylguanidine acetates against yeast and *Monilinia fructicola* has been studied[4-6]: maximum activity occurred in the C_{12}–C_{13} homologues whereas the C_{10} homologue had the greatest phytotoxicity. The fungicidal activity of dodine and similar compounds probably depends on their ability to alter the permeability of the fungus cell wall causing the loss of vital cellular components, e.g. amino acids and phosphorus compounds.

The most active member of a new group of carbamoyl hydantoin fungicides is Rovral or 3-(3′,5′-dichlorophenyl)-l-isopropylcarbamoyl hydantoin (18) discovered by Rhône–Poulenc in France and first marketed in 1976 as protectant fungicide, especially active against *Botrytis* on grapes, tomatoes, and ornamentals[15]. It is useful against fungal strains which have developed resistance to some of the benzimidazole systemic fungicides like benomyl and thiophanate methyl. Rovral (18) is also active against *Sclerotinia* and *Alternaria* but not against downy and powdery mildews. Rovral is formulated as a wettable powder and is sprayed onto the plant at a concentration of 100 g/100 1 (1 lb/100 gallons) water. The mammalian toxicity is very low: LD_{50} (oral) to rats 3500 mg/kg, and the compound does not cause environmental pollution since it is degraded in the soil to non-toxic products. (Plate 16).

(18)

Systemic Fungicides or Plant Chemotherapeutants

The idea of controlling plant diseases by the internal treatment of plants is not new (see Chapter 1, p. 7), but it is only during the last decade that commercially viable systemic fungicides have come onto the market. A systemic fungicide is a compound that is taken up by a plant and is then translocated within the plant, thus protecting it from attack by pathogenic fungi or limiting an established infection[16]. If a given candidate chemical is to be an effective systemic fungicide the following criteria must be satisfied:

1. It must itself be fungicidal, or be converted into an active fungitoxicant within the host plant. Some compounds appear to act by modifying the resistance of the host to fungal attack.
2. It must possess very low phytotoxicity. This requirement is especially important with a systemic fungicide since the chemical is brought into intimate contact with the host plant.
3. It must be capable of being absorbed by the roots, seeds, or leaves of the plant and then translocated, at least locally, within the plant. For most commercial purposes the compound will be applied as a foliar spray or a seed dressing. Unfortunately uptake from the leaves is generally much more difficult than translocation by root treatment. Consequently many materials which give promising fungicidal results by root treatment fail to exhibit comparable activity in spray tests. Conditions 2 and 3 are the major obstacles to be overcome by a candidate systemic fungicide.

The earlier protectant or surface fungicides, applied as foliar sprays, formed dried deposits on the leaves of the host plant, protecting it from fungal attack but of course the deposits are gradually removed by the effects of weathering and cannot protect new plant growth formed after spraying, or any part of the plant not covered by the spray. These disadvantages can be overcome by the use of systemic fungicides which since they penetrate the plant cuticle also offer the possibility of controlling an established fungal infection. So systemic fungicides should exhibit both protectant and eradicant activity.

Systemic antifungal action has been demonstrated in many compounds[16-18]; for instance sulphonamides, antibiotics, phenoxyalkanecarboxylic acids, 6-azauracil, and phenylthiourea, although their discovery made little impact on the large-scale control of fungal diseases because they were either too expensive, not sufficiently active under field conditions, or caused phytotoxic damage.

The development of systemic fungicides has largely arisen from the tremendous advances in the systemic chemotherapy of human diseases based on the discoveries of the antibacterial action of a *Penicillium* mould by Fleming (1929) and of Prontosil by Domagk (1935) which led to the production of antibiotics and sulphonamide drugs respectively. Plant pathologists considered that as bacteria and fungi are closely related, these materials might also show systemic activity against plant pathogenic diseases, and accordingly a number of synthetic bactericides and antibiotics were examined as potential systemic fungicides.

Sulphonamides

These are the most important class of synthetic bactericides from the viewpoint of systemic antifungal properties. Sulphonamides were first studied by Hassebrauk (1938); he showed that wheat rust could be controlled by root treatment with p-aminobenzenesulphonamide (sulphanilamide) and the systemic activity of sulphonamides has been confirmed by several workers[16,17]. Crowdy and his collaborators[17] investigated root uptake, translocation, and detoxification of sulphonamides in various plant species and showed that the behaviour of a given sulphonamide may vary from one plant to another. When a series of N^4-acyl-sulphonamides (19) was applied via the roots of wheat plants for control of wheat stem rust (*Puccinia triticina*), maximum disease control occurred with the n-butyl or pentyl derivatives (19; R = C_4H_9 or C_5H_{11}). The higher homologues were ineffective due to low aqeous solubility, or because they were so rapidly hydrolysed in the plant that severe phytotoxic symptoms developed.

$$R\,CONH-\!\!\left\langle\;\right\rangle\!\!-SO_2NH_2$$

(19)

Sulphonamides have been used mainly against rust diseases on cereals but comparatively large doses are needed and there is a danger of phytotoxic damage to the host plant. A major disadvantage is the fact that sulphonamides are fungistatic rather than fungicidal and as soon as treatment stops the fungus starts to develop[4]. The antibacterial action of sulphonamides can be suppressed by addition of *p*-aminobenzoic acid (PABA) and PABA will also reverse the antifungal action of sulphanilamide against *Trichotyton purpureum*. PABA may therefore be an essential metabolite for fungi (as well as bacteria) and the antifungal action of sulphonamides may be due to these compounds acting as antimetabolites interfering with some stage in the synthesis of folic acid[4,16,17]. The effectiveness of different sulphonamides varies: sulphanilamide enters plants better than sulphaguanidine, and although both sulphanilamide and sulphathiazole are rapidly absorbed by plants only sulphanilamide is appreciably translocated from the roots[17].

Antibiotics

Antibiotics are chemicals produced by living organisms that are selectively toxic to other organisms. The study of the biological activity of the metabolic products of microorganisms was stimulated by the successful development of penicillin for medicinal use and the properties of some 300 antibiotics have been listed[19]; more than 100 of these are produced by fungi. These substances play an important role in the biological control of soil pathogens, but the majority have very complex structures and are often too unstable for practical application as pesticides[19]. The first antibiotics examined against pathogenic fungi were those employed in human chemotherapy.

The first successful antibiotic against human diseases was penicillin (20) discovered by Fleming (1929) and first used in medicine by Chain and Florey (1940), but it has never achieved commercial significance as a systemic fungicide[9]. Gliotoxin (21), an antifungal antibiotic produced by the soil fungus *Trichoderma viride*, inhibited the growth of *Botrytis* and *Fusarium* spores at concentrations of 2 to 4 p.p.m., but the compound was too unstable for use as a soil fungicide[9].

Streptomycin (**22**), isolated from the culture filtrates of certain strains of *Streptomyces griseus*, is used for control of bacterial pathogens of plants and is especially effective against bacterial diseases of stone fruits[17]. Streptomycin (**22**) is also employed as a spray against downy mildew on hops but an interval of at least eight weeks must elapse between application and harvest[7]. Streptomycin is active too against peach blast fungus, and the copper chelate is very effective as a foliage spray against *Phytophthora infestans* on tomato[16].

Streptomycin does not appear to undergo metabolic activation in plants and this observation combined with the very low *in vitro* activity suggested that it may act indirectly by inducing a change in the tissue of the host plant[4]. The activity of plant polyphenolases may be increased, which would probably help the plant combat fungal infections. Streptomycin is easily taken up by plant roots, but unfortunately is rather phytotoxic due to its inhibition of chlorophyll synthesis, and tends to produce chlorosis. Application must therefore be at exactly the correct dosage to prevent the pathogen without causing damage to the host plant. The high cost of streptomycin is another major obstacle to its commercial development as a pesticide.

Cycloheximide or Acti-dione (**23**), an antifungal antibiotic obtained from the culture filtrates of *Streptomyces griseus*[16,17], is active against plant pathogenic fungi at concentrations of 1 to 5 p.p.m., and will eradicate cherry leaf spot and wheat stem rust[14]. Cycloheximide is absorbed by the roots of tomato plants and translocated to the leaves in sufficient quantities to be antifungal.

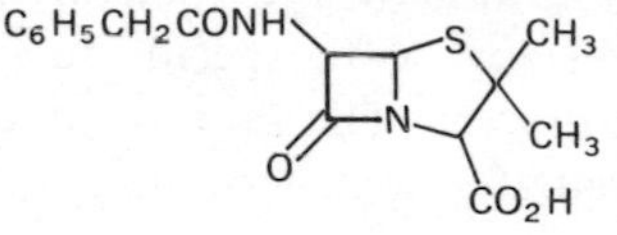

(20) 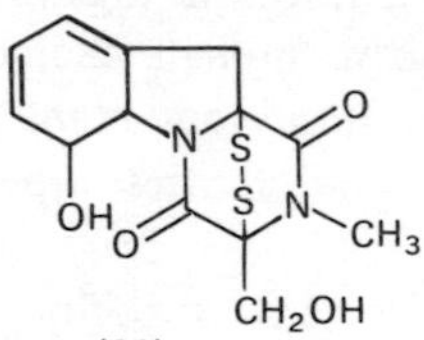   (21)

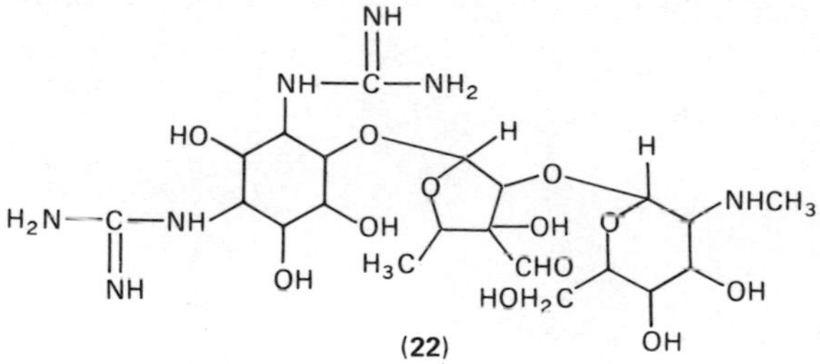

(22)

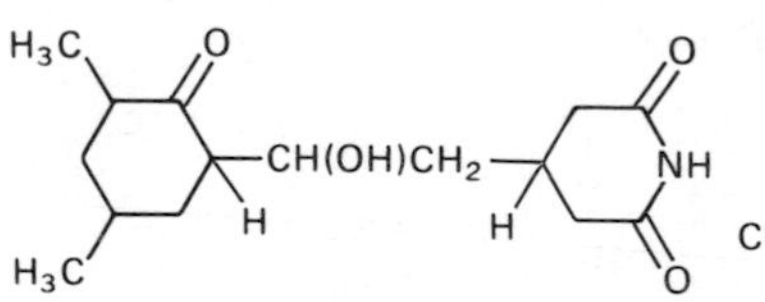

(23) 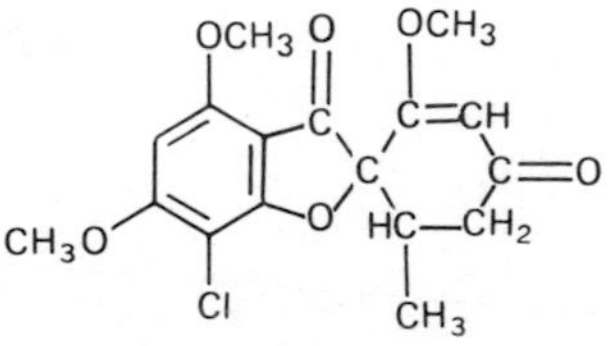   (24)

124

The control of wheat stem rust and cherry leaf spot was promising but, like streptomycin, practical application of cycloheximide is limited by the risk of phytotoxic damage to the host plant. The fungitoxicity is probably due to the ability of the antibiotic to inhibit the transfer of amino acids from aminoacyl transfer RNA to the ribosomal protein (see Chapter 3, p. 36). Cycloheximide is also toxic to animals and plants, but not to bacteria possibly due to differences in ribosomal structure[20].

Griseofulvin (24), isolated in 1939[9] from the mycelium of *Penicillium griseofulvum*, is an important antifungal antibiotic showing a wide spectrum of activity especially against *Botrytis cinerea* on lettuce and *Alternaria solani* on tomato. The fungicidal effect is generally greatest with fungi having chitinous cell walls and is associated with a characteristic distortion ('curling') of the fungal hyphae. Griseofulvin is less phytotoxic than streptomycin and cycloheximide, and is readily absorbed and translocated by many species of plants. However it does not always behave similarly; in cucumber and broad bean plants griseofulvin is absorbed and quickly translocated unchanged to the leaves. The quantity of the antibiotic moving is apparently proportional to the amount of water transpired whereas in tomatoes translocation is much slower. Griseofulvin is also promising in the systemic treatment of superficial fungal infections in animals, including man[16].

The primary fungicidal action of the antibiotic is probably interference with cell division and not with chitin synthesis, because the curling of the hyphae of treated fungi was not apparent until some 6 hours after treatment[20].

Research is being directed towards the discovery of new antibiotics which, though of no value in human chemotherapy, show specific fungicidal activity against economically important plant pathogens. The Japanese in particular have made extensive use of antibiotics, especially for control of blast and bacterial leaf blight on rice.

Thus blasticidin S (25), a pyrimidine derivative isolated from *Streptomyces griseochromogenes*, gives excellent control of rice blast[9] (*Piricularia oryzae*) and also inhibits certain bacteria, but the compound is rather phytotoxic and has been

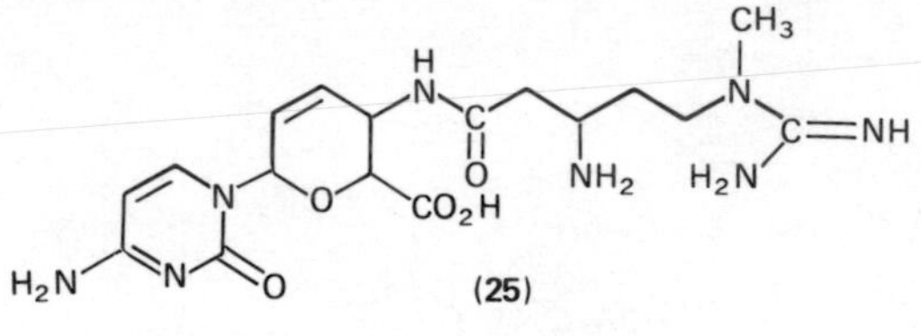

(25)

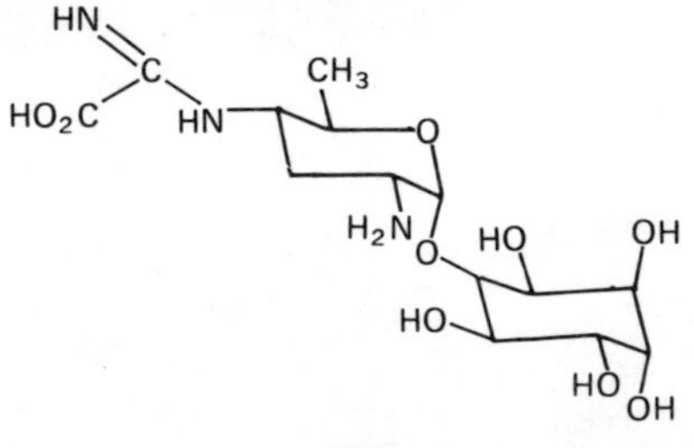

(26)

replaced to some extent by kasugamycin (26) which is both extremely effective and safer to use.

Blasticidin (25) at a concentration of 0.1 p.p.m. reduces the respiration of glucose by 50–60%, and at 1 p.p.m. completely inhibited the incorporation of radioactive glutamic acid into the protein fraction of rice blast mycelium. The synthesis of nucleic acids was not affected and the probable site of biochemical action is the transfer of amino acids to the growing protein chain on the ribosome[20]. A contributory factor is the inhibition of respiration since this limits the supply of energy to the fungal cell. Kasugamycin (26) shows selective toxicity against rice blast at low pH and also interferes with protein biosynthesis *in vivo* as well as in cell-free systems. The process inhibited is probably the binding of the aminoacyl–transfer RNA complex with the messenger RNA and ribosome to a complex[20].

The antibiotic polyoxin D (27) is another pyrimidine derivative which is toxic towards several fungi, including rice blast. Polyoxin D does not affect respiration, or nucleic acid, or protein biosynthesis, but strongly inhibits the incorporation of radioactive glucosamine into the fungal cell walls. Studies with *Neurospora crassa* showed[20] that a concentration of 100 p.p.m. of the antibiotic depressed both fungal growth and the incorporation of ^{14}C-glucosamine into chitin, an important structural material for many fungal cell walls and insect cuticles. Chitin is essentially a polymer of *N*-acetylglucosamine and is synthesized from uridine diphosphate *N*-acetylglucosamine (28) by a reaction in which *N*-acetylglucosamine units are transferred from UDP-*N*-acetylglucosamine (28) to the growing chitin chain. This is controlled by the enzyme chitin-UDP *N*-acetylglucosaminyl transferase and polyoxin D competitively inhibits the action of this enzyme in both rice blast fungus and *N. crassa* so that UDP-*N*-acetylglucosamine accumulated. The fungitoxicity of polyoxin D (27) is almost certainly due to interference with chitin synthesis, and this biochemical mode of action is consistent with the observed mycelial swelling in treated fungi and the lack of toxicity shown against mammals and plants.

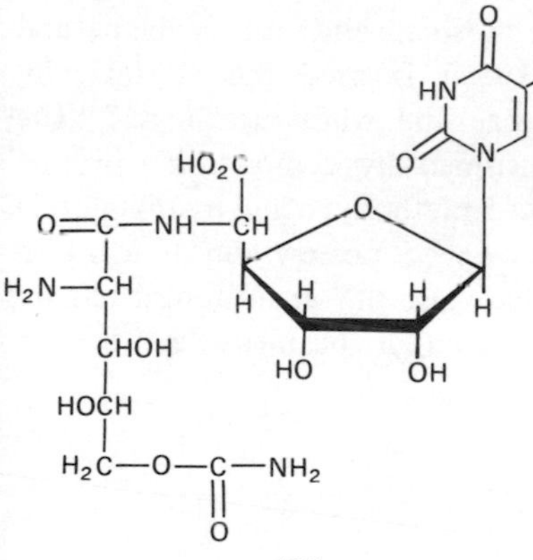

(27)

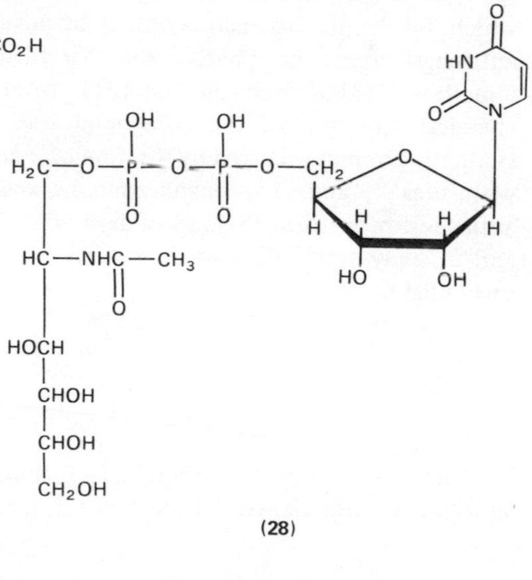

(28)

In addition to synthetic antibacterial drugs and antibiotics, several synthetic organic compounds have been discovered to show systemic antifungal properties; for instance 1-phenylthiosemicarbazide (29) and some of its derivatives, such as phenylthiourea[21−23]. Phenylthiosemicarbazide (29) has low mammalian toxicity and good systemic properties against cucumber mildew, apple scab, late blight on potatoes, and *Botrytis* and *Cladosporium* on tomatoes. On the other hand, the *in vitro* activity of phenylthiosemicarbazide against fungal spores was much less and experiments have shown[23] that it is activated *in vivo* to phenylazothioformamide (30) which is the active fungitoxicant.

$$C_6H_5NH-NHCSNH_2 \xrightarrow[(-2H)]{in\ vivo} C_6H_5N=NCSNH_2$$
$$(29) \qquad\qquad\qquad (30)$$

In contrast phenylthiourea appears to act indirectly so as to increase the resistance of the host plant to fungal infection; probably by inhibition of a pectolytic enzyme secreted by the fungus to dissolve the plant cells it is attacking[21]. Wain[8] has suggested that such inactivation of plant pectolytic enzymes may be a factor contributing to the natural immunity shown by certain plant species to fungal infection.

Phenoxyalkanecarboxylic acids are rapidly translocated from the roots to the growing points of bean plants. Most of these compounds showed little antifungal activity either *in vitro* or *in vivo*, but a few exhibited higher *in vivo* activity which is probably due to alteration of the metabolism of the host plant making it less favourable to the fungus, possibly by alteration in the level of reducing sugars present in the sap.

This leads on to a consideration of the effect of naturally occurring plant constituents, such as amino acids, phenols, and sugars on the level of resistance shown by the host plant to the invading pathogen. Unfortunately comparatively little is known about the biochemical differences between resistant and susceptible varieties of the same plant species. Indeed one route to the discovery of novel systemic fungicides is afforded by a study of naturally resistant plant varieties from which, by chemical extraction, it may be possible to isolate and identify the natural antifungal agent or phytoalexin. Virtanen (1957) isolated the phytoalexin, 6-methoxy-2(3)-benzoxazolinone (31), from maize and wheat seedlings[16]. The chemical structure of the compound was confirmed by comparison with the synthetic product prepared by reaction of 2-amino-5-methoxyphenol hydrochloride with urea[16], and 2(3)-benzoxazolinone was shown to possess almost identical antifungal properties. Extracts of garlic powder have also shown antifungal activity and in spray tests inhibited downy mildew on cucumber, bean rust, and tomato early blight.

$$CH_3O-\underset{H}{\overset{O}{\bigcirc}}C=O$$

(31)

Other methods of controlling fungi include the genetic breeding of new resistant varieties of crop plants; the resistance is generally associated with the presence of phytoalexins. In the future, it may be feasible to control plant diseases by treatment with systemic chemicals that stimulate the production of phytoalexins by the host plant. There is some evidence that injection of DL-phenylalanine enhances the resistance of apple leaves to scab, and the application of copper and mercury salts increases the production of the phytoalexin pisatin in peas.

One of the most interesting sources of natural fungitoxicity is the soil itself; the soil microbial population is exceedingly complex and soil microorganisms such as fungi and actinomycetes can synthesize appreciable amounts of antibiotics which probably diffuse into the soil solution regulating the growth of other soil organisms[18]. In sterile soil, which has been inoculated with suitable microorganisms, the formation of fungistatic antibiotics can be demonstrated easily because now there are no other organisms breaking them down. The production of fungistatic compounds has also been shown in natural soil, and the concept of a widespread soil fungitoxicity has been developed[16]. In addition, a large number of compounds, such as amino acids, can diffuse out from the roots of higher plants into the surrounding soil (the rhizosphere). These root exudates will affect the neighbouring soil microflora and this may increase or decrease the resistance of the plant roots to the attack of pathogenic soil organisms, or alter the pathogenicity of the fungus.

Bison and Novelty are two different varieties of flax; the former is resistant, while the latter is susceptible, to soil-borne *Fusarium* wilt disease. Bison exudates stimulated the growth of the soil fungus, *Trichoderma viride* which is known to produce fungistatic antibiotics (see p. 122) so the *Fusarium* wilt fungus was checked by this variety of flax; possibly in the future plant diseases might be controlled by cultivation of suitable antibiotic-producing organisms in the soil. The ideal systemic fungicide might be one which, after application to the foliage, travelled down to the roots and exuded into the rhizosphere inhibiting root parasites without interfering with naturally beneficial microorganisms, such as actinomycetes. Chemical soil sterilants, like formaldehyde and carbon disulphide, probably not only destroy the parasitic fungi, but also do not kill certain soil saprophytes so after fumigation these multiply rapidly producing a diffusible toxin that finally destroys the fungal parasites.

These ideas open up fascinating possibilities for future control of fungi, but it must be admitted that to date the discovery of commercial systemic fungicides has largely arisen from exploitation of the results from more or less random screening of synthetic organic compounds.

There are now a number of systemic fungicides on the market[16,17,23−25], which may be divided into benzimidazoles, thiophanates, oxathiins, pyrimidines, morpholines, and organophosphorus compounds.

Benzimidazoles

The most important members of this group are methyl-1-(butylcarbamoyl)benzimidazole-2-carbamate or benomyl (**32**), and 2-(4'-thiazolyl)benzimidazole or thiabendazole (**33**; X = S).

Benomyl (32), introduced in 1967, was synthesized from cyanamide and methyl chloroformate:

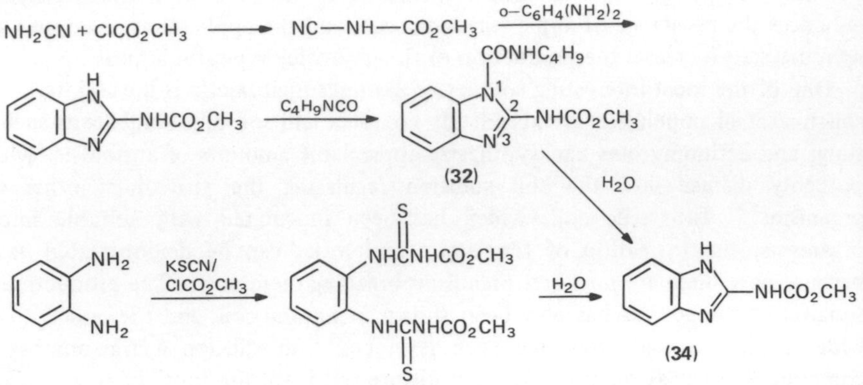

(32)

(34)

(35)

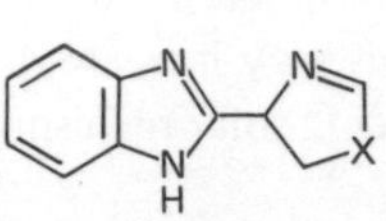

(33)

These are both wide-spectrum systemic fungicides, active against many pathogenic fungi, including powdery mildews and soil-borne pathogens, *Verticillium albo-atrum* on cotton and black spot on roses[7], but they are not effective against the *phycomycetes* group of fungi, such as potato blight, vine downy mildew, or damping off diseases. Benomyl is the more active compound and is widely applied as a foliar spray, seed dressing, or to the soil for control of grey mould (*Botrytis cinerea*), apple scab (*Venturia inaequalis*), canker and powdery mildew (*Podosphaera leucotricha*), *Gloeosporium* storage rots, leaf spot (*Cercospora beticola*) on sugar beet, the major fungal diseases of soft fruits, and some pathogens of tomato and cucumber[16,23].

Thiabendazole (33; X = S) is good against post-harvest diseases of apples and pears, and as a seed dressing against common bunt of wheat (*Tilletia caries*) when the fungicide may persist in the plants for several months. Thiabendazole has also been used against stem rot of bananas, blue and green mould on citrus fruits, and is reported[16] to be more effective than benomyl for eradication of Dutch elm disease, although neither chemical was sufficiently active to be a practical means of large-scale control[17]. Thiabendazole is translocated from either the roots or leaves of growing plants and also moves from the leaves to the roots.

Fuberidazole (33; X = O) is a valuable seed dressing against *Fusarium* diseases (e.g. of rye and peas), wheat rust (*Puccinia triticina*), and barley powdery mildew (*Erysiphe graminis*)[14].

The activity of selected benzimidazoles has been reviewed[16,17] and the fungitoxicity is clearly associated with the benzimidazole nucleus which probably accounts for the almost identical *in vitro* fungicidal spectrum shown by (32) and

(33; X = S). In aqueous solution benomyl (32) is rapidly hydrolysed to the strongly fungicidal methylbenzimidazole-2-carbamate (34) and this is probably the active fungitoxicant, since the fungicidal properties of benomyl are increased after standing in aqueous media. In this connexion, it is interesting that methylbenzimidazole-2-carbamate (carbendazim) (34) is used as a wide-spectrum systemic fungicide and may be formulated as a 50% wettable powder for control of *Botrytis*, *Gloeosporium* rots, powdery mildews, and apple scab. Carbendazim is absorbed by the roots and foliage of plants and is quicker acting than benomyl.

With thiabendazole (33; X = S) and fuberidazole (33; X = O) there is no evidence of *in vivo* metabolism. When benomyl (32) is used for control of *Verticillium* wilt of cotton and potatoes by soil treatment, the activity is enhanced by addition of surfactants probably because these improve the mobility of the metabolite (34) through the soil.

Thiophanates

These are a new group of systemic fungicides based on thiourea[23]. Thiophanate, also called Topsin or Cercobin is 1,2-bis(3-ethoxycarbonyl-2-thioureido)benzene and thiophanate-methyl or Mildothane (35) is the methyl analogue which is obtained by condensation of potassium thiocyanate, methyl chloroformate, and *o*-phenylene diamine. Both are effective against such pathogenic fungi as apple powdery mildew, apple and pear scab, sheath blight of rice, *Cercospora* leaf sport of sugar beet, and *Botrytis* and *Sclerotinia* on various crops[7,14]. They also showed a high level of persistent systemic activity by root uptake against barley and cucumber mildew. The overall fungicidal spectrum of the thiophanates resembles that of the benzimidazoles; in particular benomyl (32) and thiophanate-methyl (35) are closely similar in their antifungal properties. The latter compound, like benomyl, on treatment with water gave methylbenzimidazole-2-carbamate (34). Similarly thiophanate afforded ethylbenzimidazole-2-carbamate. The carbamate (34) has recently been isolated from plants treated with thiophanate-methyl (35) and it is concluded that the carbamates (e.g. 34) are the active fungitoxicants in both benomyl and the thiophanates. This idea is supported by the inactivity of 1,3- and 1,4-bis(3-ethoxycarbonyl-2-thioureido)benzenes which are incapable of cyclization to the benzimidazolecarbamate (34).

The thiophanates are not themselves fungicidal but are converted to the active benzimidazole derivatives (e.g. 34). The fungitoxicity of both benomyl and thiophanate-methyl appears to be related to the retention in the roots of either benomyl or, in the case of the thiophanates, of an intermediate between (34) and (35). These compounds then function as reservoirs of the active toxicant (34) which is subsequently released gradually to other parts of the plant. A similar effect is believed to operate when benomyl is applied as a foliar spray due to the effects of weathering of the deposit on the leaf surface.

Methylbenzimidazole-2-carbamate (34) probably owes its fungicidal properties to the ability to inhibit DNA synthesis or cell or nuclear division[20]. Clemons and Sisler (1971) discovered that application of a 5 μM solution of (34) to

Neurospora crassa conidia caused 85% inhibition of DNA synthesis in 8 hours. The interference with DNA synthesis by benzimidazoles is probably a consequence of the structural resemblance of methylbenzimidazole-2-carbamate to the DNA purine bases, adenine and guanine. Support for this hypothesis comes from the observation that the addition of low concentrations of purines reduces the toxicity of benomyl and thiabendazole to *Fusarium oxysporum*. These benzimidazoles do not uncouple oxidative phosphorylation, nor does it appear that inhibition of electron transfer is the primary biochemical mode of action[16,20].

The parent compounds, or some of their metabolites, may be mutagenic and benomyl is known to be mutagenic to *Aspergillus nidulans*; such mutagenicity may account for the emergence of strains of *Botrytis cinerea* on cyclamen which are resistant to benomyl[23] (see p. 138).

Oxathiins

Oxathiins are another group of heterocyclic compounds with interesting systemic fungicidal properties; examples are carboxin or Vitavax (5,6-dihydro-2-methyl-1,4-oxathiin-3-carboxanilide) (36) and the sulphone analogue known as oxycarboxin or Plantvax (37)[14]. Carboxin is prepared by reaction of α-chloro-acetoacetanilide and 2-thioethanol followed by cyclization:

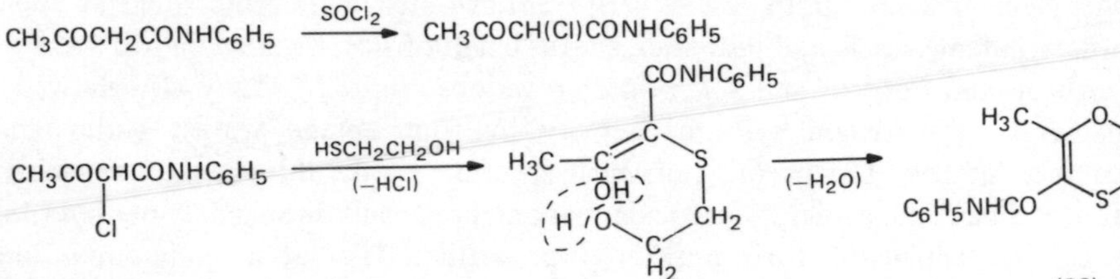

$$CH_3COCH_2CONHC_6H_5 \xrightarrow{SOCl_2} CH_3COCH(Cl)CONHC_6H_5$$

(36)

Oxycarboxin (37) is obtained by subsequent oxidation of carboxin (36) with hydrogen peroxide.

Both these compounds are primarily active against the *Basidiomycetes* class of fungi which includes such economically important pathogens as the rusts, smuts, and bunts of cereals, and the soil fungus *Rhizoctonia solani*[23]. Carboxin (36) was introduced in 1966, and is an extensively used and highly effective seed dressing for the eradication of loose smut of barley and wheat[7]. These diseases cannot be satisfactorily controlled by organomercurial seed dressings because the mycelium of the fungus is deeply within the seed. Carboxin also shows activity against oat smuts, seedling blight of wheat, leaf stripe of barley, brown foot rot of oats, and malsecco disease of lemon trees[14,17]. Oxycarboxin (37) has systemic activity against rust diseases of cereals and vegetables, and seed treatment or soil application has delayed the onset of wheat leaf and stem rust by some two months.

Carboxin is absorbed and translocated by plant roots; in water, soil and plants (barley, wheat, and cotton) the compound was oxidized to the corresponding sulphoxide, but further oxidation to the sulphone (37) was not observed, but in the roots of Pinto beans hydrolysis of the carboxanilide group also occurred.

The fungitoxicity of carboxin and eight analogues was examined in relation to their effects on metabolic pathways. All the active compounds strongly inhibited glucose and acetate oxidative metabolism and RNA and DNA synthesis, although the latter may arise from lack of cellular energy due to inhibition of respiration. Carboxin also interfered with succinate oxidation in sensitive fungi through inactivation of succinate dehydrogenase (Chapter 3, p. 29). Consequently succinate accumulated in treated fungal cells and this is probably the primary mechanism of fungitoxicity[20]. This suggestion was supported by the observation that an oxathiin-tolerant mutant fungus strain had little succinate dehydrogenase activity as compared with the normal susceptible strain.

The fungicidal properties of carboxin (36) and related carboxanilides (37—40) towards *Rhizoctonia* could be correlated generally with the oil/water partition coefficients and the uptake of carboxin by fungi is closely related to their lipid content. Electron microscopy revealed that carboxin damaged the mitochondria and vacuolar membrane of sensitive fungi causing inhibition of respiration close to the site of succinate oxidation. This probably accounts for the observed loss of cellular materials from treated hyphae of *R. solani*[17,23].

Several other substituted carboxanilides show specific fungicidal activity towards *Basidiomycetes* although the 2-methyl group appears to be essential for activity; one or both of the hetero atoms can be dispensed with. Thus 2-methyl-5,6-dihydro-pyran-3-carboxanilide (38) is rather more effective than carboxin (36) as a seed dressing against smut diseases of barley and oats and is also active against rusts and *Rhizoctonia* spp.

2-Methyl-(Mebenil) and 2-iodo-(Benodanil)benzanilide (39; $X = CH_3$ or I) are valuable systemic fungicides for control of a range of *Basidiomycetes* especially rust diseases of cereals, coffee, tobacco, vegetables, and ornamentals, and for dressing seed potatoes against *Rhizoctonia* spp. These compounds are structurally closely related to Shirlan or salicylanilide (39; $X = OH$), one of the first organic surface fungicides (see Chapter 1, p. 5).

2,5-Dimethylfuran-3-carboxanilide (40; $R = C_6H_5$) and the cyclohexylamide (40; $R = C_6H_{11}$) have been shown to have outstanding systemic fungicidal properties as seed dressings against loose smut of barley and wheat[14,15].

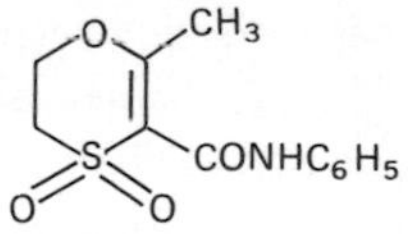

(37) 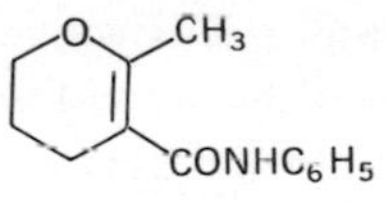 (38) 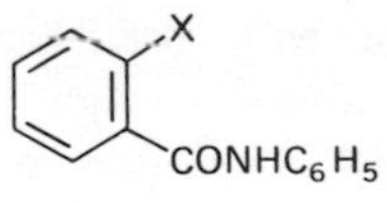 (39)

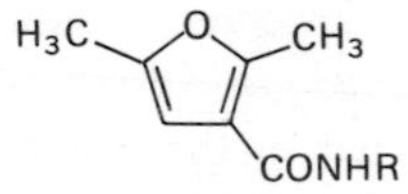

(40) 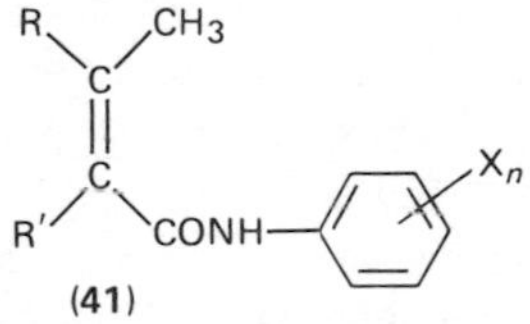 (41)

Systemically fungicidal carboxyanilides against broad bean rust (*Uromyces fabae*) and wheat brown rust (*Puccinia recondita*) were usually derived from α,β-unsaturated acids containing a methyl group attached to the β-carbon atom[17]. The double bond may form part of a planar or a non-planar ring (e.g. benzene or oxathiin) or an acyclic system, but the methyl group must be *cis* with respect to the carboxamido group which contains an imino hydrogen atom. The basic structure for activity in this series of compounds is therefore represented by (**41**; X_n = H, or one or more electron donor groups) and in these *cis*-crotonanilides, the intact molecules are probably responsible for the fungitoxicity.

Pyrimidines

Dimethirimol or Milcurb is 5-butyl-2-dimethylamino-4-hydroxy-6-methyl-pyrimidine (**42**) and is prepared by condensation of ethyl α-butylaceto-acetate with *N,N*-dimethylguanidine[14] (from dimethylamine and cyanamide):

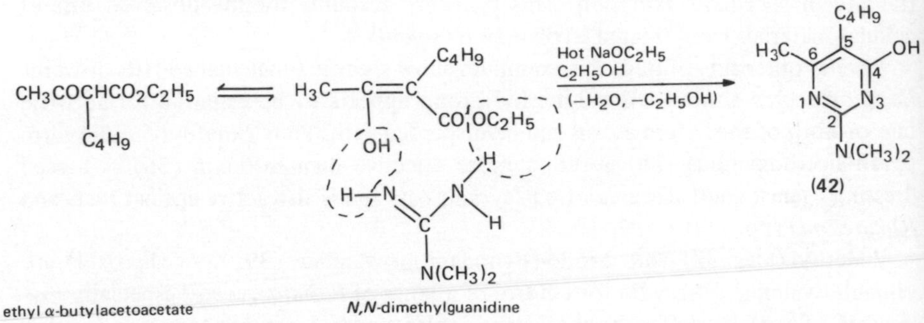

ethyl α-butylacetoacetate   *N,N*-dimethylguanidine

Dimethirimol, discovered in 1965, showed outstanding systemic activity by root application against certain powdery mildews, such as those of cucumber, melon, and some ornamentals, but only slight activity against powdery mildew of roses and vines and was not toxic to most other pathogenic fungi[24,25]. A single application by soil treatment with an aqueous solution of dimethirimol hydrochloride against cucumber powdery mildew (*Sphaerotheca fuliginea*) controlled the disease for up to eight weeks.

The closely related pyrimidine, ethirimol or Milstem, containing the ethylamino group attached to the 2-carbon atom, is effective as a seed dressing against barley powdery mildew, a major disease of barley in Europe[14,24,25]. Large-scale trials in the United Kingdom on spring and winter barley showed that seed treatment with Milstem substantially reduced the incidence of mildew and resulted in increased crop yields (Plates 12 and 13).

Both these pyrimidines are remarkably specific for powdery mildews and are absorbed by plant roots and translocated via the transpiration stream to all parts of the plant eradicating established infection and giving protection for long periods against fungal attack. They are not rapidly degraded in the soil which can thus function as a reservoir providing a slow release of the toxicant, but in plant tissue both compounds are quickly metabolized by progressive *N*-dealkylation. Dimethirimol (42) is first converted into the fungicidal monomethyl derivative, and then to the almost inactive 2-amino compound, followed by conjugation as glycosides and phosphates. Oxidation of the 5-butyl group to hydroxybutyl derivatives also occurs and these transformations afford a complex series of metabolites, some of which are fungitoxic[16,17]. Little structural variation appears possible without loss of activity. Thus the presence of a 4-oxygen function (carbonyl or hydroxy) and the 2-alkylamino group are essential, the size of the alkyl groups is critical — those in dimethirimol and ethirimol represent optimum sizes.

The precise biochemical mode of action is uncertain, but the fungitoxicity of dimethirimol (42) and ethirimol is antagonized by the addition of folic acid and riboflavin, and is slightly reduced by adenine, thiamine, and uracil. The fungicidal pyrimidines appear to be non-competitive inhibitors of enzymes related to C_1-metabolism, and experiments with powdery mildew spores suggest that they interfere with purine biosynthesis, and possibly with several pyridoxal-dependent enzymes[17,20]. These fungicidal pyrimidines are active within plant leaves at concentrations of only 0.1 μM so that a specific mode of action appears likely, and they may be antagonists to pyridoxal which is involved via C_1-metabolism in the biosynthesis of purines and amino acids[20]. This idea is supported by the observation that the fungitoxicity is partially reversed by the addition of various metabolites which are important in pyridoxal-catalysed reactions.

The N,N-dimethylsulphamate ester of ethirimol, known as bupirimate or Nimrod, has recently been developed for control of powdery mildews. It shows specific systemic protectant and eradicant fungicidal activity by spray treatment against powdery mildews on a number of crops, being particularly effective against apple powdery mildew. Bupirimate moves readily from spray deposits on the stems into the young leaves. In leaves and in aqueous solution bupirimate is degraded first to ethirimol. The mode of biochemical action has not been reported, but in view of its close structural similarity to ethirimol, probably involves inhibition of pyridoxal.

Another pyrimidine systemic fungicide is triarimol (43) obtained by condensation of 5-bromopyrimidine with 2,4-dichlorobenzophenone:

5-bromopyrimidine

(43)

Triarimol (43), unlike the hydroxypyrimidines, is a broad-spectrum fungicide showing local systemic action by spray application against a wide range of phytopathogenic fungi, including apple and barley mildew, apple scab, and cherry leaf spot[14]. Triarimol (43) was superior to dinocap and benomyl against apple mildew and is outstanding against a range of powdery mildews[23].

In the spores of *Ustilago maydis*, triarimol reduces steroid synthesis, and therefore possibly the fungicidal activity arises from inhibition of lipid biosynthesis[17,20]. Unfortunately the development of triarimol has been suspended due to undesirable toxicological effects.

Piperazine, Morpholine and Azepine Derivatives

1,4-Di-(2,2,2-trichloro-1-formamidoethyl)piperazine, or triforine (**44**) has systemic fungicidal activity as a foliar spray against mildew on cereals, apples, cucumbers, and ornamental plants, brown rot of plums (*Monilinia fructicola*), and apple scab[16,17]. In addition, as a seed dressing, it is very active against wheat leaf rust (*Puccinia recondita*) and powdery mildew[24]. Studies of the absorption and translocation of triforine (**44**) in barley plants after soil application indicated that the chemical is preferentially accumulated in the leaves via the transpiration stream in the xylem, but there is also downward flow. Translocation was stimulated by light and did not occur in the absence of photosynthetic activity, and there was no evidence for the oxidation of the formamido groups within plant tissues[17].

Two morpholine fungicides tridemorph (2,6-dimethyl-4-tridecylmorpholine, **45**) and the corresponding cyclododecyl derivative known as dodemorph (**46**) have been developed for control of *Erysiphe graminis* on barley and oats. They are eradicant fungicides with systemic action which are absorbed and translocated by the roots and leaves of barley plants giving protection from infection for 3—4 weeks; They are generally active against powdery mildews[7,14] and dodemorph has the advantage of lower phytotoxicity.

Substituted azepines of type (**47**) represent a new class of fungicide with a rather wider spectrum of activity, thus the derivative (**47**; R = *t*-butyl, R' = decyl) controls leaf spot diseases, as well as powdery mildew and rust pathogens showing limited systemic and eradicant activity. The biochemical modes of action of the piperazine, morpholine, and azepine systemic fungicides do not appear to be known[20]:

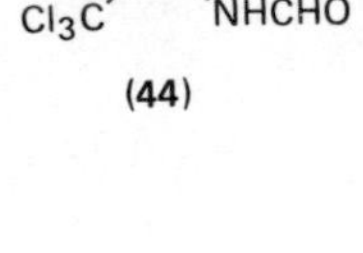

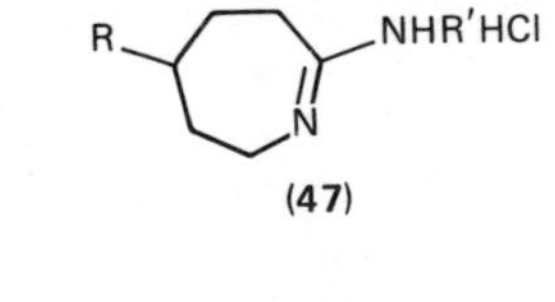

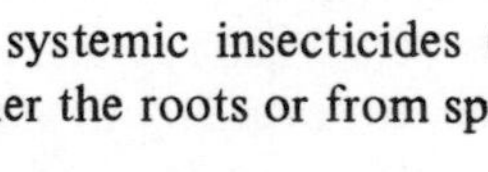

Organophosphorus Compounds

This group has produced many important systemic insecticides (Chapter 6, p. 68) which move readily in the plant from either the roots or from spray deposits on the foliage. One of the first organophosphorus fungicides was Wepsyn (48) which is obtained by condensation of the sodium salt of 3-amino-5-phenyl-1,2,4-triazole with bis(dimethylamino)phosphoryl chloride[14]:

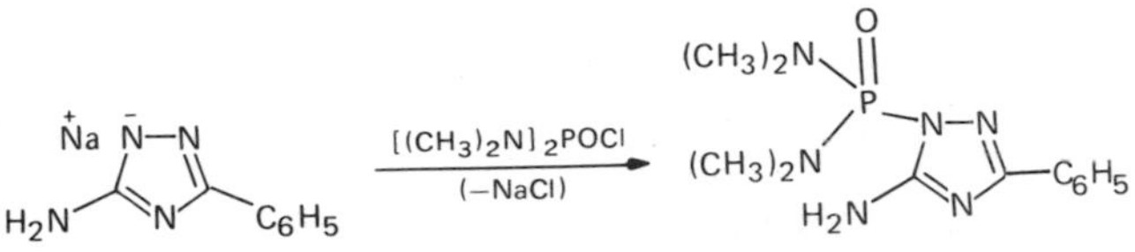

(48)

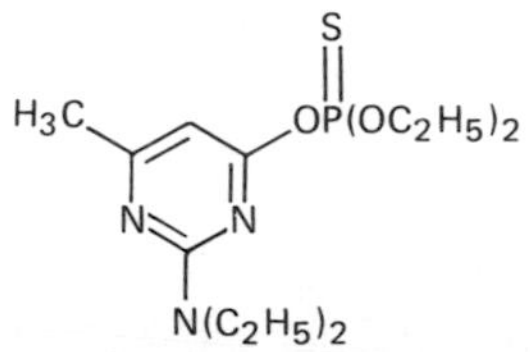

(49) 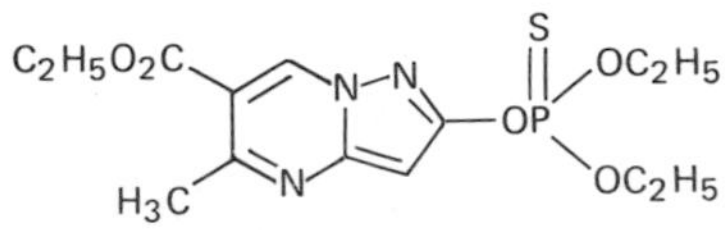   (50)

Wepsyn has been claimed (1960) to be the first commercial systemic fungicide[17,22], and has limited systemic action against a range of powdery mildews. Since then many combinations of heterocyclic ring systems and phosphorus moieties have been synthesized and evaluated in an effort to obtain more effective systemic fungicides[17,23]. Two examples are the pyrimidine derivative (49), which gave almost complete control of cucumber powdery mildew, wheat rust, and vine downy mildew and was also insecticidal[17], and Hoechst 273 or 2-(O,O-diethylthionophosphoryl)-5-methyl-6-carbethoxypyrazolopyrimidine (50), which has given promising results by foliar spraying against apple powdery mildew in field trials.

S-Benzyl O,O'-diisopropylphosphorothiolate or Kitazin P (51) is prepared by condensation of sodium benzylthiolate and O,O'-diisopropylphosphorochloridate[14,26]:

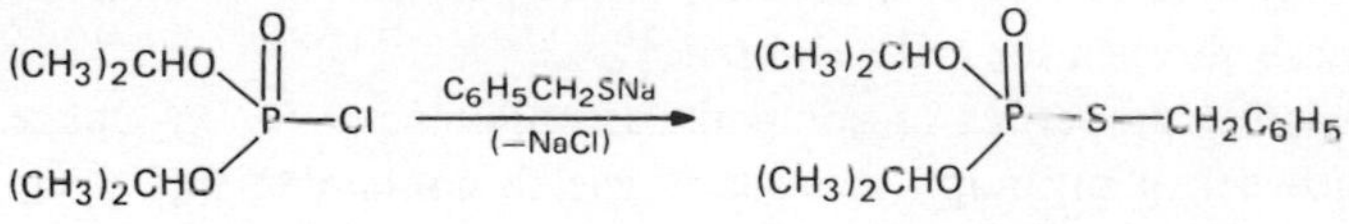

(51)

Kitazin P (51), introduced in 1968, is a systemic rice fungicide applied as granules to paddy water to control rice blast (Piricularia oryzae) and it inhibits mycelial growth in tissue. Kitazin P (51) is metabolized in rats, cockroaches, rice plants, soils and P. oryzae (Scheme 7.1) chiefly by S—C bond cleavage to give the diisopropyl hydrogen phosphorothioate. In rice plants this is converted to the phosphate. In rice blast fungus the most interesting metabolite was the m-hydroxy derivative (route ii):

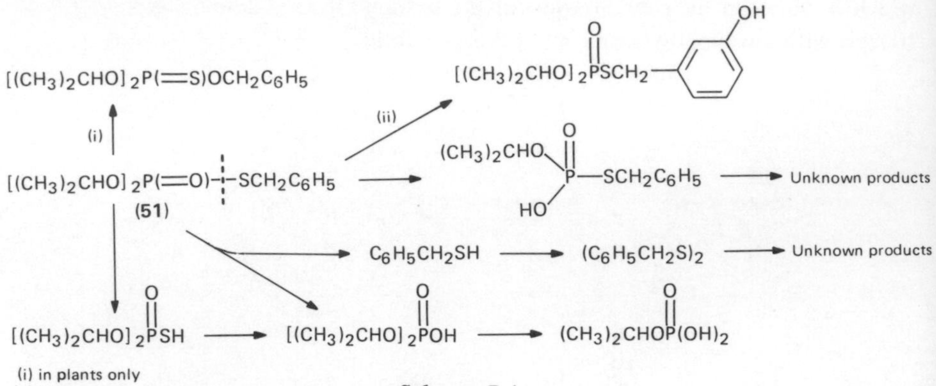

(i) in plants only
(ii) in fungi only

Scheme 7.1

Another useful fungicide against rice blast is Conen or O-butyl-S-ethyl-S-benzyl-phosphorodithioate **(52)** prepared from sodium O-butyl-S-benzylphosphorodithioate and ethyl bromide[26]:

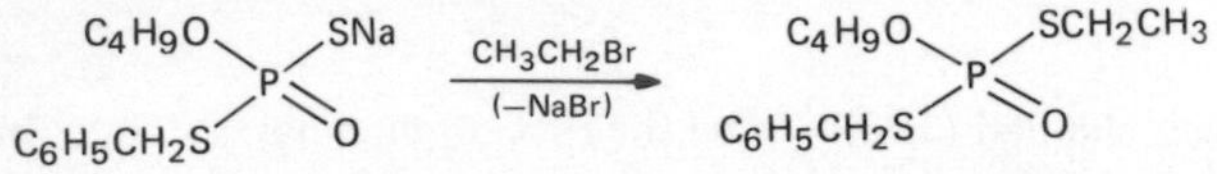

These fungicides are nearly all phosphoryl derivatives suggesting that direct attack on an enzyme system in the fungus is involved in the mechanism of fungitoxicity, unlike insects fungi cannot oxidize P=S → P=O, and hence the P=S compounds are generally inactive against fungi. To achieve penetration into the fungus, the polarity of the P=O group needs to be balanced by a large lipophilic group, such as phenyl, cyclohexyl, or butyl and many of the active fungicides, like Kitazin P and Conen, are S-benzyl derivatives. Kitazin P, like polyoxin D (p. 125) appears to owe its fungitoxicity to interference with chitin synthesis[16], although the compound may not directly inhibit the enzyme chitin synthetase since it prevents the incorporation of glucose into cell wall glucans in treated *P. oryzae* mycelium. The primary action of Kitazin may well be due to limitation of the permeation of lipid substrates for chitin synthesis through the cell membrane[20].

The fungicidal properties of compounds containing imino groups, e.g. captan, are well known. An organophosphorus fungicide containing this group is Dowco 199 or O,O-diethylphthalimidophosphorothioate **(53)**, obtained by condensation of potassium phthalimide with O,O-diethylphosphorochloridothioate[26]:

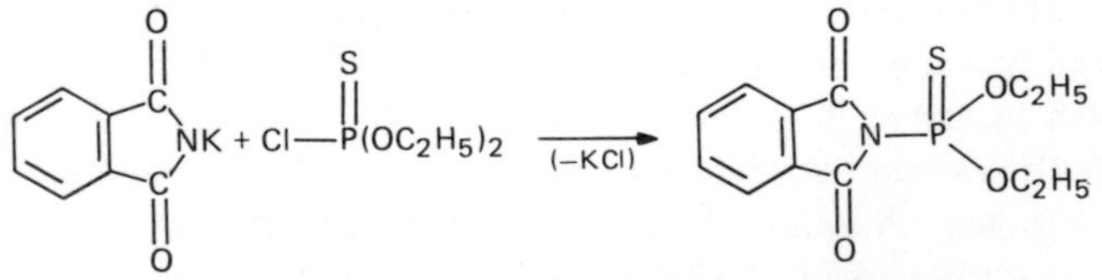

(53)

Dowco 199 (53) is extremely effective against a wide range of powdery mildews[26] and has a very low mammalian toxicity: LD_{50} (oral) to rats 4930 mg/kg.

The fungitoxicity is lost by replacement of P=S by P=O, and also if there is a bridging atom or group, e.g. $-S-CH_2-$ or $-O-$, between the nitrogen and phosphorus atoms. Certain non-phosphorus N-substituted imides like folpet (p. 115) are fungicidal and Tolkmith therefore argued that the toxophore in phthalimidophosphorothioates was not phosphorus-containing moiety but the unsaturated carboximide ring, and the activity is probably due to acylation of a vital enzyme system (H—Ez) with opening of the carboximide ring:

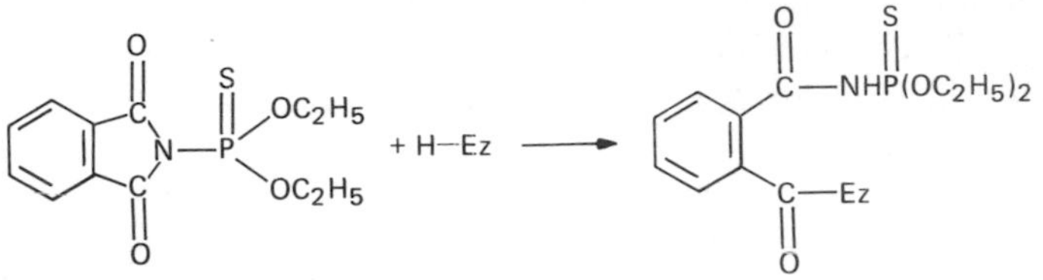

(53)

The fungicidal activity of Dowco 199 led to a systematic examination of phosphorus derivatives of other N-heterocycles having similar shape and size to the phthalimido ring. Imidazole derivatives were discovered to have good fungicidal properties against cucumber mildew and potato blight; the most active member was the compound (54; R = R' = CH₃). These compounds are obtained from phenol:

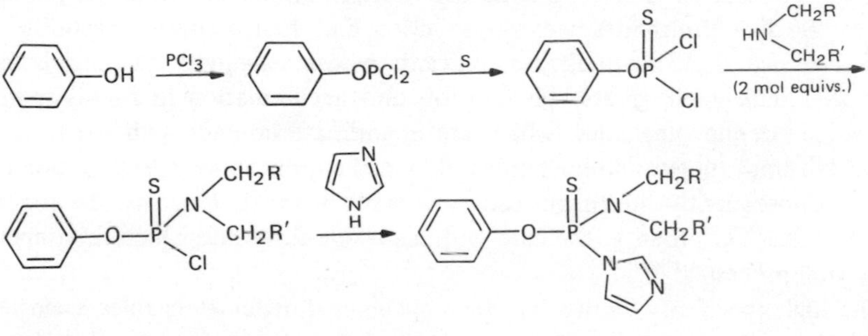

(54)

The compound (54; R = R' = CH₃) has the advantage of very low mammalian toxicity: LD_{50} (oral) to rats 1000 mg/kg[26].

Resistance of Fungi towards Fungicides

Organisms possess the capacity to adapt to changing environmental conditions; microorganisms such as fungi and bacteria reproduce extremely rapidly, so they are able to change more quickly to different conditions than higher organisms[17].

The development of strains of bacteria resistant to antibiotics was observed shortly after their introduction in human chemotherapy, and now many pathogenic protozoa can no longer be controlled by drugs formerly successful. Insects and mites are also able to adapt towards certain synthetic insecticides such as the organochlorine compounds (Chapter 5, p. 55) and this acquired tolerance has caused severe control problems.

On the other hand, there have been comparatively few examples of fungi becoming resistant to surface fungicides; for instance, although organomercurials were introduced as fungicides in 1913 only a few species of fungi have developed tolerance to them[27].

In contrast, the introduction of commercial systemic fungicides has been followed quickly by the development of resistant fungal strains[17,27]. Thus dimethirimol was introduced into Holland for control of cucumber powdery mildew in greenhouses in 1968 and by 1970 the mildew had acquired tolerance to the fungicide. Similar resistance has been noted in many greenhouses in Europe but not yet in field crops.

The fungus *Botrytis cinerea* causes large losses of cyclamen in greenhouses; treatment with surface fungicides proved ineffectual, but spraying with benomyl initially gave excellent control of the disease. However by 1971 the fungi developed resistance so that even 1000 p.p.m. of the fungicide did not kill the fungus completely, whereas the susceptible strain was eliminated by 0.5 p.p.m. of benomyl.

The resistant fungi also exhibited cross-resistance to related benzimidazole fungicides, such as thiabendazole, fuberidazole, and thiophanate methyl.

The resistance shown to systemic fungicides may be a consequence of their high selection pressure so that only naturally resistant strains in the fungal population survive. Surface fungicides were less active and hence some susceptible fungi remained; also they are multisite toxicants whose selectivity to fungi is usually associated with their greater penetration and accumulation in fungal spores. In contrast, systemic fungicides, which are in intimate contact with the host plant, would kill both fungus and host unless they had a quite specific toxic action on the fungus. Consequently the fungus can more easily adapt itself against the attack of a systemic fungicide than is the case with a surface fungicide which interferes with many vital processes[17].

The high specificity shown by some systemic fungicides enables a single-gene mutation by the fungus to give rise to a resistant strain; for instance, ultraviolet irradiation of nine strains of *Aspergillus nidulans* induced resistance to benomyl in five strains[17]. When the fungicide interferes with the metabolism of the fungus at several sites more mutations will be required for the fungus to become resistant, so surface fungicides like dithiocarbamates and copper fungicides have rarely induced resistance.

There are several possible mechanisms by which fungi can adapt themselves against fungicides. Mutation may alter the fungal cell so that the toxicant cannot reach the site of action within the cell owing either to reduced permeability of the protoplast membrane or an enhanced capacity of the fungus to detoxify the fungicide. In some cases resistance may be due to a lower affinity for the toxicant at the site of action in the fungal cell. When fungitoxicity arises from inhibition of a

vital biochemical process at a specific site in the fungus, tolerance can be gained by
the fungus modifying its metabolism so that the blocked site is bypassed[17,28].

References

1. Woods, A., *Pest Control: A Survey*, McGraw-Hill, London, 1974, p. 93.
2. Hartley, G. S. and West, T. F., *Chemicals for Pest Control*, Pergamon Press, Oxford, 1969. p. 191.
3. Martin, H., *The Scientific Principles of Crop Protection*, 6th edn., Arnold, London, 1973, p. 122.
4. Hassall, K. A., *World Crop Protection: Pesticides*, Vol. 2, Iliffe Books Ltd., London, 1969, p. 154.
5. Cremlyn, R. J. W., *Internat. Pest Control*, 5, 10 (1963).
6. Lukens, R. J., *Chemistry of Fungicidal Action*, Chapman and Hall, London, 1971.
7. Approved Products for Farmers and Growers, Ministry of Agriculture, Fisheries, and Food, 1978.
8. Wain, R. L., *Some Chemical Aspects of Plant Disease Control*, Royal Institute of Chemistry, Monograph No. 3, 1959.
9. Metcalf, R. L., 'Chemistry and biology of pesticides' in *Pesticides in the Environment* (Ed. White-Stevens, R.), Dekker, New York, 1971, p. 1.
10. Cremlyn, R. J. W., *Pest Articles and News Summaries* (B), 13, 255 (1967).
11. Rich, S., 'Fungicidal chemistry' in *Plant Pathology* (Eds. Horsfall, J. G. and Dimond, A.), Vol. 2, Academic Press, New York, 1960. p. 588.
12. Cremlyn, R. J. W., *Internat. Pest Control*, 13, 12 (1971).
13. Green, M. B., 'Polychloroaromatic and heteroaromatics of industrial importance' in *Polychloroaromatic Compounds* (Ed. Suschitzky, H.), Plenum Press, London, 1974, p. 433.
14. *Pesticide Manual* (Ed. Martin, H. and Worthing, C. R.), 4th edn., British Crop Protection Council, 1974.
15. Burgand, L., Chevril, J., Guillot, M., Marechal, G., Thiolliere, J., and Cole, R., *Proc. Brit. Insectic. and Fungic. Conf.*, Brighton, 2, 645 (1975).
16. Cremlyn, R. J. W., *J. Sci. Food Agric.*, 12, 805 (1961).
17. *Systemic Fungicides* (Ed. Marsh, R. W.), 2nd edn., Longman, London, 1977.
18. Cremlyn, R. J. W., *Internat. Pest Control*, 15 (2), 8 (1973).
19. Goodman, R. N., *Advan. Pest Control Res.*, 5, 1 (1962).
20. Corbett, J. R., *The Biochemical Mode of Action of Pesticides*, Academic Press, London and New York, 1974.
21. van der Kerk, G. J. M., *Wld. Rev. Pest Control.*, 2 (3), 29 (1963).
22. Woodcock, D., *Chem. in Brit.*, 4 (7), 294 (1968).
23. Woodcock, D., *Chem. in Brit.*, 7 (10), 415 (1971).
24. Bent, K. J., *Endeavour*, 28 (105), 129 (1969).
25. Evans, E., *Pestic. Sci.*, 2 (5), 192 (1971).
26. Fest, C. and Schmidt, K.-J., *The Chemistry of Organophosphorus Pesticides*, Springer-Verlag, Berlin, 1973, p. 144.
27. Cavella, J. F. and Price Jones, D., *The Chemist in Industry 2: Human Health and Plant Protection*, Clarendon Press, Oxford, 1974, p. 55.
28. Sbragia, R. J., *Ann. Rev. Phytopath.*, 13, 257 (1975).

Chapter 8
Herbicides

Weeds may be defined as plants growing where man does not wish them to be. As soon as man began organized agriculture which in parts of the world goes back some 10,000 years, though extensive agriculture is only about 5000 years old, weeds started to compete with the crop plants for moisture, nutrients, and light. Much of the energy expended in arable farming goes in such mechanical operations as ploughing, harrowing, and general soil cultivation, all of which both remove weeds and provide suitable conditions for the efficient growth of the crop plant. It is, therefore, somewhat ironic that the majority of our common weeds today were rare plants before man became an agriculturalist and provided conditions on which they too thrived[1]. Traditional mechanical means of weed control reached the peak of efficiency in Britain by the end of the nineteenth century when plenty of cheap agricultural labour was still available and simple horse-drawn machines kept the fields free from weeds. With the rapid growth of industrialization, the resultant drift of labour from the countryside to the factories meant a shortage of manpower on the farms and consequently agricultural wages increased and crops became weedier and weedier and different crop rotations were introduced to try and cut down weeds, not always with success. This situation provided the stimulus for the development both of more efficient mechanical means of weed control and the introduction of chemical weed killers (herbicides) (see Plates 14, 15).

The idea of controlling weeds with chemicals is not new; for more than a century chemicals have been employed for total weed control — the removal of all plants from such places as railway tracks, timber yards, and unmetalled roads. Crude chemicals such as rock salt, crushed arsenical ores, creosote, oil wastes, sulphuric acid, and copper salts were used in massive doses[2,3]. Under these conditions all plants were killed and what was much more needed were chemicals which would selectively kill the weeds, but not harm the crop plants. In the early 1900's some selective control of broad-leaved weeds in cereals was achieved by spraying with a sufficient concentration of such general plant toxicants as soluble copper salts and sulphuric acid. The selectivity was based on physical factors — the larger, rougher surfaces of weed leaves were more effectively wetted by the spray as compared with the narrow, smooth cereal leaves in which there was much greater run off of the toxicant. Copper salts are no longer used, though sulphuric acid is still applied on a limited scale for the destruction of potato haulms at the end of the season. The weed-killing properties of other inorganic compounds such as sodium chlorate, borates, and arsenic compounds (eg. sodium arsenite) have been known for a long time[3]. These compounds all function as total herbicides and treated areas remain toxic to plants for months or, in some cases, years. Sodium arsenite was formerly applied extensively as a potato haulm destroyer.

Creosote and other crude tar oils have been employed as total herbicides for a long time. More recently petroleum oils were purified by shaking them with concentrated sulphuric acid to remove unsaturated compounds, followed by fractional distillation. Such purified petroleum oils of suitable boiling points are used as contact herbicides for post-emergence control of many annual broad-leaved weeds and grasses in carrot, parsley, and parsnip, and for pre- or post-emergence weed control in forest nurseries[4].

With vegetable crops, spraying must be carefully timed, because if carried out too early there is a danger of phytotoxic damage to the crops, and if too late tainting of the crop may occur. Such hydrocarbon oils are physical toxicants. They affect photosynthesis, disrupt mitosis, and change the permeability of cell membranes; prolonged application at high concentrations causes irreversible cell damage leading to death. Consequently they are also used as insecticides and acaricides[4]

The first important discovery in the field of selective weed control was the introduction of 2,4-dinitro-o-cresol (1; R = CH_3) (DNOC or Sinox) in France in 1933. This is a contact herbicide and when sprayed onto a cereal crop it killed the majority of annual weeds infesting the crop without causing appreciable damage to the cereals. Unfortunately most perennial weeds, like couch grass and creeping thistle, were not killed because, although their top growth was desiccated, DNOC is not translocated in plants and accordingly their extensive root systems survived and in due course sent up further shoots. The related compound dinoseb (1; R = $CH_3 CH_2 (CH_3)CH-$) was mainly used for weed control in peas, beans, seedling lucerne, and cotton[3,4]. Both compounds are generally applied as high-volume sprays because even at 890 1/ha (80 gal/ac) they are usually not completely dissolved in the spray tank. These dinitro compounds played a big part in increasing food production during World War II. DNOC also finds use as a general insecticide especially as a winter wash for fruit trees (see Chapter 5, p. 51).

A serious disadvantage is that these substances are very poisonous to mammals. For DNOC (1; R = CH_3) LD_{50} (oral) to rats is 30 mg/kg, and consequently they can only be applied by operators wearing full protective clothing. They are used at fairly high concentrations of about 10 lb/acre (11 kg/ha) and they have caused considerable damage to wild life and some human fatalities[5]. However, on contact with plants or soil, they are fairly rapidly degraded to non-toxic substances and consequently do not accumulate along food chains[6].

DNOC is now little used although dinoseb (1; R = $CH_3 CH_2 (CH_3)CH-$) is still employed as a selective herbicide in peas and cotton[3]. These dinitrophenols owe their herbicidal action on plants to their ability to uncouple oxidative phosphorylation and hence inhibit ATP synthesis[7] (see Chapter 3, p. 29). This process is common to plants and mammals so it is not suprising that dinitrophenols have high mammalian toxicities.

Carboxylic Acid Herbicides

The discovery of the phenoxyacetic acid herbicides derived from the work of Kögl and his collaborators (1934) who showed that indole-3-acetic acid or IAA (2) promotes cell elongation in plants. Auxin or IAA was isolated from plants and this stimulated the search for other compounds of related structure which had growth regulating activity on plants, but which, unlike IAA, were not rapidly metabolized in plants. 1-Naphthylacetic acid was found to be active as was 2-naphthoxyacetic acid. This aroused interest in aryloxyacetic acids as potential plant growth regulators[7]. In 1942 Zimmerman and Hitchcock showed that certain chlorinated phenoxyacetic acids such as 2,4-dichlorophenoxyacetic acid or 2,4-D (3; X = Y = Cl, Z = H, n = 1) were more active than the natural growth hormone IAA (2) and furthermore were not rapidly degraded in the plant[8]. Consequently 2,4-D could be applied externally to the plant and since it was not internally regulated like IAA, produced lethally abnormal growth resulting in the death of the plant. This discovery really marked the beginning of the organic herbicide industry, since the previous herbicides had been mainly inorganic compounds. The herbicidal phenoxyacetic acids were found to be much more active against broad-leaved weeds (dicotyledons) than against cereals and grasses (monocotyledons). The selectivity may arise partly from the fact that the sprays adhere much better to the rougher surfaces of broad-leaved weeds than to the narrow waxy leaves of grasses; but despite its vital importance the selectivity of these compounds is not clearly understood[9-11]. The phenoxyacetic acids were the first really effective selective herbicides and were, like the organochlorine and organophosphorus insecticides, the product of war-directed research. They came into use at a time when the maximum home food production with a much reduced agricultural labour force was a vital factor in the war effort[2,7,8].

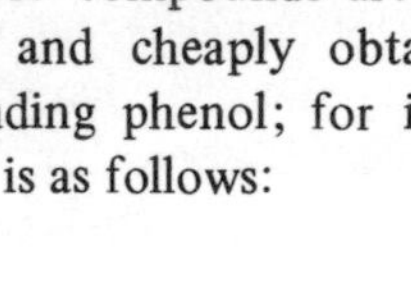

(1) 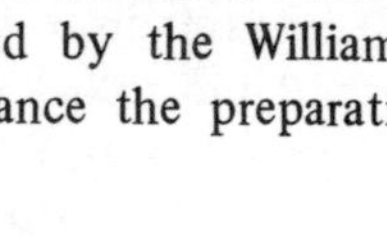 (2) 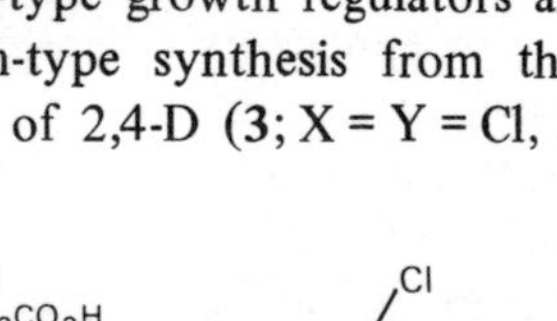 (3)

These compounds are best described as auxin-type growth regulators and are easily and cheaply obtained by the Williamson-type synthesis from the corresponding phenol; for instance the preparation of 2,4-D (3; X = Y = Cl, Z = H, n = 1) is as follows:

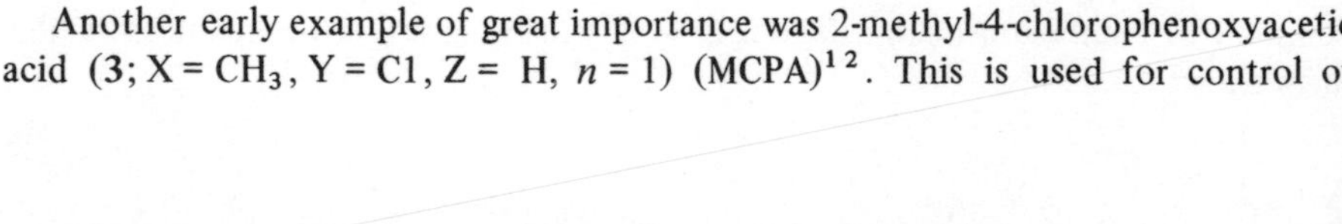

Another early example of great importance was 2-methyl-4-chlorophenoxyacetic acid (3; X = CH$_3$, Y = Cl, Z = H, n = 1) (MCPA)[12]. This is used for control of many broad-leaved weeds post-emergence in cereals, grassland, and asparagus[4,12]. Gardeners often use MCPA on lawns when the annual broad-leaved weeds die rapidly while perennial weeds such as daisies, dandelions, and plantains grow fast and grotesquely and eventually die. The hormone-type chemical induces very rapid RNA production and consequently the plant grows itself to death — in other words the growth completely outstrips the available nutrients. Plants in dry climates are more susceptible and can be effectively treated at lower dose rates than those in conditions of greater rainfall.

In Britain MCPA is used on a larger scale than 2,4-D, whereas in America the reverse is true. MCPA has the important advantage that the sodium salt is much more soluble in water than that of 2,4-D. Also local overdosage is less liable to produce phytotoxic damage to the crop, but the main reason why MCPA is more extensively used in Britain is the greater availability of the starting material o-cresol from coal tar distillation.

Another important member of this series of selective herbicides is 2,4,5-tri-chlorophenoxyacetic acid (2,4,5-T; **3**; $X = Y = Z = Cl$, $n = 1$). This translocated herbicide is employed for control of woody plants and can be used for selective weed control in conifers[4,12]; it is more persistent in soil than either MCPA or 2,4-D.

The analogous 2-phenoxypropionic acids show similar herbicidal activity but are much more effective against chickweed (*Stellaria media*) and cleavers (*Galium aparine*) in cereals[4]. Two well-known examples are mecoprop or 2'-(4-chloro-2-methylphenoxy)propionic acid (**4**) and dichloroprop which is 2'-(2,4-dichloro-phenoxy)propionic acid. The former (**4**) is obtained by condensation of 2-chloro-propionic acid with 4-chloro-o-cresol: Due to the presence of the chiral carbon atom (C*), mecoprop can exist in (+) and (−) optical isomeric forms but only the (+) isomer has appreciable herbicidal activity.

4-chloro-o-cresol   2-chloropropionic acid

(i) hot aq. NaOH
(ii) H⁺/H₂O

(4)

Research by Professor Wain of Wye College (University of London) on the herbicidal properties of a series of ω-phenoxyalkanecarboxylic acids of type (**3**; $X = Y = Cl$, $Z = H$, $n > 1$) showed that herbicidal activity alternated along the series — those members possessing an odd number of methylene groups in the side-chain were active, whereas those with an even number of methylene groups were almost inactive[13,14]. The higher homologues were shown[13] not to be themselves herbicidal, but those containing odd numbers of methylene groups owed their herbicidal activity to *in vivo* β-oxidation within the plant by the enzyme β-oxidase to give 2,4-D which was the active herbicidal entity. Biochemically the process is similar to the β-oxidative degradation of fatty acids which is a well-established metabolic route in mammals:

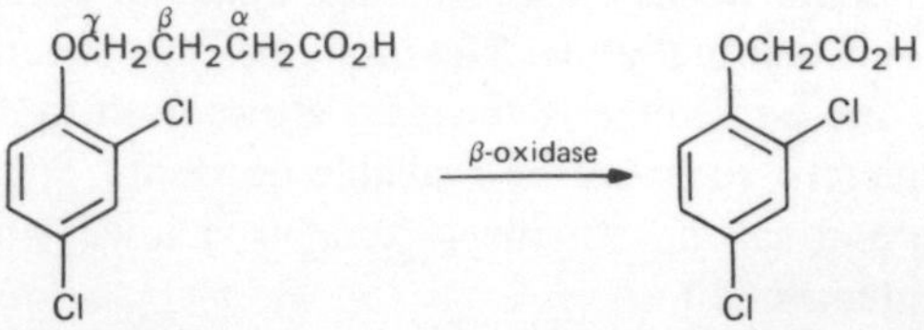

γ-(2,4-dichlorophenoxy)-
butyric acid

2,4-D (active)

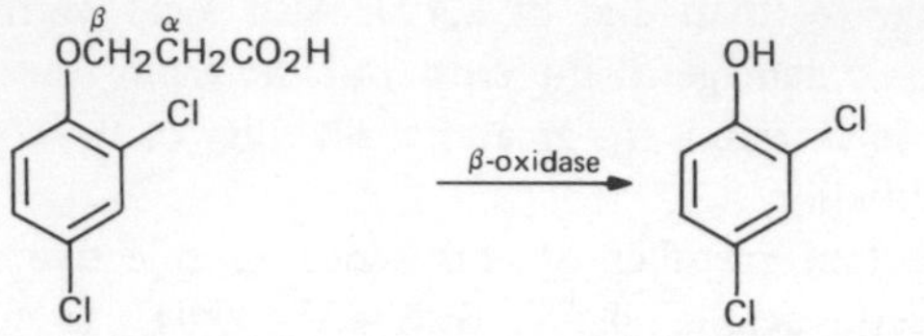

β-(2,4-dichlorophenoxy)-
propionic acid

2,4-dichlorophenol
(inactive)

Wain found[13] that different plant species vary considerably in their ability to perform this enzymic β-oxidation, which provides an additional mechanism for increasing selectivity. Several species of leguminous plants are resistant to the phenoxybutyric acid herbicides because they largely lack the β-oxidases to convert them *in vivo* to the active phenoxyacetic acid derivatives, so γ-(2,4-dichloro-phenoxybutyric acid (2,4-DB) and the related 2-methyl-4-chlorophenoxybutyric acid (MCPB) are widely used for selective post-emergence weed control in leguminous crops (e.g. peas, clover, and lucerne) and in undersown cereals where the analogous phenoxyacetic acid derivatives would damage the legumes[4,12]. The phenoxybutyric acids are synthesized by reaction of the appropriate substituted phenol with butyrolactone. Other important features of the chemical structure of herbicidal phenoxyacetic acids are as follows:

The carboxylic acid group – for high activity the molecule must generally possess either the $-CO_2H$ group or a group that is easily converted to it within the plant tissues.

The unsaturated ring system – for activity the ring appears to need at least one unsaturated bond adjacent to the side-chain; thus cyclohexanoxyacetic acid is inactive but the unsaturated derivative is active:

inactive

active

The side-chain – for a phenoxyalkanecarboxylic acid to show activity depending on cell elongation, the presence of a hydrogen atom attached to the α-carbon is necessary although this criterion does not appear to hold for the benzoic acid herbicides (see p. 146).

Substitution in the phenyl ring – in the phenoxyacetic acids, the introduction of chlorine atoms into the 2-, 4-, or 2,4-positions greatly enhances activity, although the 2,4,6-trichloro derivative is almost completely inactive. Poor activity is found for compounds containing the chlorine atoms in the 2,6- or 3,5-positions[13].

It has been argued[13] that one free *o*-position is an essential requirement for activity, but some very active compounds are known, like 2,4-dichloro-6-fluorophenoxyacetic acid in which all the *o*-positions are substituted. There is therefore some uncertainty regarding the importance of specific nuclear positions on growth regulating properties; however *at least one nuclear position must be unsubstituted.*

Optical isomers – when an alkyl group is introduced into the side-chain of a phenoxyacetic acid, the resultant asymmetric carbon atom makes resolution into (+) and (–) optical isomers possible. For instance with mecoprop (**4**) acid the (+) isomer was much more active than the (–) isomer. Generally the *dextro* isomers are more herbicidal and Wain has suggested[13] that this implies that the active stereoisomer is capable of a specific interaction with some asymmetric component in the plant cell concerned with the production of growth response.

The most satisfactory explanation of the relation between structure and activity of auxins like IAA and the benzoic acid derivatives is Thimann's theory that activity depends on the presence of a fractional positive charge situated 0.55 nm from the negative charge of the carboxyl group. Figure 8.1 illustrates how these criteria are satisfied by IAA, 2,4-D, 2,3,6-trichlorobenzoic acid, and picloram[15,17]: This hypothesis is supported by 2-chloroindole acetic acid which is more active than IAA since the chlorine atom would be expected to increase the positive charge on the nitrogen atom. In 2,4-D the two chlorine atoms increase the positive charge on the 6-carbon atom which explains why the 2,4,6-trichloro derivative is inactive since now that the 6-position is blocked no combination with the presumed site of action in the plant is possible. In the benzoic acid derivatives, auxin activity is only possible when the fractional positive charge is located at the 4-position to fulfil the spacing requirement. Presumably the correct spacing of the polar centres allows the auxins to bind to active receptor sites in the plant cell.

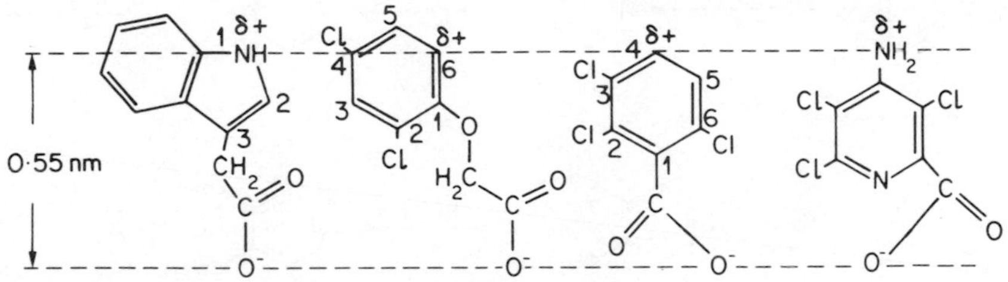

Figure 8.1

The synthetic phenoxyalkanecarboxylic acid and the benzoic acid herbicides function as persistent IAA mimics preventing normal plant growth which depends on the optimum amounts of auxins being present. They cause unrestrained growth of roots and stems eventually killing the plant. There is evidence that sublethal amounts of 2,4-D stimulate the production of ethylene (see p. 170) which is responsible for some of the observed effects of 2,4-D on plants. 2,4-D and IAA have also been shown to stimulate cell wall synthesis and possibly the formulation of RNA messengers which code for the synthesis of specific proteins[18]

The major cause of the herbicidal or auxin-like activity of phenoxyacetic acids and the various chlorinated benzoic acid derivatives is their effect on cell division. 2,3,6-Trichlorobenzoic acid or 2,3,6-TBA was originally reported (1954) as a total weed killer in America, but work by Fisons Ltd. in Britain showed that this compound can function as a translocated selective herbicide against many broad-leaved annual and perennial weeds in cereals[4,12]. The crop is not damaged provided the dosage of the chemical is sufficiently low[3]. Commercially 2,3,6-TBA is obtained from toluene[8]:

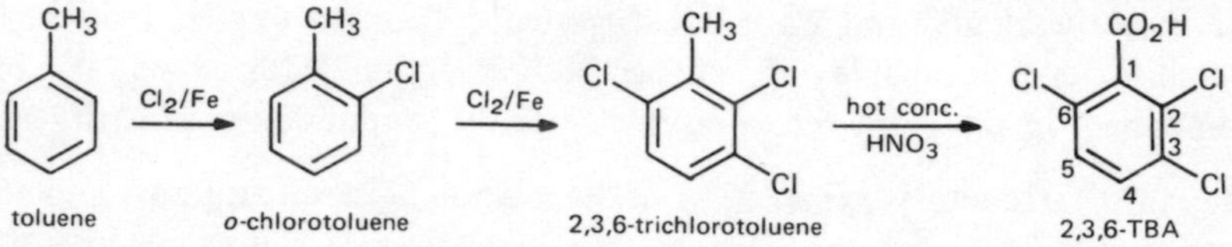

CH3 → (Cl2/Fe) o-chlorotoluene CH3, Cl → (Cl2/Fe) 2,3,6-trichlorotoluene CH3 → (hot conc. HNO3) 2,3,6-TBA CO2H

toluene   o-chlorotoluene   2,3,6-trichlorotoluene   2,3,6-TBA

o-Chlorotoluene is separated from the p-isomer by fractional distillation; further chlorination gives a mixture of isomers containing some 60% of the required 2,3,6-trichlorotoluene which is finally oxidized to the required compound. It was found that the 4-chlorobenzoic acids were almost completely inactive, but the 2,6-dichloro- and 2,3,5,6-tetrachlorobenzoic acids had a similar, but less powerful, herbicidal activity to that of 2,3,6-TBA.

2-Methoxy-3,6-dichlorobenzoic acid (dicamba)[8] finds use as a translocated herbicide against many broad-leaved weeds, especially bindweed, chickweed, and mayweed in cereals[12], and controls bracken for several years after one application. It is manufactured from 1,2,4-trichlorobenzene:

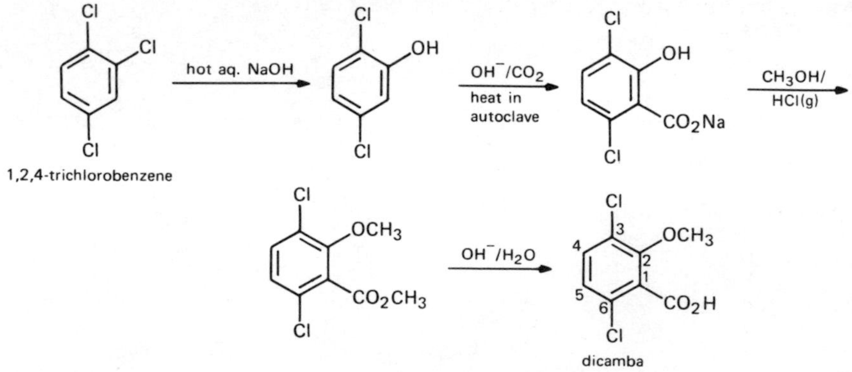

1,2,4-trichlorobenzene → (hot aq. NaOH) → (OH−/CO2, heat in autoclave) → (CH3OH/HCl(g)) → → (OH−/H2O) → dicamba

Picloram (2-carboxy-4-amino-3,5,6-trichloropyridine) is a remarkably stable compound, inspite of the free amino group, and is one of the most persistent herbicides at present available. Picloram is a translocated herbicide used for control of perennial broad-leaved weeds, such as docks and ragwort, and may be effective at concentrations as low as 2 oz per acre (0.14 kg/ha)[4]. It is also the most effective compound for killing trees by application to the bark at the base of the tree. Picloram is manufactured from 2-methylpyridine:

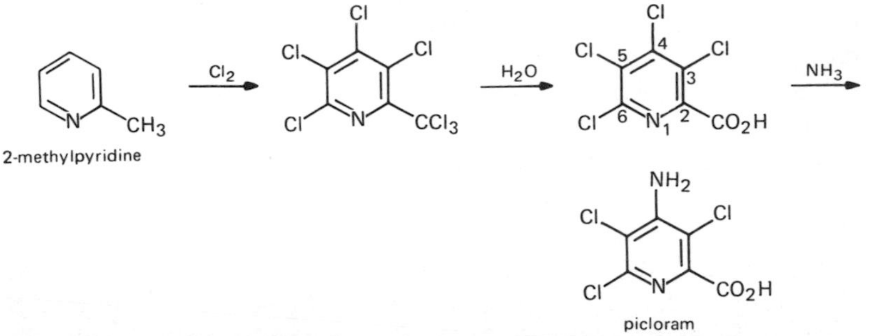

picloram

In the aromatic carboxylic acids, the optimum positions of chlorine substitution for herbicidal activity depend on the nature of the compounds; in the phenoxyacetic acids the preferred positions of substitution are 2,4- or 2,4,5-, whereas the 2,6- derivatives are not herbicidal. In the benzoic acid derivatives, on the other hand, 2,6-disubstitution appears essential, while 2,3,6-trisubstitution leads to maximum activity, but 4-substitution destroys almost all herbicidal action. These observations are in general agreement with Thimann's theory (see p. 145).

Dichlobenil (2,6-dichlorobenzonitrile) is closely related to the chlorinated benzoic acids and was introduced as a herbicide in 1960[8]. Dichlobenil is manufactured from 2,6-dichlorotoluene:

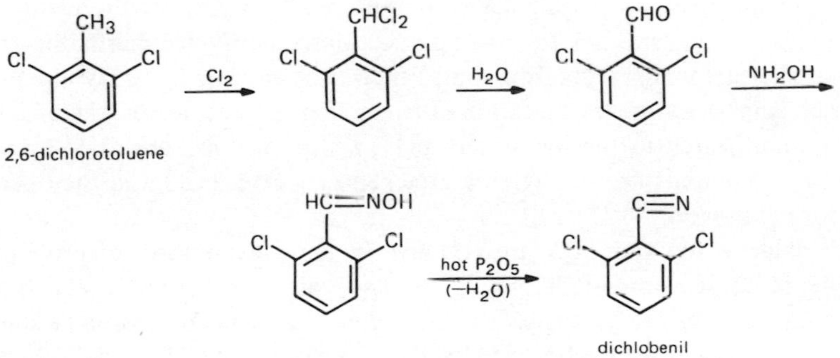

dichlobenil

Dichlobenil is a soil-acting herbicide which inhibits germination of weed seeds and will control weeds among fruit trees and bushes and the majority of annual, perennial, and aquatic weeds[4,12].

Both phenoxyacetic acid and the aromatic carboxylic acid herbicides are hormone-type herbicides and are often active at very low concentrations[9,10]. Phenoxyacetic acid derivatives generally cause contortions of leaf stalks and stems and the roots become stumpy. The benzoic acid herbicides, on the other hand, result in very narrow leaves and buds with extremely brittle stems but with no abnormal root growth. Both types of compound cause swollen stems[15,16].

Chloroaliphatic Acids

A number of chlorinated aliphatic acids have been recognized for some time to be herbicidal. The two most active members of this group are dalapon (sodium 2,2-dichloropropionate) (5) and trichloroacetic acid TCA. α-Chlorination is an essential requirement for activity, as chlorination in other positions does not lead to active compounds. Replacement of chlorine by other halogens and increasing chain length both reduce activity[16].

Dalapon (sodium 2,2-dichloropropionate) (5), introduced in 1953, is obtained by direct chlorination of propionic acid by passing chlorine gas into the hot acid until the calculated increase in weight for the replacement of two α-hydrogen atoms has been attained. Dalapon is used for control of couch and other annual and perennial grasses before planting or sowing most crops[4,12,19]. It may also be employed for the selective control of grasses in potatoes, sugar beet, and carrot, and to kill sedges and other water weeds. It ceases to affect growth 6–8 weeks after application and is readily translocated by absorption from the leaves and roots of weeds. Application is most effective when the weeds are growing vigorously. The related compound, trichloroacetic acid (TCA) (1947) shows a similar selective herbicidal action towards grasses but is less active than dalapon, since it is only a soil acting herbicide and is not appreciably translocated from foliage[20]. However TCA has the advantage of being cheaper than dalapon and is specially used against wild oats in peas and sugar beet. Both these compounds need to be applied in considerably higher concentrations (10–30 lb/acre, 11–33 kg/ha) as compared with the hormone-type herbicides at 5–25 oz/acre, 0.36–1.8 kg/ha). When dalapon is absorbed by plant roots, it collects at the growing points of the plant, and as these mature is translocated to other parts. Consequently it inhibits the growing points of the plants more than the root. Penetration appears to be favoured by low pH values, and when the transpiration rate in couch grass decreases, the quantity of dalapon translocated to the roots increased[5]. Dalapon is also absorbed by foliage, and both compounds are not readily attacked or metabolized and they persist in plants for many weeks[16,18,21].

Both dalapon (5) and TCA are effective in the precipitation of proteins; as a consequence these compounds may be toxic to all protoplasm, especially to the plant enzymes and it is suggested that they may act by combination with proteins[16,18]. Such inactivation of vital enzyme systems interferes with the plants' production of pantothenic acid, one of the B vitamins essential to plant growth and development. Support for this general mechanism of herbicidal action comes from the fact that the toxicity of dalapon to barley and oat plants was partially reversed by addition of pantothenate[16,20]. Additional inactivation of certain plant enzymes may arise from the condensation of the reactive chlorine atoms with the sulphydryl groups of enzymes. Certainly the presence of the reactive α-chlorine atoms is essential for herbicidal activity. Thus when the hygroscopic dalapon absorbs water, the resultant hydrolysis causes loss of herbicidal activity:

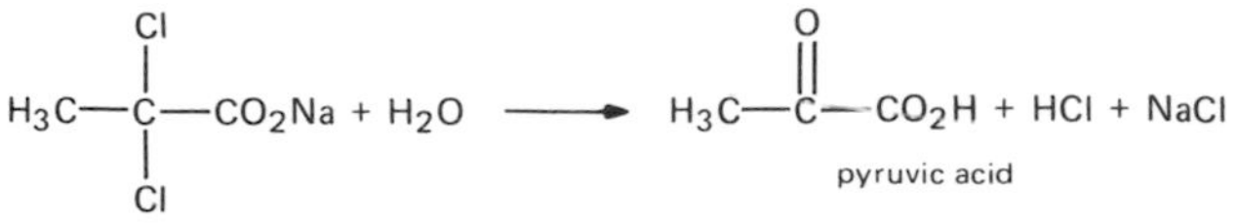

$$H_3C-\underset{\underset{Cl}{|}}{\overset{\overset{Cl}{|}}{C}}-CO_2Na + H_2O \longrightarrow H_3C-\overset{\overset{O}{\|}}{C}-CO_2H + HCl + NaCl$$

pyruvic acid

(5)

Hydrolysis at room temperature is very slow but at $50°C$ it is much faster, so it is essential to store dalapon in moisture-proof containers and to apply dalapon solutions immediately, especially in hot climates[20]. Dalapon and TCA both affect the synthesis of cuticular wax, so it is possible that the inhibition of lipid biosynthesis may be a contributory factor in their lethal action on plants[16,18].

Aromatic Carbamates and Ureas

The arylcarbamate herbicides were introduced by Imperial Chemical Industries in Britain (1946) to combat monocotyledonous weeds, such as grasses, which could not be controlled by dinitrophenols or phenoxyacetic acid herbicides. The first compound was O-isopropyl-N-phenylcarbamate or propham (6; X = H, R = $(CH_3)_2CH-$) but, although effective against seedling grasses, it was not much good against established plants and the selectivity towards grasses was not as clear cut as that shown by TCA or dalapon[3,7,10]. However, later studies of the 3-chloro derivative known as chloropropham (6; X = Cl, R = $(CH_3)_2CH-$) revealed that several crop plants are not harmed by this compound when it is applied to the soil, so it is useful as a soil-acting pre-emergence herbicide for the control of many seedling weeds, or established chickweed in bulbs, fruit trees, rhubarb, peas, and sugar beet[4,12], and to inhibit sprouting in stored potatoes. Generally chloropropham is little value against established weeds, but the acetylenic carbamate (6; X = Cl, R = $ClCH_2C\equiv C-CH_2-$), known as barban, is effective against wild oats post-emergence in wheat and barley[4,12] (Plate 17).

Certain thiolcarbamates have also been developed[16]; for instance triallate (7; R = $(CH_3)_2CH-$, R' = $Cl_2C=C(Cl)CH_2-$) for selective control of wild oats in peas, wheat, and barley[4]; also Eptam or EPTC (7; R = $CH_3CH_2CH_2-$, R' = C_2H_5) is employed as a soil-acting herbicide against some broad-leaved weeds and perennial grasses[4,12], and is especially effective against the tropical sedge.

Aromatic carbamates may be prepared by reaction of the appropriate aromatic amine and the alkyl chloroformate, or from reaction of the alcohol with the appropriate phenyl isocyanate[3,22]:

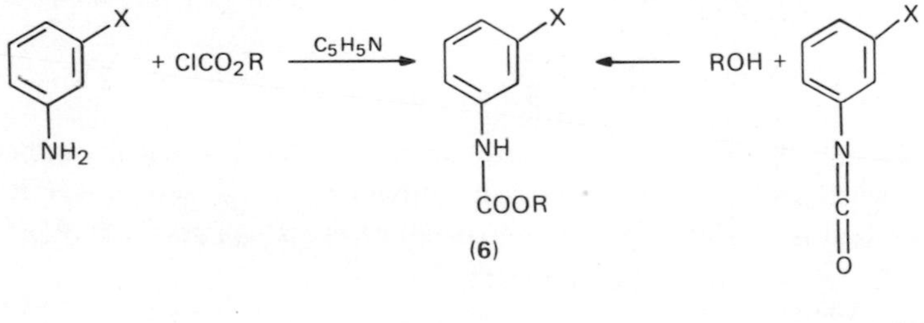

(6)

The phenyl isocyanates being obtained by condensation of the appropriate amine with phosgene or carbonyl chloride:

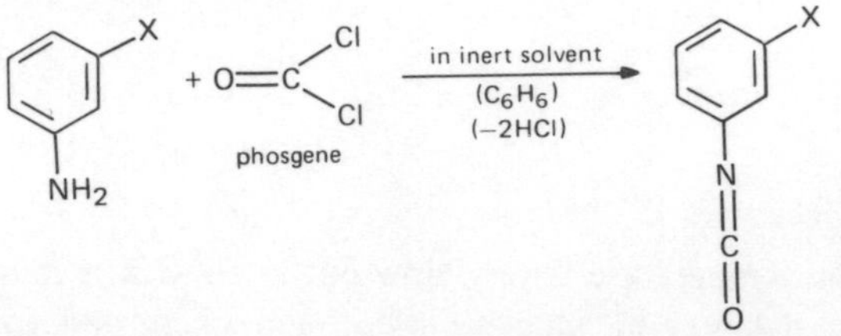

The herbicidal thiolcarbamates may be synthesized by reaction of the appropriate amine and phosgene followed by condensation with a thiol:

$$R_2NH + COCl_2 \longrightarrow R_2N{-}COCl \xrightarrow[(-NaCl)]{R'SNa} R_2N{-}COSR'$$

(7)

It is interesting that several O-arylcarbamates are insecticides (Chapter 6, p. 96) but, not surprisingly, they suffer from being slightly phytotoxic, and some alkyldithiocarbamates are valuable surface fungicides (Chapter 7, p. 111).

The herbicidal N-phenylcarbamates affect photosynthesis by inhibition of the Hill reaction of photosynthetic electron transport, but, in most cases, the effect does not appear to be marked and probably the major cause of their herbicidal activity is their effect in inhibiting cell division or mitosis (Chapter 3) which directly interferes with plant growth[18]. Biochemical studies[16,18] with thiolcarbamates (7) indicate that these compounds inhibit the formation of plant waxes and so it is possible that their herbicidal action arises from interference with lipid biosynthesis.

The sulphonyl carbamate asulam (Asulox), introduced by May and Baker Ltd. in 1965 for control of docks in grassland and against bracken[4,22] (Plates 1, 2, and 18), is also believed[18] to act by inhibition of cell division since the shoots of treated plants do not behave normally. Asulam is prepared from aniline according to Scheme 8.1.

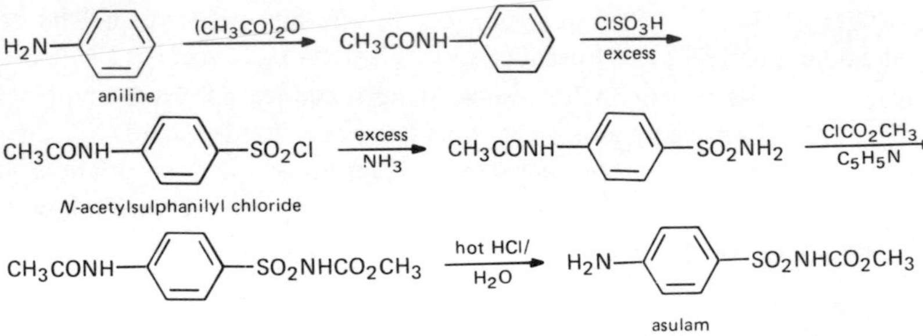

Scheme 8.1

In 1951 C. W. Todd of DuPont described the herbicidal properties of 3-(p-chlorophenyl)-1,1-dimethylurea or monuron (8; R = Cl, X = H, R' = CH$_3$) and emphasized its toxicity against annual and perennial grasses[16]. DuPont initially developed (1952) the following substituted arylureas as commercial herbicides: monuron; fenuron $(8; R = X = H, R' = CH_3)$; diuron $(8; R = X = Cl, R' = CH_3)$; and neburon $(8; R = X = Cl, R' = -(CH_2)_3CH_3)$. In neburon the substitution of one of the methyl groups by butyl reduced herbicidal activity but increased selectivity so that, while neburon is not useful like diuron as a total herbicide, it is more selective for weed control in cereals. A similar effect was noted with linuron $(8; R = X = Cl, R' = OCH_3)$ in which one of the methyl groups is replaced by methoxy.

These ureas are persistent herbicides and are valuable as total weed killers, but later work[16] showed that they have considerable potential for selective application in agriculture; thus chlortoluron or Dicurane $(8; R = CH_3, X = Cl, R' = CH_3)$ was developed by Ciba-Geigy (1969) for selective control of grass and broad-leaved weeds in cereals[4,22] (Plate 19). Currently some 25 different ureas are important commercially and they have become the most widely used herbicides after the phenoxyacetic acid derivatives[3,8]; thus they can be used for pre-emergence control of a wide range of annual weeds in naturally resistant crops, e.g. potatoes, leeks, carrots, and parsnips. (Plate 21) Linuron $(8; R = X = Cl; R' = OCH_3)$ shows better post-emergence weed control than the other areas[4,12].

The arylureas (8) are manufactured from the appropriate arylamine:

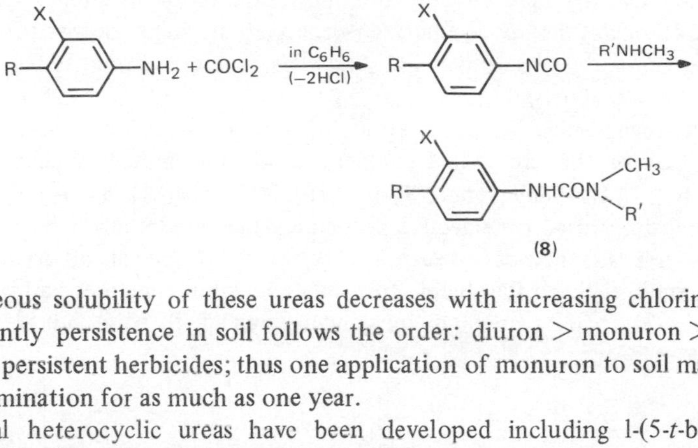

(8)

The aqueous solubility of these ureas decreases with increasing chlorination and consequently persistence in soil follows the order: diuron > monuron > fenuron. They are persistent herbicides; thus one application of monuron to soil may prevent weed germination for as much as one year.

Several heterocyclic ureas have been developed including 1-(5-t-butyl-1,3,4-thiadiazol-2-yl)-1,3-dimethylurea or Spike which was introduced by Eli Lilly Ltd. (1974) as a

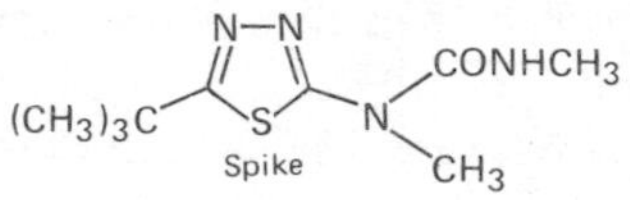

Spike

persistent broad-spectrum herbicide for control of herbaceous and woody plants. When sprayed at a rate of 6–8 lb/acre ≡ 6.4–9 kg/ha it can selectively control woody plants in pastures, and at higher rates can be used for total vegetation control.

The herbicidal activity of ureas is due to their inhibition of the Hill reaction in photosynthetic electron transport, and it may therefore be concluded that they kill plants by interference with photosynthesis which deprives the plant of food (starvation); a contributory factor is probably the irreversible damage which they cause to photosynthetic processes resulting in a permanent lack of food production[15,16,18]. The majority of the active compounds contained small 1-alkyl groups, e.g. they were 1,1-dimethyl-3-phenylureas. Also the introduction of a chlorine atom into the *para* position of the phenyl ring often enhanced activity[21].

Ureas are easily absorbed from the soil by the plant roots and are rapidly translocated to the stems and leaves in the transpiration stream. In the soil diuron and other ureas are microbiologically degraded by stepwise dealkylation and deamination as illustrated by the metabolism of diuron:

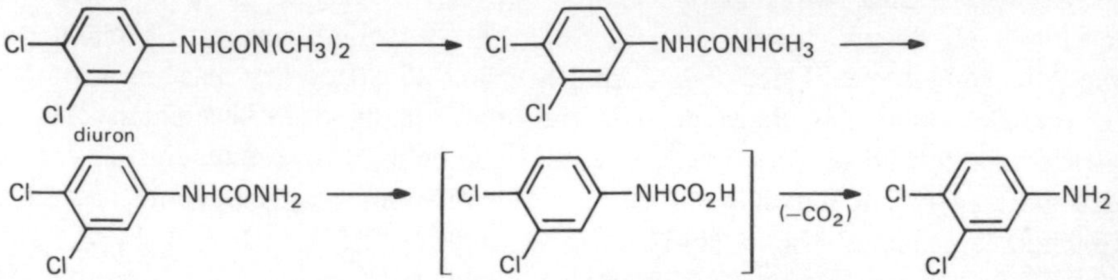

The aromatic chloroamines are finally rapidly decomposed probably to carbon dioxide, ammonia, and halogen. In plants the formation of the amines appears less common, although the preliminary *N*-dealkylation steps occur to give the monoalkylurea, and it is concluded that this is either oxidized or becomes conjugated to carbohydrates[16].

Substituted amides are another group of herbicides whose activity is closely related to that of the ureas[21]. Examples include the anilides such as propanil (9; R = Cl, R' = C_2H_5) and solan (9; R = CH_3, R' = $-CH(CH_3)(CH_2)_2CH_3$). Propanil is extensively used for selective post-emergence weed control in rice fields[4]; the rice is resistant because it can deactivate the chemical by hydrolysis to 3,4-dichloroaniline. Solan is valuable for weed control in tomatoes and potatoes[3]. These amides are readily prepared by condensation of the acid chloride with the appropriate amine:

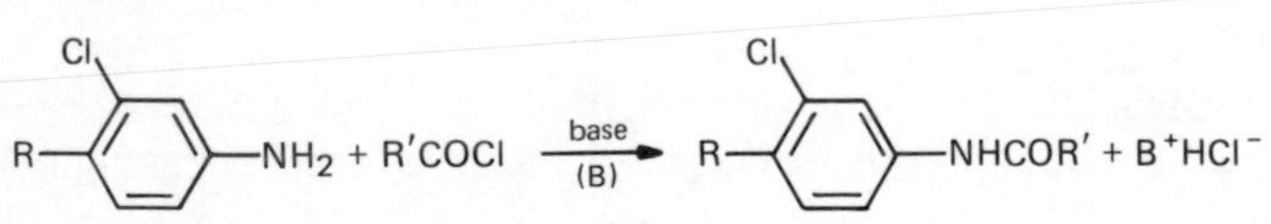

(9)

N-(1-Naphthyl)phthalamic acid or naptalam is used for pre-emergence control of weed seedlings and acts as root growth inhibitor (see p. 169).

Several acetamide derivatives are useful herbicides: allidochlor or CDAA is *N*,*N*-diallyl-2- chloroacetamide and is obtained by condensation of diallylamine and chloroacetyl chloride:

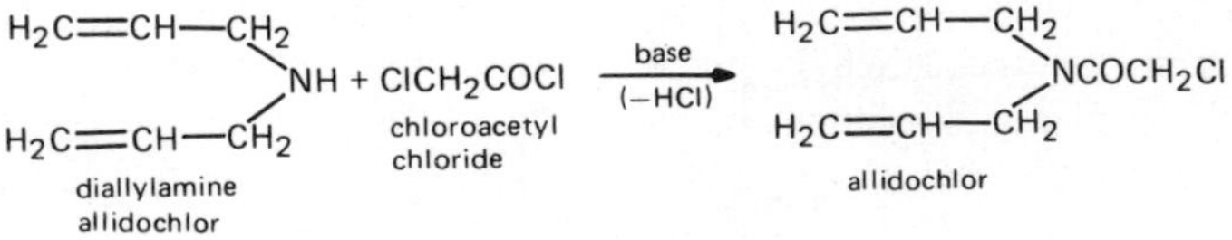

This is a selective pre-emergence herbicide against grasses and is relatively short-lived in soil and is degraded within 3–5 weeks[21,22]. Other chloroacetamides are alachlor (**10**; $R' = CH_3O$, $R = H$, $X = C_2H_5$) and CGA 24705 (**10**; $R' = CH_2OCH_3$, $R = CH_3$, $X = CH_3$); the former is used for pre-emergence

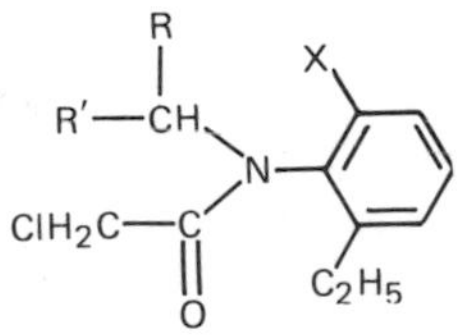

(10)

control of grasses and broad-leaved weeds and remains effective for some 10–12 weeks[22]. The latter introduced by Ciba–Geigy (1974), is effective for inhibiting germination of grasses selectively in maize, soybeans, and peanuts at 1.0–2.5 kg/ha. Diphenamid or Dymid (**11**) is prepared by condensation of diphenylmethane and dimethylcarbamoyl chloride:

$(C_6H_5)_2CH_2 + (CH_3)_2NCOCl$
diphenylmethane   dimethylcarbamoyl
chloride

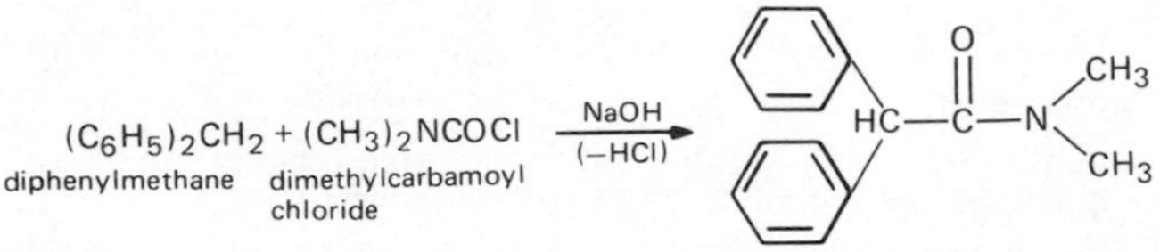

(11)

Dymid is used as a selective pre-emergence soil herbicide against annual grasses and certain broad-leaved weeds[22]. A related compound is benzoylprop-ethyl or Suffix (**12**) introduced by Shell Research (1969) for post-emergence control of wild oats in wheat[4,22] at dosages of

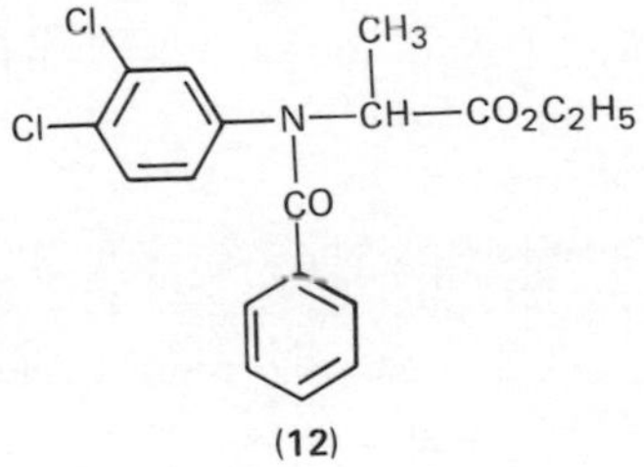

(12)

1–1.5 kg/ha. Wild oats can present a serious problem in spring and winter wheat. The increasing extent of cereal monoculture encourages wild oat infestation which has been helped by the use of the hormone selective weed killers like MCPA which remove the competing broad-leaved weeds.

The amide herbicides, like the ureas, owe their activity to inhibition of the Hill reaction in photosynthesis[15,18,21]. Several 2,6-dinitroanilines are very valuable herbicides[16]; these can be obtained from 1-chloro-4-trifluoromethylbenzene as indicated:

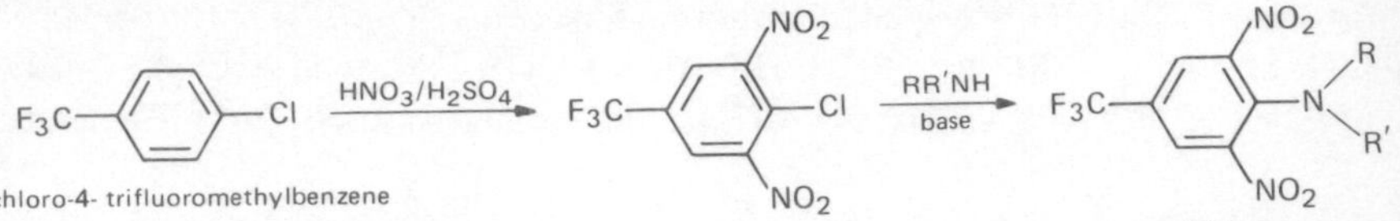

1-chloro-4- trifluoromethylbenzene

(13)

Examples are trifluralin $(13; R = R' = CH_3CH_2CH_2-)$; Balan $(13; R = CH_3CH_2, R' = CH_3CH_2CH_2-)$, and Sonalan $(13; R = CH_3CH_2, R' = CH_2=C(CH_3)CH_2-)$. These compounds were introduced by the Eli Lilly Company (1960) as selective pre-emergence herbicides most effective when incorporated into the soil[4,22]. These dinitroanilines prevent germination of susceptible weed seeds and also stop weed development by inhibition of root growth. Sonalan is specially valuable for selective control of annual grasses and broad-leaved weeds in cotton and soybeans at rates of 0.5 to 1 kg/ha.

The dinitroaniline herbicides are considered[16,18] to owe their activity primarily to inhibition of plant nuclear and cell division, in contrast to the dinitrophenols (p. 141) which act by uncoupling oxidative phosphorylation.

Heterocyclic Compounds

The herbicidal properties of substituted s-triazines were discovered by the Swiss firm J. R. Geigy Ltd. in 1952[16]. Two well-known examples are simazine $(14; R = R' = C = C_2H_5)$ and atrazine or Gesaprim $(14; R = C_2H_5, R' = (CH_3)_2CH-)$. Like the ureas, these are persistent soil-acting herbicides which can be applied in large concentrations (5–20 kg/ha) as total weed killers in industrial sites or on paths etc., but in lower concentrations (1–4 kg/ha) can be used for the selective control of many germinating weeds in a variety of crops (beans, maize, asparagus, straw-berries) and around fruit bushes[4,12,22] (Plate 22). They are taken up by the roots of the emergent weed seedlings causing them to turn yellow and die, but owing to their low aqueous solubility they do not appreciably penetrate to lower levels of the soil, and consequently they have little effect on deep-rooted crops, such as tree and bush fruits (see Chapter 2, pp. 23).

Many varieties of maize and sugar cane are resistant to the herbicidal action of triazines, such as atrazine $(14; R = C_2H_5, R' = (, = (CH_3)_2CH-)$ and simazine $(14; R = R' = C_2H_5)$ because these plants contain an enzyme which detoxifies the compounds by hydrolysis in the plant tissues:

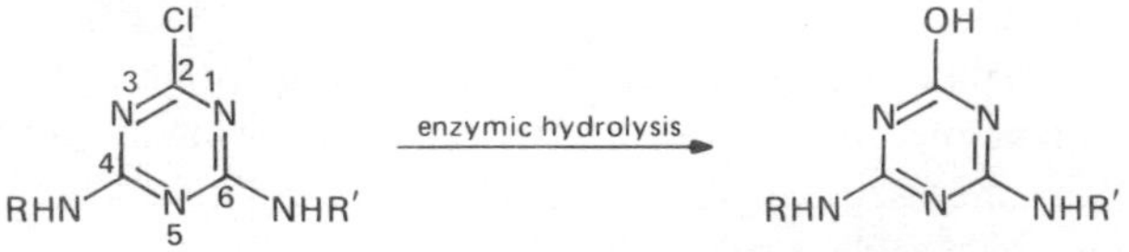

enzymic hydrolysis

(14)

The products of this enzymic hydrolysis are not herbicidal, so these triazines are very valuable for selective weed control in these crops[21,22]. Owing to their low aqueous solubility atrazine and simazine are more effective when they are applied to damp soil. Several other triazines in which the 2-chlorine atom has been replaced by the methylthio group have been developed for specific uses[16]: for instance prometryne $(15; R = R' = (CH_3)_2 CH—)$ is a translocated and soil-acting herbicide for post-emergence control of a wide range of annual weeds in carrot, celery, potatoes, and parsley[4,22]; while desmetryne $(15; R = CH_3, R' = (CH_3)_2 CH—)$ is used against annual broad-leaved weeds in brussels sprouts, cabbage, and kale[4,22].

The triazine herbicides are obtained by reaction of cyanuric chloride with the appropriate nucleophilic reagents. The chlorine atoms may be successively replaced, since as each is substituted further replacement of the remaining chlorines becomes progressively more difficult[23]. The general synthesis of triazines from chlorine and sodium cyanide is illustrated in Scheme 8.2:

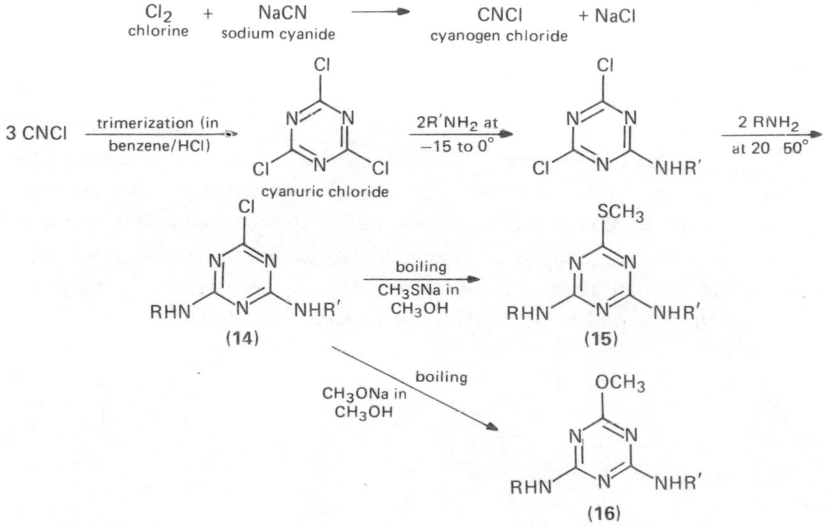

Scheme 8.2

Certain methylthiotriazines (15) are also effective as algicides.

It was also possible to obtain useful herbicidal s-triazines by replacing the 2-chlorine atom of (14) by the methoxy group; an example is prometon $(16; R = R' = (CH_3)_2 CH—)$ used as a total weed killer. Attempts to substitute the 2-chlorine with groups other than methoxy and methylthio did not lead to effective herbicides. However considerable variations in the N-alkylamino side-chains are possible and these extend the range of persistency and biological activity. For instance, terbuthylazine $(14; R = C_2 H_5, R' = C(CH_3)_3$ is more persistent than simazine and is used as pre-emergence herbicide in peas and potatoes[22] (Plate 23); and cyanazine $(14; R = C_2 H_5, R' = —C(CH_3)_2 CN)$ was introduced by Shell (1971) as a relatively short-persistence pre- and post-emergence herbicide against weeds in peas, maize, and beans[4,22]. The corresponding 2-methylthiotriazine $(15; R = C_2 H_5, R' = —C(CH_3)_2 CN)$ known as Aqualin was developed by Shell (1974)[24] for control of aquatic weeds with minimal ecological damage. With the 2-methylthiotriazines (15) an alkoxyalkylamino group may be introduced; an example is methoprotryne (15; R = $-CH(CH_3)_2$, R' = $(-CH_2)_3OCH_3$); altogether some 20 s-triazines are of commercial importance as herbicides.

Triazines are absorbed by clay minerals in the soil which reduces their concentration in the soil solution (see Chapter 2, p. 21). They are readily absorbed by plant roots from the soil solution and generally the 2-chlorotriazines (14) have lower aqueous solubilities than the corresponding 2-methylthio- (15) and 2-methoxy- (16) triazines.

Triazines are metabolized in plants and in the soil by both chemical and microbiological processes. Thus when ^{14}C-side-chain labelled simazine (14; R = R' = C_2H_5) was added to a nutrient solution containing corn, cotton, or soybean plants appreciable amounts of [14]CO_2 were liberated and N-dealkylated metabolites were identified. N-Dealkylation has been established as a principal degradation mechanism in both resistant and susceptible plants[16]. The monodealkylated triazines retain their phytotoxicity, but become inactive when the 2-chloro group is hydrolysed to the hydroxy. Hydrolysis of the 2-methylthio- and 2-methoxy-triazines was also observed in plants. In soils, triazines similarly suffer N-dealkylation and hydrolysis of the 2-substituent, then the free amino groups are replaced by hydroxy groups and ultimately there is cleavage of the triazine ring with liberation of carbon dioxide. Certain soil microorganisms can utilize trazines as sources of carbon and nitrogen. Biguanides are possible intermediates in the ring cleavage, being formed after hydrolytic removal of the 2-carbon atom as carbon dioxide, and these would be susceptible to further hydrolysis[16,23]:

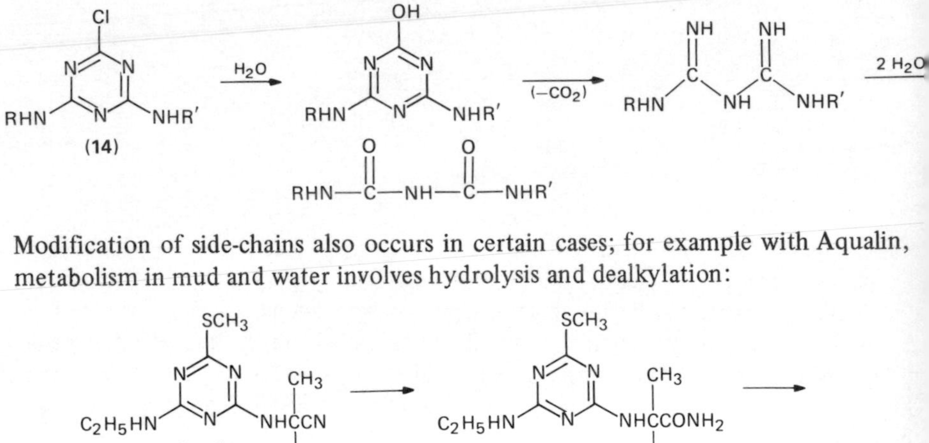

Modification of side-chains also occurs in certain cases; for example with Aqualin, metabolism in mud and water involves hydrolysis and dealkylation:

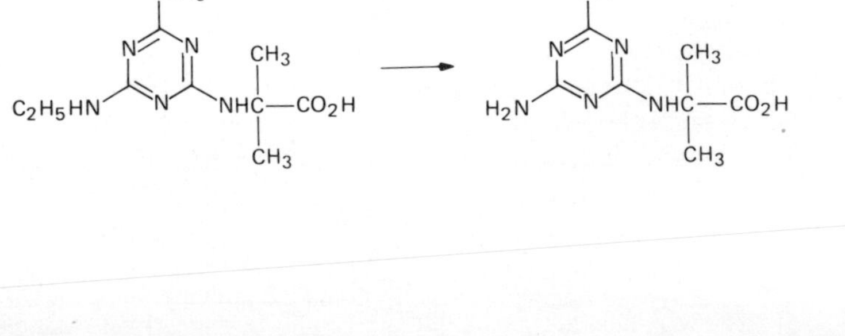

Sulphoxides and sulphones were not detected, and in aquatic plants no degradation products were observed two weeks after treatment[24].

Triazines kill plants by interfering with photosynthesis and it seems clear that, like the urea and amide herbicides, the primary site of action is inhibition of the Hill reaction of photosynthetic electron transport[10,15,16,18]. Triazines are potent inhibitors of the Hill reaction in isolated chloroplasts; all the herbicides inhibiting the Hill reaction possess the common structural feature:

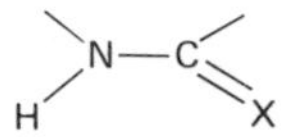

in which X is an atom possessing a lone pair of electrons (either N or O) and it is possible that this grouping represents the essential toxophore in these herbicides and is responsible for binding them to a vital enzyme involved in the Hill reaction and thus preventing the photolysis of water and so depriving the plant of its energy supply.

Several other heterocyclic nitrogen compounds are herbicidal; the following are some examples:

Triazoles

The best known member of this group is amitrole (3-amino-1,2,4-triazole) (17), introduced as a herbicide and growth regulator in 1954, and prepared by condensation of aminoguanidine and formic acid[16]:

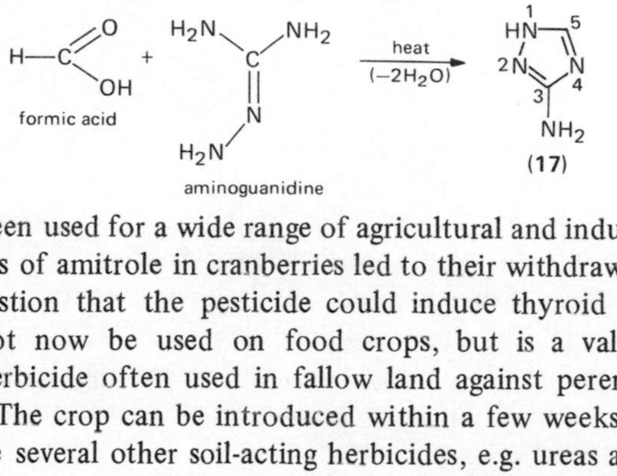

Amitrole has been used for a wide range of agricultural and industrial purposes, but in 1959 residues of amitrole in cranberries led to their withdrawal from the market after the suggestion that the pesticide could induce thyroid tumours in rats[16]. Amitrole cannot now be used on food crops, but is a valuable translocated, non-selective herbicide often used in fallow land against perennial weeds such as couch grass[22]. The crop can be introduced within a few weeks of treatment since amitrole, unlike several other soil-acting herbicides, e.g. ureas and triazines, is not very persistent. The action of the pesticide is synergized by addition of ammonium thiocyanate.

Low concentrations of amitrole stimulate plant growth, but higher concentrations cause chlorosis and death of plants[21]. Amitrole is readily absorbed by the roots and leaves of plants and is translocated in both xylem and phloem.

The primary site of the herbicidal action of amitrole is concluded[15,16,18] to be interference with carotenoid biosynthesis. Is the absence of carotenoids, chlorophyll in the plant leaves is destroyed by photooxidation giving rise to the characteristic symptoms of chlorosis noted in plants after treatment with amitrole.

A number of N-phenylpyridazinones show herbicidal properties; the most active member is pyrazon or 5-amino-4-chloro-2-phenylpyridazin-3-one (18) which is obtained by condensation of phenylhydrazine and mucochloric acid:

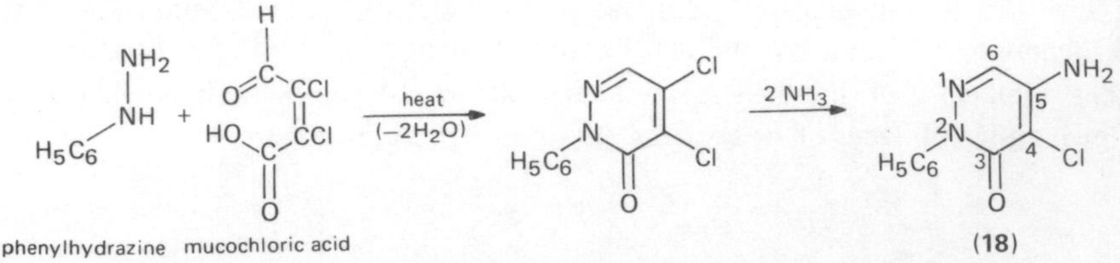

phenylhydrazine mucochloric acid   (18)

Pyrazon (18) is a soil-acting herbicide used for pre- or post-emergence weed control in sugar beet[4,22], and is most effective when incorporated into the soil before sowing the crop. It can be used selectively in beets, because the crop plant is able to detoxify the chemical by metabolism to the inactive aminoglucosyl derivative. Pyrazon is a potent inhibitor of the Hill reaction in photosynthesis which accounts for the herbicidal action[18]:

Some substituted uracils are also herbicidal (Plate 25); two which have attained commercial importance are bromacil or 5-bromo-6-methyl-3-(sec-butyl)uracil (19) and the corresponding 5-chloro-3-t-butyl derivative or terbacil (20):

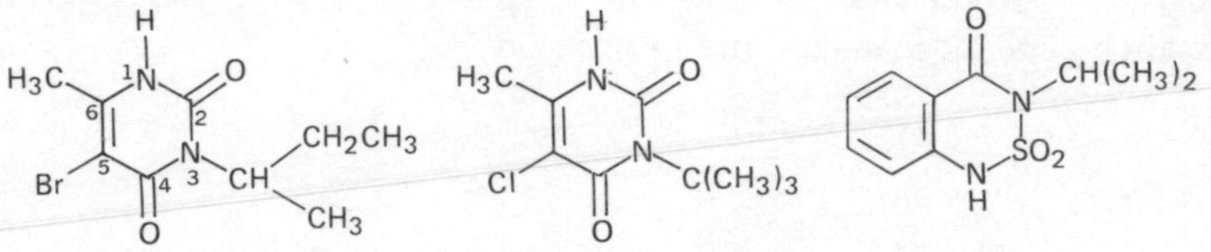

(19)   (20)   bentazon

Bromacil (19) is used as a total weed killer on industrial sites at rates of 20 lb/acre (21 kg/ha), or for selective weed control in cane fruit at 2–6 lb/acre (2.2–6.6 kg/ha)[4,16,22]. Terbacil (20) is used for selective control of annual and perennial weeds in sugar cane, fruit orchards, asparagus, and strawberry at rates of 1–4 lb/acre (1.1–4.4 kg/ha)[4,22]. These are both soil-acting herbicides which were developed by DuPont Ltd. (1966); the primary mode of herbicidal action is again inhibition of the Hill reaction in photosynthesis[16,18].

Bentazon, a member of the thiadiazine group, containing nitrogen and sulphur hetero atoms, was developed by BASF AG (Germany) (1968) for post-emergence control of dicotyledonous weeds in cereals, peas, and beans; also for weed control in rice, maize, soybeans, and peanuts[4,22]. It acts primarily as a contact herbicide, but does have some residual action.

Endothal

This provides an example of a herbicidal oxygen heterocycle which is prepared from maleic anhydride and furan by the Diels–Alder reaction[3,7,22]:

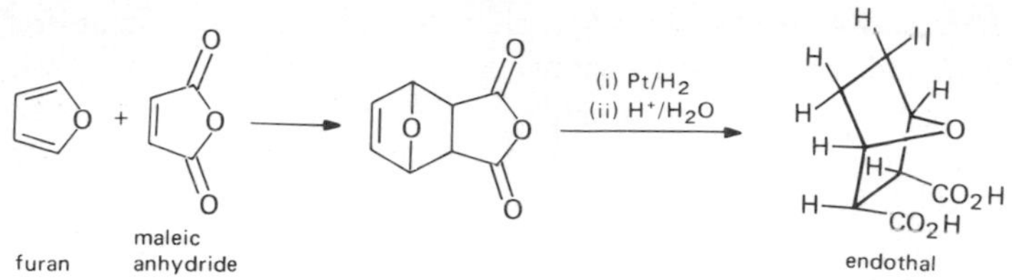

The product is mainly the *exo-cis* isomer, and this is the most active of the three stereoisomers. Endothal as the disodium salt is quite soluble in water and the aqueous solution is readily absorbed by plant roots. It was developed (1951) as a defoliant for cotton, soybeans, and legumes, and can also be used for pre- or post-emergence weed control in turf, beets, and spinach, against aquatic weeds, and as a potato haulm desiccant[21,22]. A disadvantage is the fairly high mammalian toxicity: LD_{50} (oral) to rats for the acid 51 mg/kg, although it has only low toxicity to fish.

The mode of action is not definitely known, but in certain cases this chemical inhibits both lipid and protein biosynthesis[18].

Bipyridylium Herbicides

These compounds were developed from the observation that quaternary ammonium germicides, like cetyl trimethylammonium bromide, will desiccate young plants[21]. The two most important examples diquat (21) and paraquat (22) were introduced by the Plant Protection division of Imperial Chemical Industries Ltd. in 1958. These herbicides are synthesized from pyridine according to Scheme 8.3.

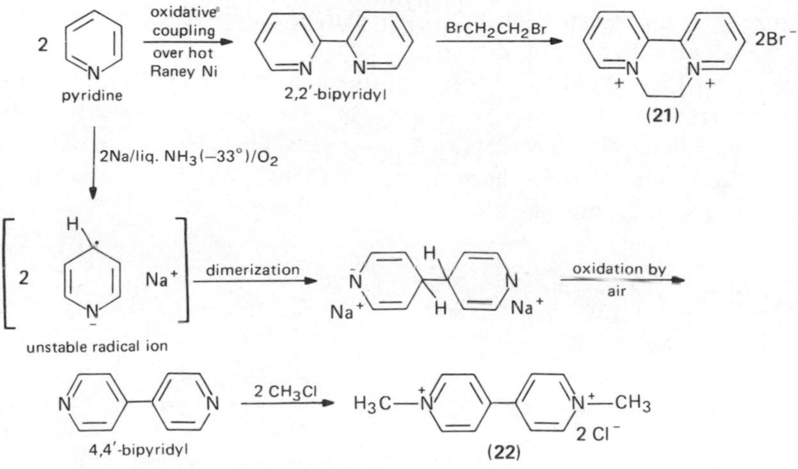

Scheme 8.3

The formation of paraquat provides an interesting example of a commercial synthesis based on a free-radical reaction.

These bipyridylium herbicides are rapidly translocated from the foliage, but not from plant roots because they are immediately deactivated on contact with the soil due to ion exchange with clay constituents which bind the compounds firmly to the soil particles, so they act as contact herbicides rapidly killing all green plant growth on which they fall.

Both paraquat and diquat are widely employed as plant desiccants; paraquat is rather more effective against grasses than diquat[4,22] and can be used to kill weeds before sowing crops. The rapid deactivation by the soil enables paraquat to be used in 'chemical ploughing' in which the seed is sown after spraying with paraquat without the need for ploughing. This technique is of special importance in areas where soil erosion is a serious problem and is becoming increasingly attractive because it reduces energy costs. Paraquat is useful in renovating degenerate pastures, while diquat is widely applied for potato haulm destruction and against aquatic weeds. Both compounds are easily soluble in water since they are salts and paraquat is very useful to the gardener because of its safety in application only killing those plants directly hit by the spray. Paraquat shows only moderate mammalian toxicity: LD_{50} (oral) to rats is 150 mg/kg (diquat is considerably less toxic), but when large amounts are ingested they cause proliferation of human lung cells leading to respiratory failure and ultimately death. At present there appears to be no adequate antidote[21]. The bipyridylium herbicides are almost inactive in the dark, but are extremely active in sunlight when photochemical decomposition products can be detected in treated leaves[5].

Extensive studies[15,18,25,26] have demonstrated that in solution these compounds are almost completely dissociated into ions, and in chloroplasts during the process of photosynthesis the positive ion of paraquat is reduced to relatively stable, water-soluble, free radicals. In the presence of oxygen the free radicals are reorganized to the original ion and hydrogen peroxide which is probably the ultimate toxicant destroying the plant tissue.

Support for this suggestion comes from the discovery that paraquat (22) can be chemically reduced by sodium dithionite or zinc dust to intensely coloured solutions containing resonance-stabilized radical ions in which the odd electron can be delocalized over all twelve nuclear carbon atoms with partial removal of the positive charges on the nitrogen atoms:

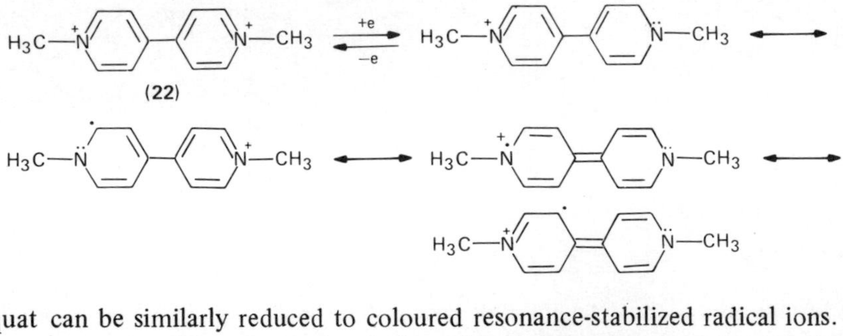

Diquat can be similarly reduced to coloured resonance-stabilized radical ions. The reduction can be effected in isolated plant chloroplasts and plays an essential part in the herbicidal activity of these compounds, since only those quaternary bipyridyls which can be reduced by zinc dust to coloured solutions showed appreciable herbicidal properties[25,26]. Measurement of the redox potentials for the reduction of various bipyridylium salts showed that the majority of herbicidal derivatives had redox potential values within the range -300 to -500 mV. In other words only those bipyridylium compounds in which the two heterocyclic rings are essentially coplanar are herbicidal. Any twisting causes loss of activity. Consider the derivatives of diquat and paraquat shown below:

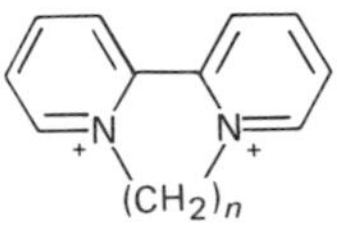

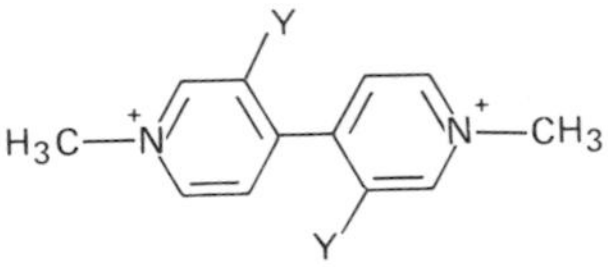

In diquat where $n = 2$ the rings are coplanar and the compound was an active herbicide, but when $n = 3$ the compound was much less active and with $n = 4$, the substance was almost completely inactive. In paraquat with $Y = H$, the compound was very active, but when Y was methyl or other alkyl radicals there was almost no activity. In all these cases, apart from paraquat and diquat, ultraviolet spectroscopy showed that the rings are not coplanar, hence the formation of resonance-stabilized free radicals on biological reduction is no longer possible.

Photosynthesis is basically an oxidation—reduction process coupled to ATP formation, and so photosynthesis could provide sufficient reduction potential to reduce diquat and paraquat to the stabilized free radicals; this reduction by isolated plant chloroplasts in the absence of oxygen has been demonstrated[26]. Bipyridylium herbicides are only active in the presence of both light and oxygen, because in the absence of oxygen the free radicals are extremely stable and are unlikely to participate in radical reactions leading to phytotoxic damage. On the other hand, in the presence of oxygen diquat or paraquat present in chloroplasts is quickly reduced to the radical ions which are subsequently reoxidized with continuous generation of hydrogen peroxide. Hydrogen peroxide is generally considered to be the active phytotoxic agent in these compounds, although possibly other transitory radical intermediates may be implicated in the herbicidal action[15,18,26].

Hydrogen peroxide kills the plant by lipid peroxidation involving a chain reaction which destroys the cellular membranes. Paraquat and diquat are total herbicides and later research has investigated the possibility of introducing some degree of selectivity into this class of herbicides[25].

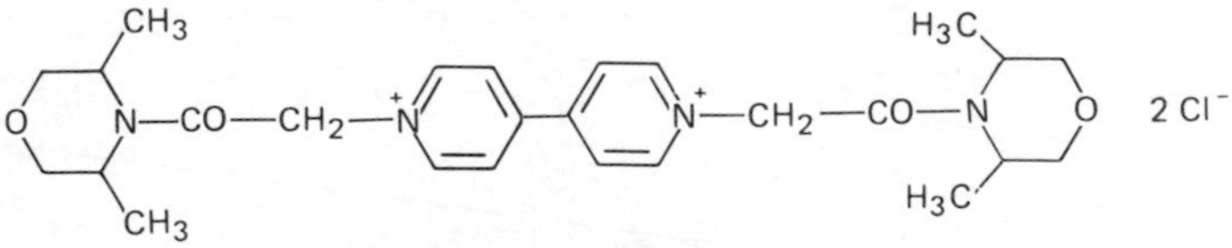

(23)

It has been discovered[25] that certain carbamoyl derivatives of paraquat, such as morfamquat (23), while being very active against broad-leaved weeds, are non-injurious to grasses and so are useful for post-emergence weed control in cereals[22]. This selectivity may be due to differential movement of the compound from the cytoplasm into the chloroplasts of the susceptible plant species[18].

Miscelleneous Compounds

Benzonitriles

Several hydroxybenzonitriles are herbicidal; two of the best known examples of this group are bromoxynil (24; X = Br) and ioxynil (24; X = I). These contact herbicides were independently discovered in 1963 by Professor Wain of Wye College and by May and Baker Ltd. and can be prepared from p-hydroxybenzaldehyde[3,21,22]:

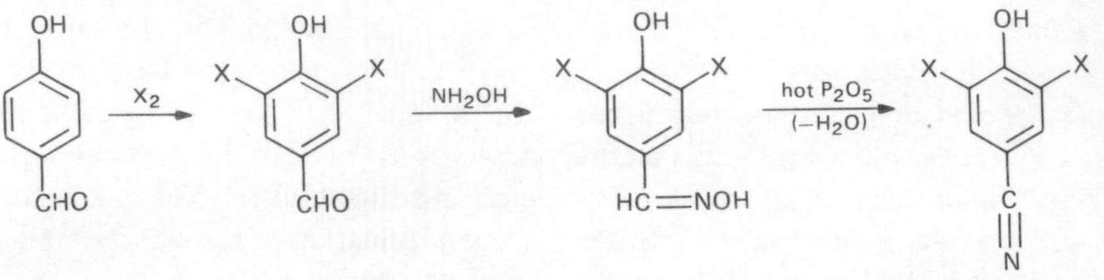

Ioxynil is chiefly used for general weed control in turf, onion, and leek, while bromoxynil functions as a contact herbicide for the post-emergence control of many broad-leaved weeds especially chickweed and mayweed in cereals at rates of 4–12 oz/acre (0.3–0.9 kg/ha)[4,12] (Plate 24). The mode of action of the hydroxybenzonitriles is complex, since, like dinoseb (p. 51), they both uncouple oxidative phosphorylation and inhibit photosynthesis. It is not certain which action is most important in their herbicidal activity, but the primary cause of activity is generally considered to be their interference with photosynthetic electron transport[18].

Another useful herbicide is 2,6-dichlorobenzonitrile or dichlobenil (26) and the hydrogen sulphide adduct, chlorthiamid (25), which slowly loses hydrogen sulphide in the soil:

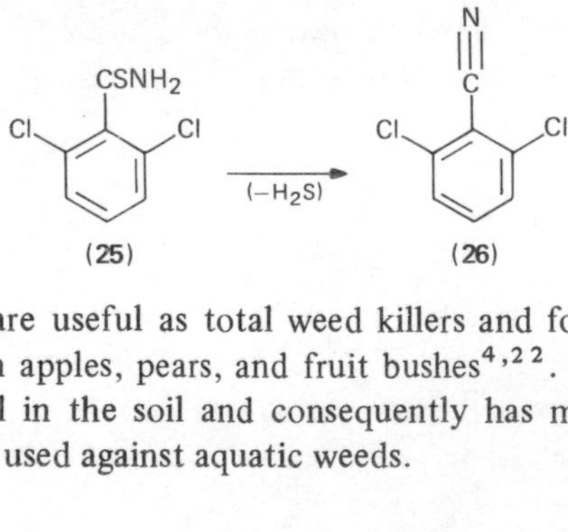

These compounds are useful as total weed killers and for selective pre-emergence control of weeds in apples, pears, and fruit bushes[4,22]. Chlorthiamid (25) slowly liberates dichlobenil in the soil and consequently has more residual action; both compounds are also used against aquatic weeds.

Diphenyl Ethers

A number of substituted diphenyl ethers (27) have been developed as herbicides[26]. They are generally prepared by the Williamson synthesis from *p*-chloronitrobenzene and the appropriate substituted sodium phenolate:

Y—⟨ ⟩—ONa + Cl—⟨ ⟩—NO$_2$ $\xrightarrow[(-NaCl)]{heat}$ Y—⟨A⟩—O—⟨B⟩—NO$_2$

(27)

The best-known example is nitrofen (27; X = Y = Cl), introduced by the Rohm and Haas Company (1964). Nitrofen is a valuable pre-emergence herbicide against several annual broad-leaved and grass weeds in brassica (e.g. kale, rape, cabbage, Brussels sprouts) and winter-sown wheat[4,22]. The pesticide is applied as a thin layer on the surface of the soil at a rate of 3–4 lb/acre (3.4–4.5 kg/ha); herbicidal activity is however quickly lost when the chemical is mixed with the soil.

Another example is fluorodifen (27; X = NO$_2$, Y = CF$_3$), developed by Ciba-Geigy Ltd. (1968). This is applied as a contact pre- or post-emergence herbicide, and is promising for pre-emergent weed control in soybeans or pre- or post-emergence application in rice at 3–4 lb/acre (3.3–4.4 kg/ha). On the majority of crops, other than rice, activity remains for 8–12 weeks especially in dry soil[22].

The Japanese firm Nihon Nohyaku Ltd. introduced 2,4-dichlorophenyl 3'-methoxy-4'-nitrophenyl ether (27; X = Y = Cl, 3'-OMe) for weed control in rice[22].

The herbicidal diphenyl ethers can be divided into two main groups[26]:

(a) those containing at least one *ortho* substituent on the benzene ring; these compounds are only active in the presence of light;
(b) those without *ortho* substituents which are active even in the dark.

The majority of the diphenyl ethers belong to the first group, the favoured substitution pattern being 2,4-disubstitution in ring A with a 4'-nitro group in ring B, although the *para*-nitro groups can sometimes be replaced by a chloro or cyano group. The second group is illustrated by 3-methoxy-4'-nitrodiphenyl ether.

Diphenyl ether herbicides are formulated either as granules or wettable powders. In plants, diphenyl ethers are rapidly taken up by the roots and degraded in plant tissue – the main metabolites of fluorodifen (27; X = NO$_2$, Y = CF$_3$) are shown in Scheme 8.4. The unknown compounds I and II may be amino acid conjugates of the phenols[26].

The diphenyl ethers (27) inhibit the Hill reaction in photosynthesis, but they also reduced the respiration of plant mitochondria. The herbicidal activity could, therefore, result from interference with the production of ATP by either of these routes. However, the major cause of herbicidal action is probably[18] the interference with photosynthesis because nitrofen (27; X = Y = Cl) prevented the growth of *Chlorella* in light, but not that of the shoots and roots of higher plants in the dark. In the case of those diphenyl ethers, without *ortho* substituents, which

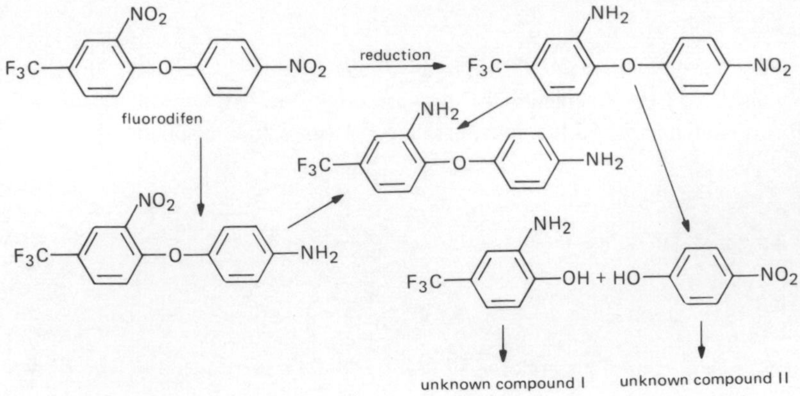

Scheme 8.4

were active in the dark, herbicidal action may depend on the inhibition of cell division[26].

Organophosphorus Compounds

The value of organophosphorus compounds as systemic insecticides is well established, especially since several members of the group show remarkable selective toxicity to insects (Chapter 6). It is only comparatively recently that organophosphorus compounds have become important as herbicides, but currently this is a rapidly expanding area of pesticide research.

A wide range of different types of organophosphorus compounds show some herbicidal properties but the major groups are phosphorothiolates, phosphoramidates, and phosphonates.

a) Phosphorothiolates. Bensulide (**28**), an effective pre- and post-emergence herbicide with 4–12 months persistence of action for brassicas, lettuce, and cotton[22,27], was introduced by Stauffer Chemical Company (1964). Bensulide (**28**) was prepared by condensation of sodium *O,O*-diisopropylphosphorodithioate with the appropriate sulphonamide:

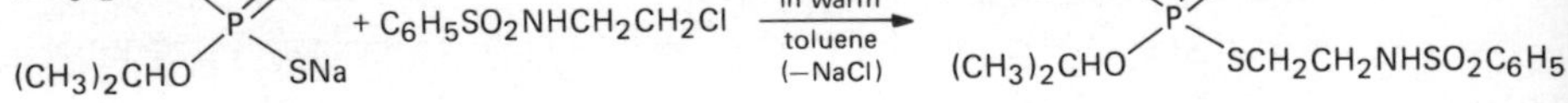

(**28**)

This has a low mammalian toxicity: LD_{50} (oral) to rats 770 mg/kg, and possibly owes its herbicidal properties to the inhibition of cell division in root tips.

Defoliants play an important role in cotton growing since their application facilitates mechanical harvesting of the crop. Folex (**29**) is a valuable defoliant, which is obtained by reaction of finely divided white phosphorus with dibutyl disulphide:

$$P + C_4H_9S\text{---}SC_4H_9 \xrightarrow[\substack{\text{under atmosphere} \\ \text{of } N_2}]{\text{in } (CH_3)_2SO} (n\text{-}C_4H_9S)_3P$$

$$\underset{\text{dibutyl disulphide}}{} \qquad \qquad \qquad \qquad (29)$$

The corresponding phosphine oxide, known as DEF (30), is also a useful defoliant and inhibits new growth of green plants. DEF (30) is made by oxidation of Folex, or from butyl thiol and phosphorus oxychloride:

$$3\ C_4H_9SH + POCl_3 \xrightarrow[(-3HCl)]{\text{base}} (C_4H_9S)_3PO$$

$$(30)$$

Another interesting phosphine oxide is the tris-(1-aziridinyl) derivative or tepa (31), a growth inhibitor which is claimed[27] to increase the sugar content of corn, sugar beet, and sugar cane.

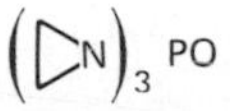

$$\left(\triangleright N\right)_3 PO$$

$$(31)$$

Such aziridinyl compounds are alkylating agents which interfere with cell division and are important as chemosterilants (Chapter 13, p. 200). Tepa (31) cross-links cellulose in fabrics and its stunting effects on plants may arise from the *in vivo* cross-linking of cellulose. The phytotoxicity may depend on inhibition of cellulose biosynthesis, since the cell wall constituents were reduced in treated plants[18].

(b) Phosphoramidates DMPA or Zytron (32) was introduced by the Dow Chemical Company (1959) and is prepared from 2,4-dichlorophenol:

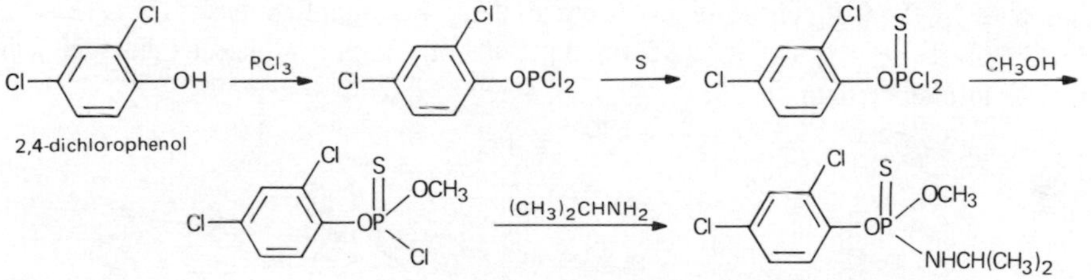

DMPA (32) is used as a germination inhibitor and for control of crabgrass in turf; a related phosphoramidate cremart is used for pre-emergence control of a range of annual weeds in cereals, cotton, beans, and carrots[27].

(c) Phosphonates Glyphosate (33), introduced by the Monsanto Chemical Company (1971), is obtained from glycine and chloromethylphosphonic acid[22,27]:

$$\underset{(HO)_2\overset{\displaystyle O}{\overset{\|}{P}}CH_2Cl}{} + NH_2CH_2CO_2H \xrightarrow[(-HCl)]{OH^-} (HO)_2\overset{\displaystyle O}{\overset{\|}{P}}CH_2NHCH_2CO_2H$$

$$(33)$$

Glyphosate (33) acts as a broad-spectrum translocated post-emergence herbicide for controlling couch and other annual and perennial weeds before sowing cereal crops[4,22]. The herbicidal action may be due to glyphosate interfering with the biosynthesis of phenylalanine[27]. A related phosphonic acid is Ethrel which acts as a plant growth regulator (see p. 170).

Plant Growth Regulators

In higher plants the control of growth and development is exceedingly complex and depends on the presence of various chemical plant growth hormones or regulators. In 1932 observations showed that ethylene and acetylene promoted flowering in pineapples, and in 1934 indole acetic acid (IAA) (34; $n = 1$) was shown to enhance the growth of certain plant tissues, e.g. roots. IAA was soon isolated from a number of plants and it gradually became clear that all the remarkable physiological changes in plants are controlled by hormones. These chemicals control growth, initiate flowering, cause blossoms, fruit, and leaves to fall, induce setting of fruit, control initiation and termination of dormancy, and stimulate root development. Plant hormones fall into five main types[17,28].

Auxins

The only naturally occurring plant auxin is indole acetic acid (IAA) (34; $n = 1$) which probably occurs in all plant tissue. Auxins have the characteristic effect of promoting cell elongation in shoots and assisting the rooting of cuttings. The biochemical mode of action is not clearly understood[18], but it probably involves loosening the cell wall to facilitate cell enlargement and is related to the action of the phenoxyacetic herbicides, many of which have auxin-type activity[9,10,15,17] (see p. 142). Useful synthetic analogues of IAA are 1-naphthylacetic acid (35) and 3-indolylbutyric acid or IBA (34; $n = 3$), both of which are used to enhance root growth in plant cuttings:

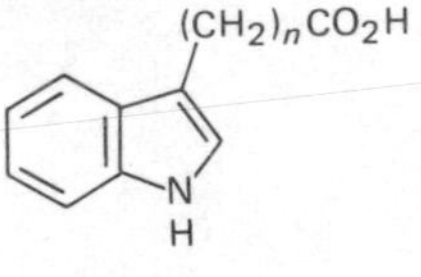

(34) 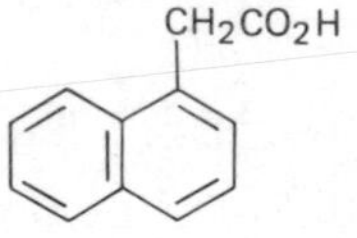   (35) 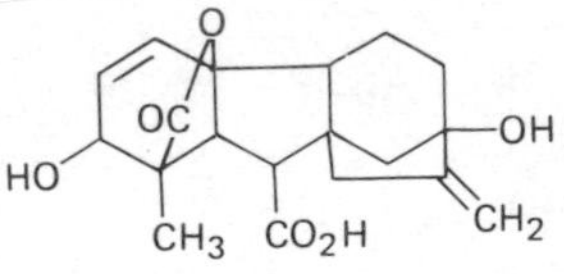   (36)

Gibberellins

This group of plant growth hormones (over 30 occur naturally) have varied morphological effects which differ from plant to plant. They stimulate cell division or cell elongation or both. Gibberellins may possibly act by modifying auxin levels in plant tissues[18]. In 1957 gibberellic acid or GA3 (36) was introduced for a variety of purposes, the most important of which were inducing germination in barley during brewing, and ending dormancy of seed potatoes. Gibberellic acid (36) was originally isolated from the fungus *Gibberella fujikuroi* and caused fantastic growth in many species of plants. Since then more than 30 gibberellins have been isolated known as GA1, GA2, etc., with gibberellin action. They do not affect root growth but influence the dormancy of buds and seeds. The most characteristic effect of gibberellin treatment is the stimulation of plant growth as indicated by exceptionally long stems. Gibberellins are believed to be important in mobilizing sugars in certain plant organs and some of their effects may be due to this property.

Cytokinins

These compounds control cell division and hence can exert a decisive effect on differentiation and initiation of roots and buds. The majority of cytokinins are adenine derivatives which were discovered in 1955. Their practical potential lies in their ability to extend the storage life of green vegetables, cut flowers, and mushrooms. Two examples of natural cytokinins are zeatin (37) and 6-(γ,γ-dimethylallylamino)purine (38) while 6-benzylamino- and 6-furfurylaminopurine (39;

$R = C_6H_5CH_2-$ or 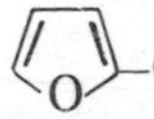 $-CH_2-$ respectively) are examples of synthetic cytokinins:

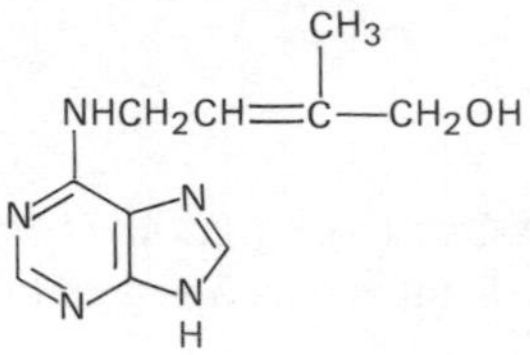

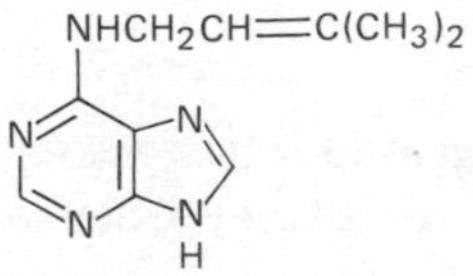

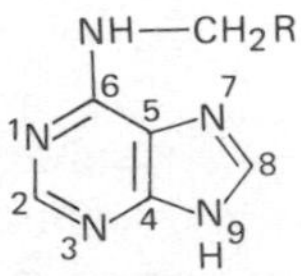

Cytokinins also affect leaf growth, light response, and ageing. Their outstanding effect is the induction of cell division. The mechanism of action is not definitely known but they apparently act by becoming incorporated in the cellular nucleic acids.

Inhibitors

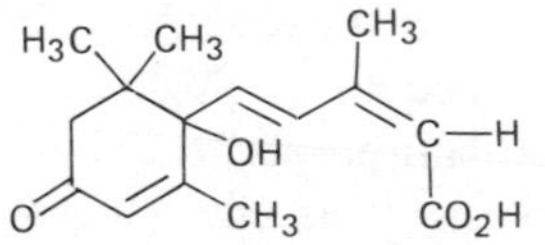

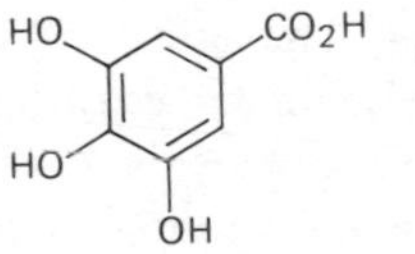

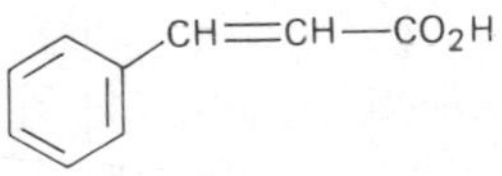

Natural growth inhibitors exist in cotton, sycamore, and yellow lupin. The cotton growth inhibitor abscisic acid (40) also occurs in sycamore leaves, yellow lupin, and other plants. Abscisic acid induces dormancy in cotton, birch leaves, plum twigs and reduces germination in rose hips. The inhibitory effects can be relieved by treatment with gibberellin and cytokinins[18]. Adenosine-3',5'-cyclic monophosphate (cyclic AMP) may possibly function as an intercellular messenger for auxins, gibberellins, and cytokinins. In plant tissue the specific effect exerted by each hormone is reflected by a local change of cyclic AMP. Other natural growth inhibitors include gallic acid (41) and cinnamic acid (42) and such compounds may be used to prevent sprouting of stored onions, potatoes and root crops, retard sucker development on tobacco plants, and shorten plant stems.

A number of synthetic herbicides act by modifying plant growth — important examples are the phenoxyalkanecarboxylic and benzoic acid herbicides already described (p. 142) which possess auxin-type activity.

Other synthetic growth inhibitors include the following compounds:

Maleic hydrazide (MH) (43) has been used since 1949 and is obtained by condensation of maleic anhydride and hydrazine[21]:

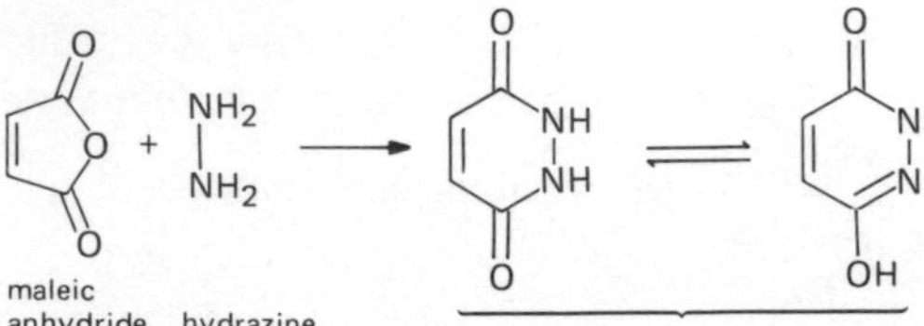

maleic anhydride   hydrazine

(43)

MH is used as a stunting agent to reduce the rate of growth of grass and some other weeds on verges, suppressing root and sucker growth on tobacco and certain trees, and preventing sprouting of potatoes and onions in storage[3-5,21].

Maleic hydrazide (43) inhibits cell division in the actively growing tissues of treated plants, but apparently does not affect cell enlargement[18]. It is an isomer of uracil, one of the pyrimidine bases in RNA, so it is suggested[18] that the effect on mitosis may arise from MH becoming incorporated into the RNA molecule.

One of the most important synthetic growth regulators is chlormequat chloride (44), introduced by the American Cyanamid Company (1959) and prepared from ethylene dichloride and trimethylamine[21]:

$$ClCH_2CH_2Cl + N(CH_3)_3 \longrightarrow ClCH_2CH_2\overset{+}{N}(CH_3)_3Cl^-$$

(44)

Chlormequat (44) produces compact plants with short stems and reduced internodes, and is used commercially for reducing the height of cereals to prevent the crop bending over under its own weight so facilitating mechanical harvesting and allowing more intensive use of fertilizers[4]. It may make the crop more resistant to attack by insects and fungi. The effect of chlormequat on higher plants is competitively inhibited by gibberellin and growth retardant properties are probably due to the compound inhibiting gibberellin biosynthesis[18], as chlormequat does prevent production of gibberellin by the fungus *Gibberella fujikuroi*[18].

Daminozide (45) is prepared by reaction of succinic anhydride and *N,N*-dimethylhydrazine[21]:

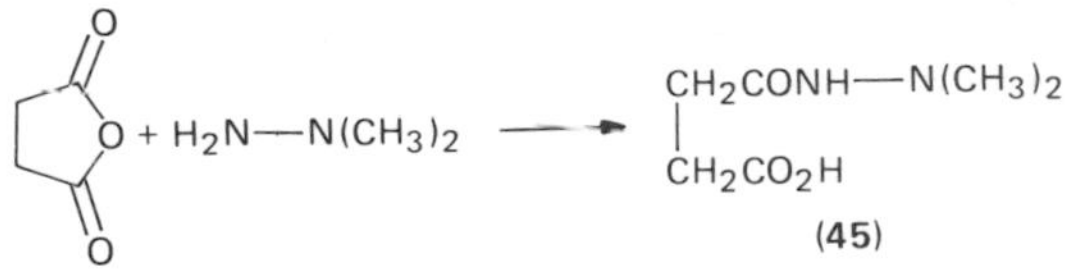

succinic
anhydride   *N,N*-dimethylhydrazine

Daminozide is used to control the growth of fruit trees and the shape and height of ornamentals, e.g. to stunt the growth of chrysanthemums making them more suitable for indoor cultivation[4]. Daminozide affects plant growth similarly to chlormequat and probably also owes its action to interference with gibberellin biosynthesis. The pyrimidine derivative, ancymidol (46), was introduced by the Eli Lilly Company (1971) as a plant growth retardant. It reduces internode elongation and is effective on a large range of plant species by soil or foliar application. The inhibitory effect of ancymidol (46) can be reversed by treatment with gibberellin, suggesting that again this compound acts by interference with gibberellin biosynthesis. It has delayed flower maturation in various greenhouse flowering plants (e.g. chrysanthemums, poinsettias, and Easter lilies) by up to 5 weeks. It is interesting that the related *N*-arylphthalamic acids also have growth regulating properties; an example is the α-naphthyl derivative or naptalam (47) used as a selective pre-emergence herbicide for vegetable crops, soybeans, potatoes, and groundnuts[21,22].

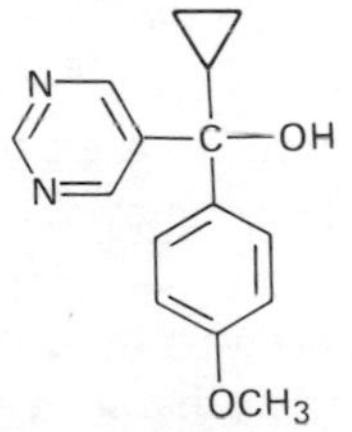

(46) 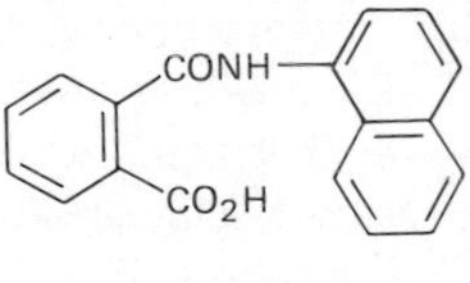   (47) 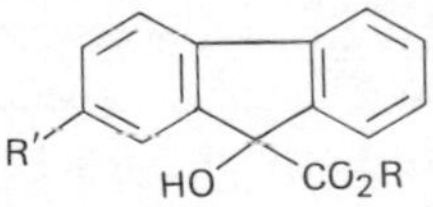   (48)

Naptalam (47) has the characteristic effect of abolishing the normal curvature of plant roots towards the ground or of shoots towards light. Certain fluorene derivatives known as flurecols (48) show similar effects; examples are chlorflurecol (48; R' = Cl, R = H), chlorflurecol-methyl (48; R' = Cl, R = CH$_3$), and flurecol-butyl (48; R' = H, R = C$_4$H$_9$). These compounds are mainly used in conjunction with phenoxyacetic acid herbicides for weed control in cereals[7,22]. The response of plant organs to light or gravity may be ascribed to the redistribution of auxin to the lower or to the dark side, and it is therefore possible that both naptalam and the flurecols owe their influence on plant growth to interference with indole acetic acid transport[18].

Some organophosphorus compounds regulating plant growth include Phosfon (49) which is likewise believed[18] to owe its activity to inhibition of gibberellic acid biosynthesis, and is a useful stunting agent for various ornamental plants[22,27].

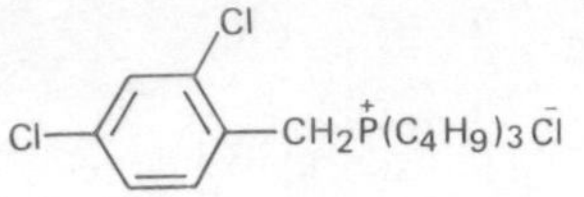

(49) 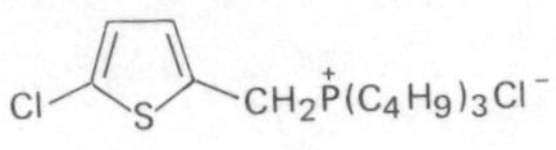 (50)

In contrast, the corresponding thienyl phosphonium chloride (50) stimulated the growth and yield of snap beans[27].

The di- and triphenoxyphosphites (51) and (52) behave as selective herbicides against broad-leaved weeds in cereals due to *in vivo* oxidative hydrolysis to 2,4-D, so extending the residual action of the herbicide[18]:

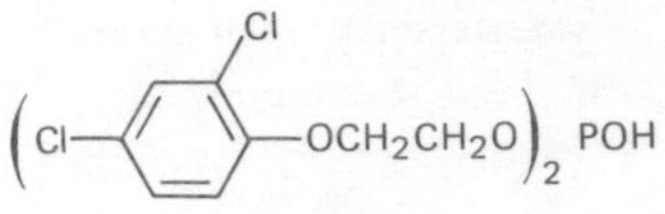

(51) 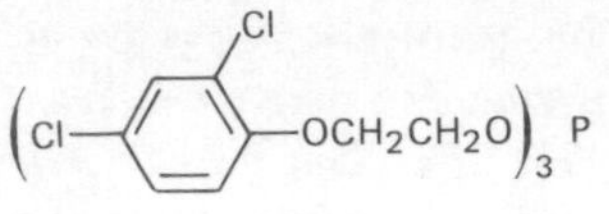 (52)

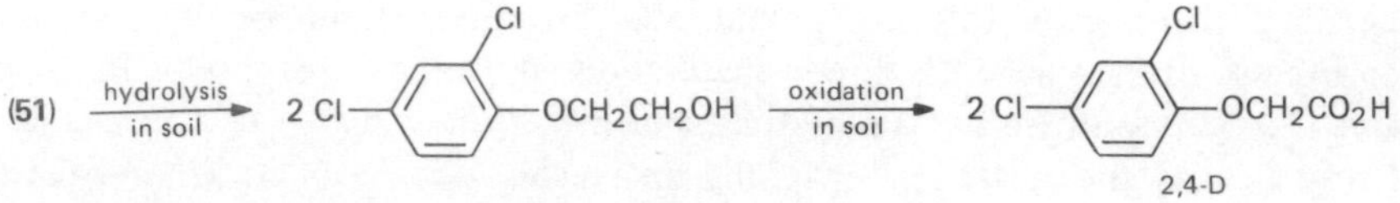

2,4-D

Ethylene

Ethylene is the simplest plant growth hormone and has several effects on plants. It can inhibit growth, accelerate leaf and flower drop (abscission) and the ripening of fruit, and flowering, depending on the stage at which the plant is exposed to the chemical. Recently chemicals have been developed which when sprayed onto plants release ethylene and hence control different phases of the plant's development. Thus in orchards, Ethrel (53) is sprayed onto the trees to promote ripening and loosening of the fruit. Ethrel acts by slowly decomposing in the plant generating ethylene. Ethrel (53) is obtained from phosphorus trichloride and ethylene oxide:

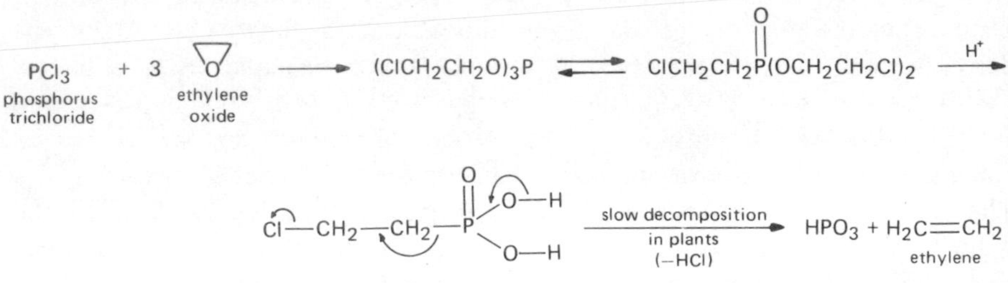

PCl_3 + 3 (ethylene oxide) ⟶ $(ClCH_2CH_2O)_3P$ ⇌ $ClCH_2CH_2P(OCH_2CH_2Cl)_2$ ⟶ H^+

phosphorus trichloride   ethylene oxide

(53)

slow decomposition in plants (−HCl) ⟶ $HPO_3 + H_2C=CH_2$ ethylene

Ethrel (53) is capable of forming ethylene in solutions with pH higher than 5, and hence can be used to regulate various phases of plant metabolism, growth, and development[18].

Bis-benzoyloxy-2-chloroethyl-methylsilane (54) was introduced (1972) by

Ciba–Geigy as a chemical thinning agent for peaches and other fruits. This novel type of chemical structure, like Ethrel, gradually decomposes in the presence of water with evolution of ethylene:

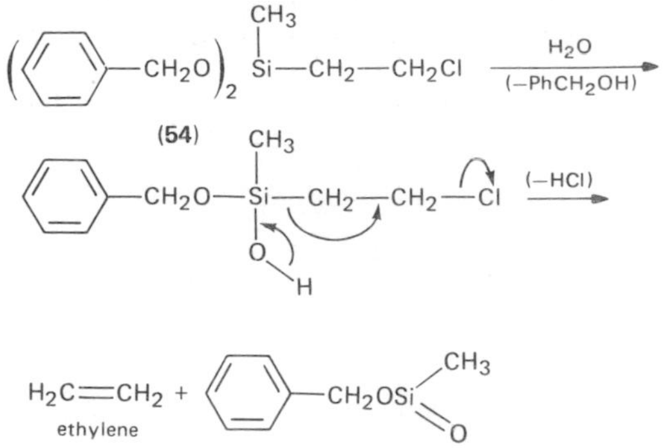

The driving force for the reaction being the formation of the very strong Si=O bond.

Ethylene is now recognized as a natural plant growth hormone whose main action is to accelerate the ageing process; the effect is fairly specific since related compounds like propylene are much less active.

Resistance to Herbicides

, With the large-scale use of certain herbicides, such as the phenoxyacetic acid derivatives, for weed control in cereals, the development of major weed strains that were resistant to these chemicals would be a major disaster to cereal production. It is likely to develop since insect resistance to insecticides is now widespread (Chapter 6, p. 101) and reported examples of resistant fungi are rapidly increasing since the introduction of commercial systemic fungicides (Chapter 7, p. 137). Luckily, however, there are at present comparatively few examples of resistant weed species. Resistance is most likely to develop in short-lived annual weeds which may go through several generations in one season, and have been frequently subjected to a particular chemical herbicide. Chickweed *(Stellaria media)* has been shown[19] to acquire tolerance to treatment with 2,4-D by the third generation, and this chemical has also ceased to be effective against *Erechtipes hieracifolia,* a common annual weed, in sugar plantations[3].

References

1. Salisbury E., 'Weeds and aliens', *The New Naturalist,* No. 43, Collins, London, 1961.
2. Whitten, J. L., *That We May Live,* Van Nostrand, Princeton, U.S.A., 1966. p. 31.
3. Hartley, G. S. and West, T. F., *Chemicals for Pest Control,* Pergamon Press, Oxford, 1969.

4. Approved Products for Farmers and Growers, Ministry of Agriculture, Fisheries and Food, 1978.
5. Hassall, K. A., *World Crop Protection: Pesticides*, Vol. 2, Iliffe Books Ltd., London, 1969.
6. Mellanby, K., *Pesticides and Pollution*, Collins, London, 1967.
7. Martin, H., *The Scientific Principles of Crop Protection*, 6th edn., Arnold, London, 1973.
8. Green, M. B., 'Polychloroaromatics and heteroaromatics of industrial importance' iin *Polychloroaromatic Compounds* (Ed. Suschitzky, H.), Plenum Press, London, 1974, p. 419.
9. *Physiology and Biochemistry of Herbicides* (Ed. Audus, L. J.), Academic Press, London, 1964.
10. Crafts, A. S., *Chemistry and Mode of Action of Herbicides*, Interscience, London, 1961.
11. Crafts, A. S., and Robbins, W. W. *Weed Control*, 3rd edn., McGraw-Hill, London, 1962.
12. *Weed Control Handbook* (Eds. Fryer, J. D. and Evans, S. A.), 5th edn., Blackwell, Oxford, 1968.
13. Wain, R. L., *Advan. Pest Control Res.*, 2, 263 (1958).
14. Wain, R. L., 'Chemistry and crop protection', *RIC Lecture Ser.*, No. 3, 1965.
15. Ashton, F. M. and Crafts, A. S., *Mode of Action of Herbicides*, Wiley, New York, 1973.
16. Kearney, P. C. and Kaufmann, D. D., *Herbicides – Chemistry, Degradation and Mode of Action*, 2nd edn., Vol. 1, Dekker, New York, 1975.
17. Audus, L. J., *Plant Growth Substances: Chemistry and Physiology*, 3rd edn., Vol. I, Hill, London, 1972.
18. Corbett, J. R., *The Biochemical Mode of Action of Pesticides*, Academic Press, London and New York, 1974.
19. Woods, A., *Pest Control*, McGraw-Hill, London, 1974.
20. Klingman, G. G., *Weed Control as a Science*, 2nd edn., Wiley, New York, 1963.
21. Metcalf, R. L., 'The chemistry and biology of pesticides' in *Pesticides in the Environment* (Ed. White-Stevens, R.) Vol. 1, Part I, Dekker, New York, 1971.
22. *Pesticide Manual* (Eds. Martin, H. and Worthing, C. R.), 4th edn., British Crop Protection Council, 1974.
23. Harris, C. T., Kaufmann, D. D., Sheets, T. J., Nash, R. G., and Kearney, P. C., *Advan. Pest Control Res.*, 8, 1 (1968).
24. Roberts, T. R., Proc. E.W.R.C., *4th Int. Sympos. on Aquatic Weeds*, Wien, 1974.
25. Calderbank, A., *Advan. Pest Control Res.*, 8, 127 (1968).
26. Kearney, P. C. and Kaufmann, D. D., *Herbicides – Chemistry, Degradation and Mode of Action*, Vol. 2, Dekker, New York, 1975.
27. Eto, M., *Organophosphorus Pesticides: Organic and Biological Chemistry*, CRC Press, Cleveland, Ohio, U.S.A., 1974, p. 325.
28. Thimann, K. V., *Plant Physiol.*, 54, 450 (1974).

Chapter 9
Fumigants

The process of fumigation is of great antiquity, thus the burning of aromatic resins, herbs, and incense has been widely practised for centuries as a means of disinfection. In this way, the foul smell of disease and putrefaction could at least be disguised, if not cured!

In the absence of special formulation or apparatus for application, if chemicals are to be effective fumigants they must have high volatility at room temperature. Chemical fumigants, therefore, will be gases or volatile liquids of comparatively low molecular weight, their volatility enabling the chemical to penetrate the material to be protected.

Fumigation is an important method for soil sterilization as a means of killing insects, nematodes, weed seeds, and fungi, which is widely used in glasshouses. It is also useful for protecting crops in store against attack by pests[1–3], and for killing rodents e.g., rats, and also moles and rabbits, by treatment of their burrows. In the treatment of stored products, it is of paramount importance that the fumigant should not leave any toxic residues.

Chemical fumigants, because of their volatility require an enclosed space (e.g. greenhouses, food stores, warehouses) if they are to be effective, otherwise too much of the chemical will be lost to the atmosphere. For soil treatment, the surface of the soil has to be first covered with polythene or other plastic sheeting to cut down loss of the toxicant from the soil surface; citrus trees can be effectively fumigated by enclosing the tree in an air-tight tent before introducing the fumigant.

The earliest commercial chemical fumigant was probably hydrogen cyanide used against scale insects on citrus trees in California in 1886 (Chapter 1, p. 4). The hydrogen cyanide is usually generated by the action of acid on sodium or calcium cyanide; although the gas can cause phytotoxic damage but this is reduced if fumigation is done at night[3].

Hydrogen cyanide is the most effective fumigant for empty buildings or containers but has poor powers of penetration and is therefore little used in the treatment of stored grain and is useless for soil fumigation. It is, however, widely employed by quarantine authorities for treatment of plant materials which might carry infectious diseases. Hydrogen cyanide is a very dangerous acute poison; a concentration of 1000 p.p.m. quickly causes human death; accordingly handling the concentrate requires proper equipment and great care. Although some people can detect the gas by its slight bitter almond smell others cannot which increases the danger. It is now general practice to add chloropicrin or cyanogen chloride to the concentrate so that people are warned of the presence of the gas by the effects of the added chemicals on the nose, throat, and eyes.

Carbon disulphide is a very volatile and cheap organic liquid; it is inflammable and hence is generally used in non-flammable mixtures with for instance carbon tetrachloride. These are valuable for fumigation of grain stores[1,2] and carbon disulphide is particularly good against fungi.

Methyl bromide is a widely used non-flammable general fumigant for fumigation of soil, stored grain, and buildings (see Chapter 11, p. 185) but has high mammalian toxicity. Ethylene dibromide is also used as a general fumigant[2].

Sulphur dioxide has only about one-tenth of the toxicity of hydrogen cyanide towards most pests, but is safer and cheaper. Like hydrogen cyanide, sulphur dioxide has poor penetrating properties and is only useful for fumigation of empty buildings[1].

Ethylene oxide is a gas (b.p.t. $12°C$) and is marketed as a liquid in pressure cylinders; it is mainly employed as a fumigant in grain stores, empty buildings, and for plant disinfection – but is rather toxic to seeds and hence should not be used in seed grain. Propylene oxide is used in fumigation of dried fruit in packets. Phosphine, generated by the action of moisture on aluminium phosphide tablets, is useful to control insects in grain and as a rodenticide (Chapter 10, p. 177).

Trichloronitromethane or chloropicrin (1) is used as a nematicide (Chapter 11, p. 185) and is also one of the best chemicals for disinfection of stored grain, but after fumigation adequate ventilation must be provided otherwise the treated grain becomes tainted. Chloropicrin (1) is a valuable general fumigant controlling insects, fungi, nematodes, and weed seeds in soil[4].

$$Cl_3C-N\overset{\displaystyle O}{\underset{\displaystyle O}{}}$$

(1)

Formaldehyde generally used as a 40% aqueous solution (formalin) is mostly used as a surface dressing for seed beds to combat soil fungi; for instance those associated with 'damping off' diseases which often cause high losses of seedling plants. Formaldehyde has poor penetration in soil which makes it rather ineffective in normal soil sterilization[1]. Methyl formate is mainly used for fumigation of animal furs in store[1].

Soil fumigation is generally performed before introduction of the crop so complete sterilization is most satisfactory and little selectivity is required; this is also true for chemicals used in fumigation of buildings and grain stores. If fumigation is to be successful, there must be rapid diffusion of the fumigant into the soil or mixing with air in the space to be treated which must be sealed off from the outside atmosphere. In this respect, the practice of fumigation has been substantially assisted by developments in the plastics industry – field clamps of vegetables, houses, and even food factories can be sealed off by plastic covers enabling effective fumigation to be carried out. Agricultural land can be covered by polythene sheeting held down by being embedded in soil all round the perimeter which permits effective use of methyl bromide for fumigation. This chemical is so volatile that too much would be lost to the atmosphere without the plastic cover. In the fumigation of grain, initial treatment with the chemical fumigant under slightly reduced pressure helps to draw the chemical into the interior of the store more effectively than when diffusion only is relied on. Grain is especially sensitive to fungal attack when stored under conditions of high moisture content and the moisture content also alters the effect of the chemical fumigant on the viability of the seed. The dry seed being much more resistant to the chemical, the moisture concentration should not exceed 16% during fumigation and storage.

Other fumigants include methyl isothiocyanate (MIT) (2) and sodium N-methyldithiocarbamate dihydrate (metham-sodium or SMDC (3))[4]:

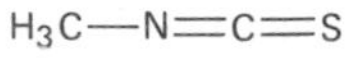

$$H_3C-N{=}C{=}S \qquad H_3C-NH-C\!\!\nearrow^{\!S}_{\!\searrow SNa}\;\;.2\,H_2O$$

(2)

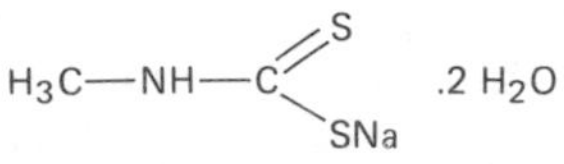

(3)

MIT (2) is applied in soil sterilization and is highly toxic to germinating weed seeds and nematodes; it is often used as the water-soluble SMDC (3) which gradually evolves the active fumigant MIT (see Chapter 11, p. 187). Tetrachloroethane is valuable as a greenhouse fumigant against whitefly[5], but its use is limited by phytotoxicity.

The original fumigants were highly volatile, low molecular weight molecules, but recently the application of the method of fumigation has been considerably extended by volatilization of chemicals from thermostatically controlled heaters or by their formulation as pyrotechnic mixtures[5].

In these ways compounds of comparatively low vapour pressure at room temperature can be successfully used as fumigants. Examples are HCH, DDT, dichlorvos and azobenzene.

References

1. Hartley, G. S. and West, T. F., *Chemicals for Pest Control*, Pergamon Press, Oxford, 1969, p. 274.
2. Woods, A., *Pest Control*, McGraw-Hill, London, 1974, p. 92.
3. Fletcher, W. W., *The Pest War*, Blackwell, Oxford, 1974, p. 95.
4. Ware, G. W., *Pesticides: An Auto-Tutorial Approach*, W. H. Freeman and Co., San Francisco, 1975, p. 120.
5. Martin, H., *The Scientific Principles of Crop Protection*, 6th edn., Arnold, London, 1973, p. 294.

Chapter 10
Rodenticides

In certain circumstances many vertebrates, even elephants, can rank as pests, but among mammals by far the most serious are the rodents (e.g. squirrels, rats, and mice). The majority of poisons effective against vertebrate pests are also very toxic towards man and domestic animals, although there are a few examples of specific pesticides in this field. However, even with generally toxic chemicals, they can often be applied in such a way that they are not taken by man or domestic animals[1,2].

Three rodent species which are major pests of agriculture, stored produce, and present a potential danger to public health throughout the world are the common brown or Norway rat *(Rattus norvegicus),* a major pest in temperate regions; the black or ship rat (*Rattus rattus*) a dominant pest in warmer climates; and the house mouse *(Mus musculis).* The latter is a less important pest economically but tends to live in close association with man in his dwellings to a greater extent than rats and therefore demands effective control.

The rat is the traditional enemy of man, often living in his homes and consuming and fouling the produce of his labour. Rats are carriers of several deadly diseases (Chapter 1, p. 8). In Britain black rats, the main carriers of diseases, are largely confined to ports and large towns but the brown or common rat is much more numerous adapting itself to both town or country habitats and does extensive damage especially to stored foodstuffs such as grain. The brown rat follows man all over the world and probably came to Britain from the East on trading ships about 1740. Rats breed very rapidly and the female produces up to 50 young per year although the majority die before becoming adults and most of the adults do not live more than a year[3].

The high rate of reproduction and turnover makes rats a difficult pest to control. They will eat almost anything, all kinds of vegetables, and animals. In many countries, including Britain, the black rat has been driven out by the brown rat and has now become quite rare.

Rats are very destructive to the fabric of buildings and furniture where this is sufficiently soft to be gnawed in their efforts to gain access. Wooden buildings are particularly vulnerable to attack by rats; the most important defence against these pests is a sound building and perhaps the greatest contribution of modern technology has been the manufacture of Portland cement.

There is often a public outcry against the use of chemical pesticides against birds like pigeons, sparrows, and bullfinches or against grey squirrels; however rats are so universally detested that no one objects to the use of poisons to control them.

In Britain, as in many other countries, the use of poisons against rodents and other pests is controlled by the Protection of Animals Acts 1911 to 1927. These permit the laying of poisons against 'insects, rodents, and other small ground vermin' but at present prohibit laying of poison against rabbits, birds, or coypus, although fumigation of rabbit burrows and the controlled use of baits containing stupefying narcotic substances against certain bird pests is allowed.

In special situations (e.g. warehouses, ships' holds) rats and other animals and insects can be eliminated by fumigation with such poisons as sulphur dioxide, hydrogen cyanide, or methyl bromide (Chapter 9). More generally rodenticides are laid down as baits to control rodent pests in buildings and on farms. If the bait is to be safe and effective the following criteria must be satisfied:

(a) it must not be unpalatable to the target pests;
(b) it must not induce bait shyness which often happens if more than one dose is needed to kill, and the first feed results in painful symptoms;
(c) the poison should make the rats go out into the open to die, otherwise the rotting corpses create health hazards;
(d) it should have a much lower toxicity to domestic animals, especially cats and dogs, which often eat the poisoned rats; sometimes this problem can be reduced if the pesticide is rapidly degraded to less toxic substances in the corpse.

In many ways, the most satisfactory method of control would be by introducing an infectious disease specific to rats. At one time this appeared to have been achieved with 'Liverpool virus' — a *Salmonella* bacterium presented as an infected meat bait, but unfortunately the rats quickly developed an immunity to this not very specific disease so the procedure had to be abandoned[2]. A very successful illustration of this general method was introduction of the myxomatosis virus to control the rabbit population in 1953. During the period 1954–55 some 90% of the total rabbit population in Britain was killed by the spread of this specific disease. More recently the virulence of the virus seems to have decreased with a consequent partial recovery of the rabbit population and so trapping, shooting, and chemical methods have again to be used for their control.

The most effective method is by the introduction of a powdered mixture of sodium cyanide, magnesium carbonate, and anhydrous magnesium sulphate (Cymag) into the rabbit burrows, when the powder gradually liberates the lethal hydrogen cyanide gas in contact with moist soil[3]. Aluminium phosphide pellets evolving poisonous phosphine gas are useful to control mole rats in paddy fields. Both sodium cyanide and aluminium phosphide are also effective against moles by placement in their burrows.

Rodenticides may be classified either as acute (single-dose, rapid-acting) poisons or as chronic (multiple-dose, slow-acting) poisons. Chronic poisons are taken at present only to include the anticoagulants and have advantages in efficacy and safety, but are slower to use and more expensive as regards labour and bait material; also resistance to anticoagulants is becoming an increasing problem.

Acute poisons have the advantage of rapid kill, but may be too hazardous or not effective for use in many circumstances. Ideally a candidate rodenticide needs the three attributes of toxicity, acceptability, and safety in use[4].

Early rodenticides included strychnine, now only important for controlling moles in their burrows, white arsenic, yellow phosphorus, squills, and cantharides. By 1939 however, only red squill and arsenic remained much in use. The natural product extracted from the bulb of the red squill plant[5] was a safe poison because it was rapidly detoxified in the rat's body and, if directly consumed by other animals, quickly acts as an emetic and so they do not ingest sufficient quantities of the poison to harm them. The specific toxicity of the product to rats therefore depends on the rat's inability to vomit.

Barium carbonate, zinc phosphide (Zn_3P_2), and arsenious oxide came into prominence later; although moist zinc phosphide emits an unpleasant smell which repels most animals it is surprisingly well accepted in baits by the generally suspicious rat.

Thallium sulphate (Tl_2SO_4) has been used for a long time as a rodenticide and also to control some bird pests. It is a general cellular toxin and inactivates a number of vital enzymes, such as those containing thiol groups. Thallium sulphate has a very high mammalian toxicity and the toxicant is readily absorbed through the skin. It has cumulative and secondary, as well as acute toxicity hazards, and consequently its application is severely restricted. Like barium sulphate, poisoning with thallium sulphate causes rats to seek water and consequently they die in the open. Death arises from respiratory failure and thallium sulphate must be used only with the greatest care by expert operators. A variety of antidotal treatments have been developed, the most recent being Prussian blue[4].

The first synthetic organic rat poison was α-naphthylthiourea or Antu (1) reported in 1945, and obtained by reaction of α-naphthylamine and ammonium thiocyanate[5]:

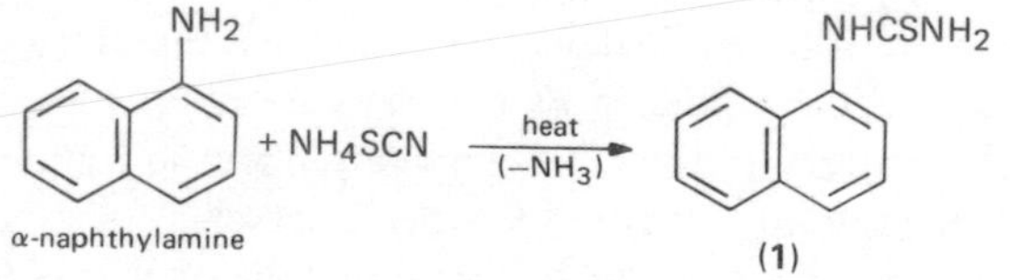

Antu (1) is a somewhat specific poison for adult Norway rats; the toxic dose is 6–8 mg/kg and it is less effective against other species of rats. It suffers from the disadvantage that it induces extreme bait shyness in those rats which do not consume a fatal dose at the first feed and tolerance develops in rats by repeated administration of sublethal doses[1,2]. Antu has been banned in Britain and some other countries because of the carcinogenicity of occasional impurities such as β-naphthylamine[4].

Fluoroacetic acid obtained by heating together sodium fluoride and chloroacetic acid at high temperature and pressure was found (1945) to be an effective rodenticide either as the sodium salt (sodium fluoroacetate) or as the amide (fluoroacetamide)[2,5]. The toxicity of these compounds arises from their role in the biosynthesis of fluorocitrate which blocks the action of the enzyme aconitase in the tricarboxylic acid cycle[6]. Both compounds are highly toxic to rats and most other animals and there is no recognized antidote. The flesh of poisoned animals remains toxic for some time and there have been several instances of secondary poisoning. Welsh sheepdogs have been killed as a consequence of eating oatmeal baits containing fluoroacetate[2,4]. Owing to their hazardous nature, the use of fluoro-acetates is restricted but they are still employed against rats in sewers, ships, and port areas. The demand for other rat poisons was almost completely eliminated by the introduction of warfarin or 3-(α-acetonylbenzyl)-4-hydroxycoumarin (2) which seemed to make other rat poisons obsolete. It was discovered by the Wisconsin Alumini Research Foundation (1944) and has remained the most successful of a number of hydroxycoumarin derivatives. Warfarin is prepared by condensation of 4-hydroxycoumarin with benzalacetone[2,5]:

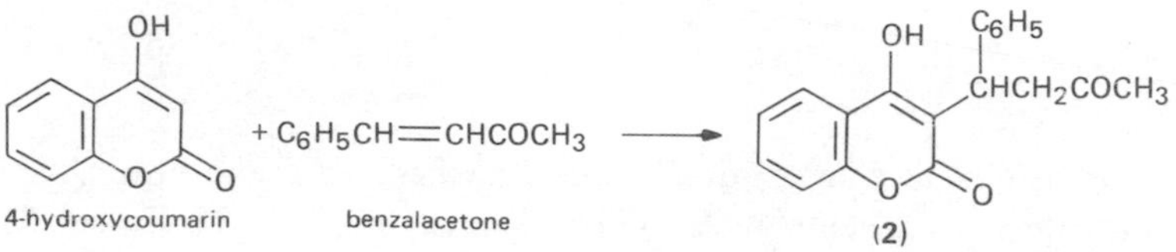

| 4-hydroxycoumarin | benzalacetone | (2) |

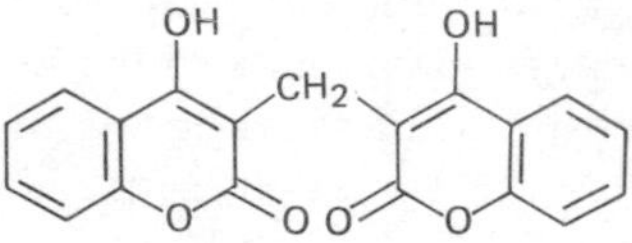

(3)

Spoiled sweet clover has long been known to be highly poisonous to cattle and in 1941 Dr. K. P. Link and his colleagues at Wisconsin found that the active ingredient in the sweet clover was dicoumarin (3) which interferes with the action of vitamin K in the body reducing the formation of prothrombin, the blood clotting factor, so that a minor injury can cause a fatal haemorrhage[7,8]. A range of hydroxycoumarin derivatives was examined and the forty-second on their list was selected as the most promising rat poison; this became known as warfarin or WARF 42 (2)[5]. Warfarin, and other similar compounds, function as anticoagulants so that the rats eventually die from internal haemorrhage after coming into the open in search of water. Rats are very susceptible to this toxicant and do not develop bait shyness; a dose of 1 mg/kg for five days is lethal, whereas other animals are generally less affected, although the safety is mainly dependent on use of a suitable bait placement. Warfarin has also been the main weapon to control the spread of the grey squirrel but they need a higher dosage than rats.

Normally these anticoagulant (chronic) rodenticides control rodent pests effectively within 2 to 5 weeks and are relatively safe largely because other animals rarely have access to the baits over the prolonged period needed to accumulate a lethal dose. Another advantage is that in cases of accidental poisoning, the anti-coagulating effect can be readily reversed by treatment with vitamin K.

Several substituted indandiones also interfere with blood coagulation; an example is pindone (4) prepared by condensation of diethyl phthalate and pinacolone[5, 8]:

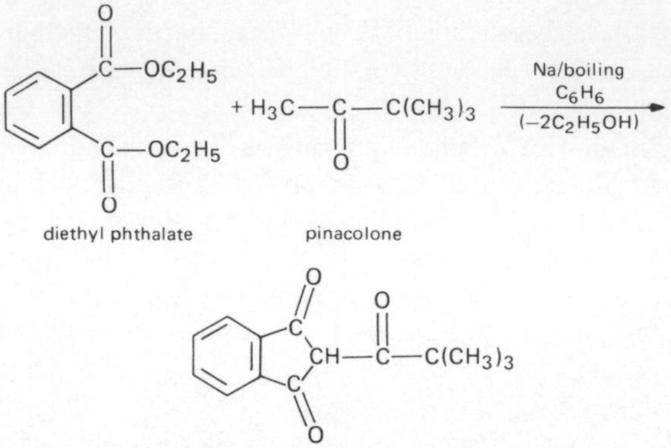

diethyl phthalate   pinacolone

(4)

Pindone can be used as a substitute for warfarin when rats become bait shy.

Unfortunately a resistant strain of rats has appeared in the United Kingdom which is practically immune to warfarin and related anticoagulant poisons. There are now resistant or 'super' rats in many parts of the world which are multiplying alarmingly in areas largely dependent on warfarin for rat control. A resistant strain of mice has also appeared so there is currently a need for new selective, quick-acting, rodenticides.

In 1964 Roszkowski and coworkers[9] reported a new rodenticide called norbormide or Raticate (5) obtained by reaction of the condensation product of cyclopentadiene and 2-benzoylpyridine with maleimide:

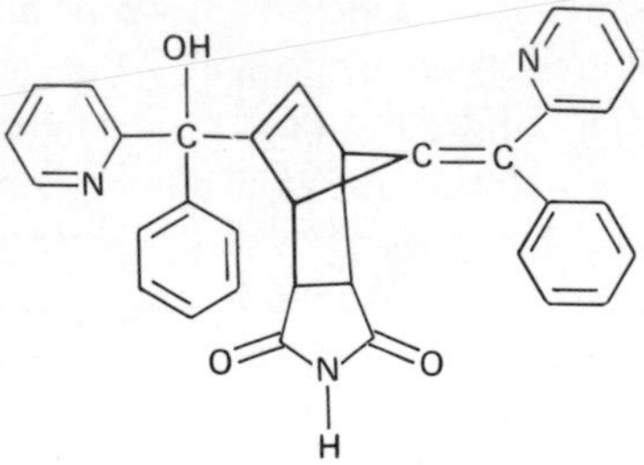

(5)

Norbormide (5) is selectively toxic to rodents of the genus *Rattus* and particularly the Norway rat *(Rattus norvegicus)* to which it has an oral LD_{50} value of about 11 mg/kg[4,5,9]; dosages of 1000 mg/kg have produced no ill effects on cats, dogs, monkeys, or mice[8]. The compound acts by constriction of the smaller peripheral blood vessels leading to a catastrophic increase in blood pressure. Norbormide initially appeared extremely promising as a rodenticide but, unfortunately, it suffers from a low acceptability to rats and easily induces bait shyness so that non-selective, more efficacious poisons are generally to be preferred for rat control.,

The widespread resistance developed by rodents to the anticoagulant chemicals has encouraged the development of new chemical rodenticides; one of considerable promise is alphachloralose or chloralose (6) prepared by condensation of chloral (trichloroacetaldehyde) with glucose[5] :

(6)

Chloralose probably has the chemical structure (6) and exists as two stereoisomers: (α- m.p. 187°C, and β- m.p. 227—230°C) both isomers have low aqueous solubility. Baits are usually formulated by incorporating about 1.5% of the active ingredient as a dust onto moist grain[2] to make birds eating the grain sleepy. If they are left alone the birds completely recover[5]. The farmer can kill the pest birds, e.g. pigeons, sparrows, and starlings while the other birds are released. Use of large grain such as field beans makes the operation more selective. Chloralose and other narcotics can also be applied to control birds in food stores without any danger of contamination of the food by either the poison or dead birds[3].

Chloralose (6) is primarily a sleep-inducing narcotic drug; the hypnotic phase of its action is often preceded by a convulsive or hyperactive phase. Death of the animal generally results from hypothermia and therefore use of the compound as a rodenticide becomes unreliable at temperatures above 15°C. It is now widely employed as baits containing 4% of the active ingredient for control of mice, and has been permitted for use in stupefying baits against pigeons and house sparrows since 1959. There are about 5 million wood pigeons in Britain and they now rank as one of the commonest and most destructive agricultural pests in Europe. Pigeons dig up and eat newly sown grain and in winter attack brussels sprouts, cabbages, and other green crops — it has been estimated that pigeons cost British agriculture about £2 m per annum.

There are at present no alternatives to the anticoagulant rodenticides, e.g. warfarin, which have the same degree of effectiveness coupled with safety, so the occurrence of resistant rodent strains has posed a severe problem to rodent control operators.

New chemicals being developed include the organophosphorus compound Gophacide (7). Although many organophosphorus compounds are highly toxic to rodents few of these have proved useful for their control, possibly because the onset of the cholinergic effects by the majority of these compounds is so rapid that ingestion of a lethal dose is prevented.

Gophacide (7) introduced by Bayer-AG as a selective rodenticide has an oral LD_{50} for a male rat of 7.5 mg/kg, and is prepared by condensation of

O,O'-di-p-chlorophenylphosphorochloridothioate with acetamidine[10]:

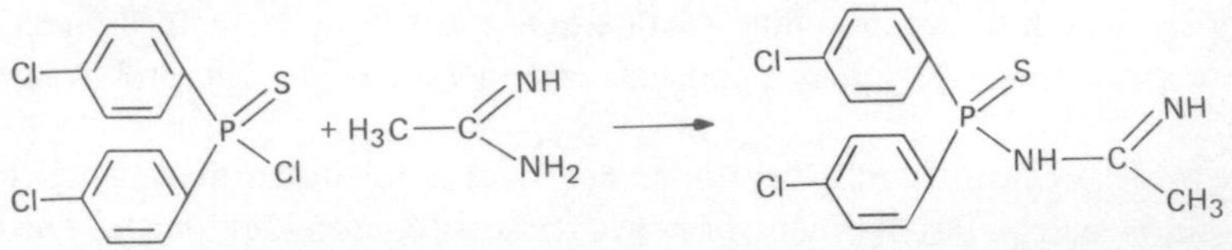

(7)

Gophacide does not possess anticholinesterase activity *in vitro* but is slowly activated *in vivo*, which results in a slow development of symptoms probably accounting for its acceptability to rodents as a bait, permitting them to consume a lethal dose before becoming bait shy. Gophacide appears promising for the control of a range of rodent pests but may prove to be rather hazardous to non-target species[4,8].

Silatrane, 1-(p-chlorophenyl)-2,8,9-trioxa-5-aza-1-silabicyclo(3,3,3)undecane is a very quick-acting rodenticide but is unstable in the presence of moisture[4]. This makes it difficult to formulate as a bait, but allows rapid detoxification of poisoned rodent corpses and any residual bait, and the compound has considerable potential for large-scale agricultural usage. Silatrane has special promise for large-scale agricultural application as compared with control of domestic infestation.

The chlorinated heteroaromatic compound, 2-chloro-4-dimethylamino-6-methylpyrimidine or crimidine (8) has been a widely used rodenticide since the 1940s, but resistant strains of rats have appeared which has reduced its effectiveness[5,8]. Crimidine is manufactured by condensation of urea and ethyl acetoacetate, followed by chlorination and dimethylation[5]:

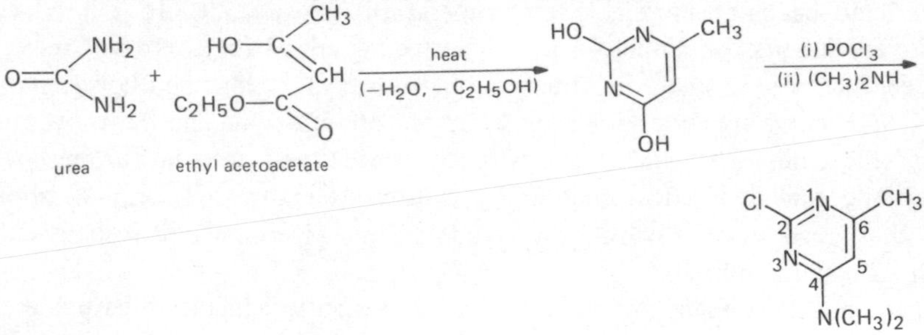

Crimidine (8) is an intense poison to mammals, rapidly producing convulsions and death: LD_{50} to rats 1.25 mg/kg. It is however metabolized rapidly to non-toxic materials, hence poisoned rats are not toxic to predators

References

1. Woods, A., *Pest Control: A Survey*, McGraw-Hill, London, 1974, p. 105.
2. Hartley, G. S. and West, T. F., *Chemicals for Pest Control,* Pergamon Press, Oxford, 1969, p. 138.

3. Fletcher, W. W. *The Pest War,* Blackwell, Oxford, 1974, p. 103.
4. Greaves, J. H., *Repts. Progr. of Appl. Chem.,* **56,** 465 (1972).
5. *Pesticide Manual* (Ed. Martin, H.), 4th edn. British Crop Protection Council, 1974.
6. Corbett, J. R. *Biochemical Mode of Action of Pesticides,* Academic Press, London and New York, 1974, p. 21.
7. Martin, H., *The Scientific Principles of Crop Protection,* 6th edn., Arnold, London, 1973, p. 351.
8. Metcalf, R. L., 'Chemistry and biology of pesticides', in *Pesticides in the Environment,* Vol. 1, Part 1, Dekker, New York, 1971, p. 131.
9. Roszkowski, A. P., Poos, G. I., and Mohrbacher, R. J., *Science N.Y.,* **144,** 412 (1964).
10. Eto, M., *Organophosphorus Pesticides: Organic and Biological Chemistry,* CRC Press, Cleveland, Ohio, U.S.A., 1974, p. 332.

Chapter 11
Nematicides

Unsegmented eelworms belong to a group of animals termed nematodes of which certain species attack the roots of crop plants doing considerable damage. The last thirty years or so have seen a substantial increase in the number of harmful species of nematodes coupled with a greater awareness of the damage they inflict on crops. The increasing trend towards large areas of monoculture has created an environment particularly conducive to the increase of nematode populations[1]. Nematodes generally live in the soil feeding on plant roots, although some species also invade the plants above ground, e.g. the species feeding inside chrysanthemum leaves causing them to die progressively up the stem; another eats the stems and bulbs of narcissi.

However, the most economically damaging is the root eelworm of potato which can survive in a hard encysted form for several years in the absence of the host crop[2]. The nematode cysts can remain in a dormant state in the soil until some chemical substances secreted by the growing potato roots potentiates them to hatch releasing their eggs. The eelworm hatching factor has been shown to be a complex oxidized sugar, which is, unfortunately impossible to manufacture and add to the soil.

If a suitable synthetic chemical hatching factor could be produced it might be possible to hatch the cysts in the absence of the host and so starve them to death. Some host plants have their own built in defences against the attack of parasitic eelworms, and recently potato varieties have been bred which are resistant to potato eelworm. When feeding on normal potato roots, the newly hatched eelworm can become either a male or female adult and the larvae develop into approximately equal numbers of both sexes. However, when eating the roots of the resistant varieties, the larvae nearly all become males; the attack on the potato in the first season is therefore not reduced but in subsequent seasons the nematode population cannot expand and the number of cysts gradually diminishes. The potato growing areas of Britain, until recently, had become heavily infested with eelworm cysts and so much of the land had to be planted with other, less economically valuable crops. It is now generally recommended that an interval of at least three years should be observed between planting potato crops on the same piece of land as a precaution against building up the eelworm population, although in some parts of the country a very early crop of potatoes can be set which may be harvested before the eelworms reach maturity.

Many species of nematodes only feed on one host plant or closely related plants, while others will attack almost any plant, but the latter are fortunately not very damaging. It is however difficult to assess the damage arising from nematodes, of which there are a large number of different varieties, because of the effect of soil fungi some of which do not invade the plant unless it has been previously damaged by eelworm attack. Some plants, while not greatly affected by either eelworms or soil-borne fungi alone, are seriously damaged when both organisms are present.

Plant parasitic eelworms are controlled by pesticides known as nematicides[2-4]. Chemicals are particularly effective against the more vulnerable stem and leaf eelworms and are also widely used against eelworms invading expensive greenhouse crops, e.g. tomatoes, where the value of the crop is sufficiently high to justify the cost of the massive doses of nematicide often required for effective control of the pest. Other crops often protected by chemical treatment include tobacco, especially in seed beds, and citrus fruits. Nematodes are covered with an impermeable cuticle which protects them. Effective nematicides therefore need the ability to penetrate the lipophilic cuticle.

Chemical control can be achieved by treatment of the soil with fumigants; compounds of sufficient volatility to penetrate through the upper layers of the soil where most of the insects and nematodes are to be found (Chapter 9)[5]. Some nematodes have seasonal migratory habits and, in these cases, soil fumigation must be carried out when they are not residing in deep layers of the soil. Damage caused by eelworms is most serious if they attack a newly transplanted plant or seedling so elimination of eelworms from the initial rooting zone of soil may often achieve satisfactory results.

In order to achieve efficient nematode control the chemical must be well distributed in the soil[3], and fairly long lasting in its nematicidal action. Chemicals, such as dichloropropene or ethylene dibromide, that release toxic gases slowly are therefore particularly effective.

The first soil fumigant used was carbon disulphide in Germany (1871) against sugar beet eelworm.

Chloropicrin, the tear gas of World War I and obtained by nitration of chloroform, was used as a fumigant in Britain (1919) and later extensively in the United States of America. D-D, a mixture of the *cis* and *trans* isomers of 1,3-dichloropropene was introduced as a soil fumigant by Shell Ltd. (1943). It is an extremely valuable nematicide — injection into the soil at rates of 220—670 kg/ha killed both eelworms and their cysts. D-D has been the most widely used nematicide; an additional benefit arising from its application is an increase in the availability of nutrient nitrogen due to the effect of the chemical on soil microogranisms[3].

Ethylene dibromide (1,2-dibromoethane) has also been used in soil fumigation and this led to the examination of other bromine compounds. Methyl bromide, which is highly volatile, was developed for soil sterilization in glasshouses, but is too volatile for general soil treatment. Methyl bromide is most effective as nematicide when used to treat infested seed or plants in an air-tight chamber. Shell Ltd. introduced (1958) 1,2-dibromo-3-chloropropane or Nemagon (1) as a soil fumigant and nematicide. Nemagon is prepared by liquid-phase addition of bromine to allyl chloride:

$$H_2C\!\!=\!\!CH\!-\!CH_2Cl \xrightarrow{\;Br_2\;} BrH_2C\!-\!CH(Br)\!-\!CH_2Cl$$

allyl chloride   (1)

Nemagon is effective against many varieties of nematodes at rates of 10–125 kg/ha and it is less phytotoxic than D-D and therefore can be used for nematode control on established crops under carefully controlled conditions[2,4,6]. Thus Nemagon has been applied for eelworm control in citrus orchards in California; the chemical was added to the soil on one side of the row of trees one year and the other side of the row was treated the next year. This allowed the tree roots to recover from any damage inflicted by the first treatment before the second application was made.

The effectiveness of soil fumigation against nematodes depends on several factors, the most important being soil type, conditions, and temperature. A warm soil is needed so that the volatile toxicant can disperse effectively through the soil layer. Even dispersion is aided by a well-worked fine soil with a reasonable amount of moisture present. Different types of soil can vary considerably in their capacity to absorb fumigants[7], thus soils rich in organic matter often deactivate a high proportion of the toxicant so that control of the nematodes is unsatisfactory, whereas sandy soils give much better results, e.g. on sandy soils application of D-D at 60 gal/acre controls citrus nematodes but on clay soils 250 gal/acre are required for comparable effect. Of course, one can have too rapid a dispersion of the fumigant resulting in considerable loss of the material from the top layer of the soil (down to about 7 cm) making it difficult to maintain sufficient of the toxicant to destroy the eelworms in this region. The situation can often be improved by covering the soil with a layer of polythene or using a sealant such as xylenol. The lower layers of the soil are not reached by the fumigant, so these untreated areas eventually cause recolonization of the sterilized soil by both harmful and beneficial organisms. This situation might result in a transient enhancement of some pest species, but luckily the beneficial saprophytic fungi are the most resistant species present and so these usually reinfest the treated soil first and cause some further delay in the spread of the parasitic fungi[2,5].

The majority of nematicides are relatively simple molecules which are the by-products of large-scale chemical industry which often have other major uses. D-D is a special product obtained for use as a nematicide by distillation of petroleum cracking products during chlorination, but it can only be manufactured economically as part of a larger chemical programme.

The fumigation of soil is a costly operation. With volatile materials, such as those previously mentioned, the chemical is injected into the soil in a grid pattern at about 30 cm spacing and the soil often needs covering. The high expense generally prohibits such treatment with the majority of field crops, but is a routine procedure with many varieties of glasshouse crops. When the toxicant is relatively non-volatile, distribution in the soil is achieved mechanically. For instance, the nematicide formulated as a dust is well stirred into the soil to the required depth by a powerful rotary hoe. Another method is by soil drenching which can only be effective on well-broken, good draining soils, so that the toxicant is washed down rapidly and reasonably uniformly. It is also important that the chemical is not appreciably adsorbed onto the soil colloids (Chapter 2, p. 22).

In 1959 methyl isothiocyanate (m.p. 35°, b.p. 119°) was introduced as a nematicide and is a general soil fumigant for control of soil fungi, insects, nematodes, and weed seeds (Chapter 9, p. 175). Methyl isothiocyanate is phytotoxic and, as with the majority of nematicides, treatment must be carried out sufficiently early to allow the toxicant to decompose before the crop is planted[6]. Methyl isothiocyanate can generally be more effectively applied as metham-sodium or Vapam (2)[5], which slowly decomposes liberating methyl isothiocyanate:

$$H_3C-N=C=S + SH^-$$

methyl isothiocyanate

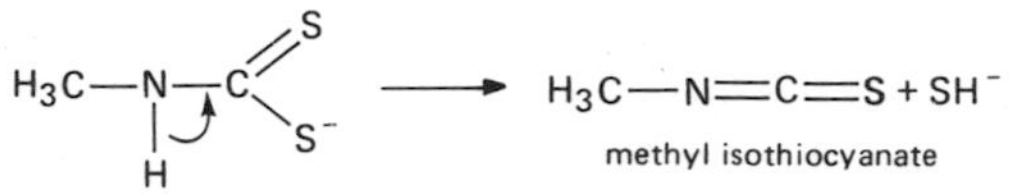

(2)

Metham-sodium (2) is prepared by reaction of methylamine with carbon disulphide in the presence of sodium hydroxide:

$$CH_3NH_2 + CS_2 + NaOH \longrightarrow$$

methylamine

$$+ H_2O$$

(2)

Another useful soil sterilant is the thiadiazine-2-thione or dazomet (3)[4,6] which is prepared by reaction of carbon disulphide, methylamine, and formaldehyde. Dazomet has been developed by BASF under the name Basamid and can be applied to the soil as granules at rates of 40–50 g/m^2 for control of soil nematodes, fungi, insects, and weeds[5]. Dazomet, like metham-sodium, owes its activity to decomposition in the soil to give methyl isothiocyanate – the rate of decomposition depending on the type of soil, the humidity, and temperature:

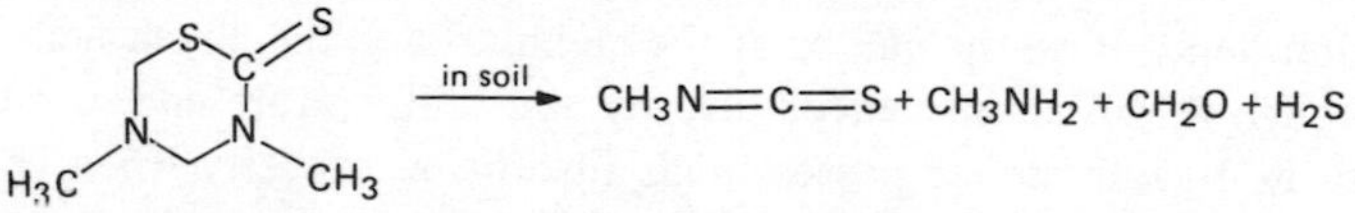

$$\xrightarrow{\text{in soil}} CH_3N=C=S + CH_3NH_2 + CH_2O + H_2S$$

(3)

$$H_2C=CH-S-CH_2-CH_2N=C=S$$

(4)

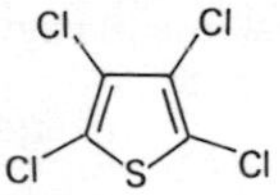

(5)

Several alkyl isothiocyanates, such as the vinyl thioethyl derivative (4) have useful nematicidal properties as does tetrachlorothiophene (5).

The majority of volatile commercial nematicides are saturated or unsaturated halides, e.g. methyl bromide, ethylene dibromide, chloropicrin, D-D, Nemagon, and their activity probably depends on reaction with a nucleophilic site, e.g. OH, SH, or NH_2 groups, in a vital enzyme system in the nematode:

$$Ez—\overset{\frown}{S}H + \overset{\frown}{R}—Hal \longrightarrow Ez—SR + H—Hal \quad (S_N2)$$

Ez = remainder of the enzyme structure

This is an S_N2-type reaction and there is also the possibility of a competing S_N1 hydrolysis by the large excess of water present in the soil which will deactivate the nematicide:

$$R—Hal + H_2O \longrightarrow R—OH + H—Hal \quad (S_N1)$$

So compounds having high reactivity in S_N2 reactions and low reactivity in S_N1 reactions should be the most active nematicides[4,5]. Nematicidal aliphatic halides induce a period of hyperactivity in nematodes, followed eventually by paralysis and death; mortality being dependent on the product of concentration and time of exposure. The alkyl isothiocyanates, including metham-sodium (2) and dazomet (3) owe their nematicidal properties to the ability of isothiocyanates to react with nucleophilic centres, e.g. thiol groups, in vital enzymes in nematodes by a similar type of S_N2 reaction to that operating with the aliphatic halogeno nematicides[4]:

$$R—N{=}C{=}S + Ez—SH \longrightarrow RNH—C\overset{\displaystyle S}{\underset{\displaystyle SEz}{}}$$

Recent developments have centred on the introduction of new, non-volatile nematicides[8]. Volatile nematicides, such as alkyl halides, are injected into the soil and their vapour penetrates by soil diffusion; if there is to be effective control of nematodes, they must be well distributed in the soil. With non-volatile nematicides, the distribution has to be achieved either by mechanical mixing of the toxicant, formulated usually as granules, with the soil, or by application as aqueous drenches around the roots of the crop plants. Movement of non-volatile chemicals in soil is limited by the degree of adsorption onto soil colloids (Chapter 2, p. 22). Adsorption depends on the nature of the chemical and the physical character of the soil, although it will be influenced also by soil temperature and moisture content. Soils rich in organic matter possess high adsorption capacity when chemicals tend to be strongly adsorbed on the soil organic colloids and removed from the soil solution and therefore unavailable to kill soil nematodes. The partition coefficient (Q) of a given chemical in soil is given by the equation.

$$Q = \frac{[\text{chemical in soil organic matter}]}{[\text{chemical in soil solution}]}$$

For good nematicidal activity the values of Q should be fairly low[8]. Two important groups of non-volatile nematicides are organophosphates and carbamates which are incorporated into the soil at rates of 2–10 kg/ha as compared with 150–1150 kg/ha needed for control of nematodes by volatile aliphatic halides. The carbamates generally have values of Q of approximately 10 (cf. 150 for the organophosphates), so carbamates tend to be more effective against soil nematodes.

Organophosphorus compounds[9,10] have found extensive application as insecticides (Chapter 6), and several of these have nematocidal action such as parathion, good against leaf nematodes. Phenyl N,N'-dimethylphosphorodiamidate or Nellite (6) was introduced as a nematicide by the Dow Chemical Co. (1962)[6]; it is systemic in plants and hence effective against nematodes that have entered the root system.

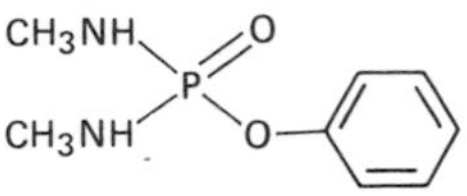

(6) 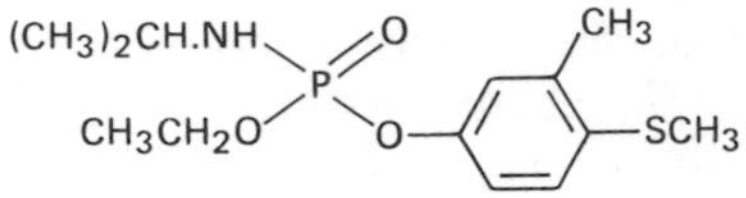   (7)

Nellite (6) is a very hydrophilic molecule with low adsorption onto soil organic matter and is very effective as a soil drench against root-knot nematode larvae. A related phosphoramidate is phenamiphos or Nemacur (7) which is active against nematodes by broadcast application at 5–20 kg/ha with or without soil incorporation. A disadvantage is the high mammalian toxicity: LD_{50} (oral) to rats 17 mg/kg[6].

Several phosphorothioates, especially those containing heterocyclic groups have been developed as nematicides: some examples are provided by thionazin (8), a valuable

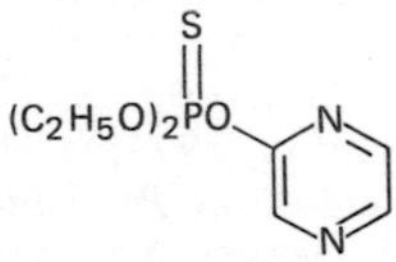

(8) 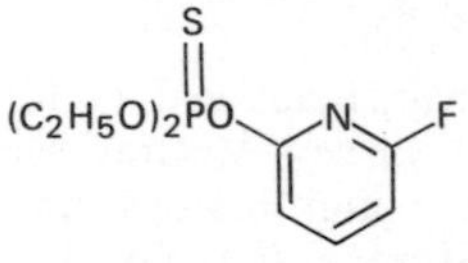   (9) 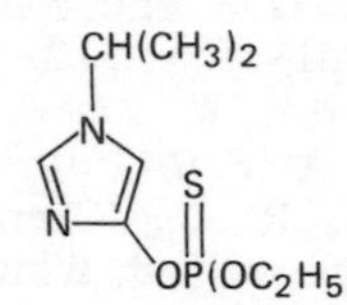   (10)

soil nematicide; since it is systemic this is especially useful against species of nematodes attacking bulbs[6] although it has a very high mammalian toxicity: LD_{50} (oral) to rats 12 mg/kg. Other useful soil nematicides of this type are the compounds (9) and (10).

Several carbamates (Chapter 6, p. 96) are also highly nematicidal and examples include the N-methylcarbamate, carbofuran (65, p. 98) effective as a soil nematicide when broadcast at 6–10 kg/ha[6], and has a short residual life so it is useful on food crops, and the oxime carbamates aldicarb (11) and oxamyl (12) are both highly active against soil nematodes but suffer from very high mammalian toxicities[8].

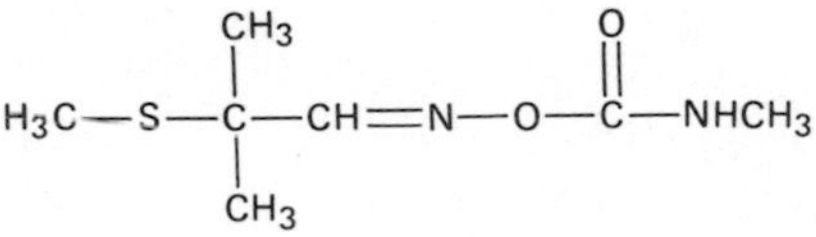

(11)

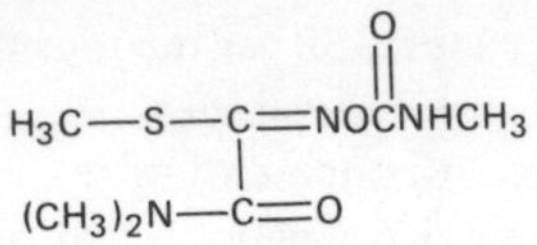

$$H_3C-S-C=NOCNHCH_3$$
$$(CH_3)_2N-C=O$$
(12)

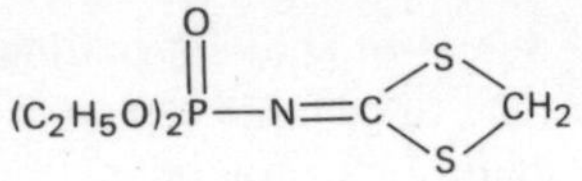

$$(C_2H_5O)_2P-N=C\begin{smallmatrix}S\\S\end{smallmatrix}CH_2$$
(13)

The dithioiminocarbonate (13) has been recently introduced $(1975)^{11}$ by the American Cyanamide Co. as a broad-spectrum nematicide effective against plant nematodes by soil incorporation; the activity is probably related to its decomposition in soil to the thiocyanate ion.

The organophosphorus and carbamate nematicides probably owe their activity to poisoning of the enzyme cholinesterase by phosphorylation or carbamoylation; the P=S derivatives being activated by conversion to the corresponding P=O compounds by the action of soil microorganisms or within the nematode (cf. Chapter 6, p. 84).

A severe disadvantage of the non-volatile nematicides is that they often have long-term residual effects which restrict their application, especially since many of these chemicals possess high mammalian toxicities. The traditional volatile nematicides which leave no residues will therefore continue to play an important role in nematode control for many years to come.

References

1. Hartley, G. S. and West, T. F., *Chemicals for Pest Control*, Pergamon Press, Oxford, 1969, p. 112.
2. Fletcher, W. W., *The Pest War*, Blackwell, Oxford, 1974, p. 60.
3. Woods, A., *Pest Control: A Survey*, McGraw-Hill, London, 1974, p. 102.
4. Metcalf, R. L., 'Chemistry and biology of pesticides' in *Pesticides in the Environment* (Ed. White-Stevens, R.), Vol. 1, Part I, Dekker, New York, 1971, p. 125.
5. Moje, W., *Advan. Pest Control Res.*, **3**, 181 (1960).
6. *Pesticide Manual* (Ed. Martin, H.), 4th edn., British Crop Protection Council, 1974.
7. Allen, M. W., 'Nematicides' in *Plant Pathology* (Eds. Horsfall, J. G. and Dimond, A. E.), Vol. II, Academic Press, New York, 1960, p. 603.
8. Hague, N. G. M., *Proc. 8th Brit. Insectic. and Fungic. Conf.*, Brighton, **3**, 837 (1975).
9. Fest, C. and Schmidt, K. -J., *The Chemistry of Organophosphorus Pesticides*, Springer-Verlag, Berlin, 1973.
10. Eto, M., *Organophosphorus Pesticides: Organic and Biological Chemistry*, CRC Press, Cleveland, Ohio, U.S.A., 1974.
11. Whitney, W. K. and Aston, J. L., *Proc. 8th Brit. Insectic. and Fungic. Conf.*, Brighton, **3**, 625 (1975).

Chapter 12
Molluscicides

Molluscs include land slugs and snails which do a considerable amount of damage by eating vegetable seedlings and mature green vegetables such as lettuce and cabbages; they also severely attack autumn-sown wheat[1]. Molluscs do not directly harm mammals but are alternate hosts for some species of parasites; for instance the liver-fluke which sometimes kills its mammalian host (generally sheep) spends an essential phase of its life cycle in small *Limnaea* snails which are taken in with the grass by sheep. Lungworms, causing 'husk' in cattle, are transmitted in a similar manner, and several different species of aquatic snails act as alternate hosts for the *Schistosoma* parasites, the causal agent for the debilitating human disease of bilharzia, which is widespread in tropical countries.

Barnacles which attach themselves to the hulls of ships severely reduce the power to speed ratio, an increasingly vital factor in these times of high energy costs. Chemicals specifically designed to combat these various molluscs are termed molluscicides.

The most effective chemical for controlling slugs and snails is metaldehyde or Meta (1), which is obtained by polymerization of an ethanolic solution of acetaldehyde in the presence of sulphuric acid at a temperature $< 30°C$:

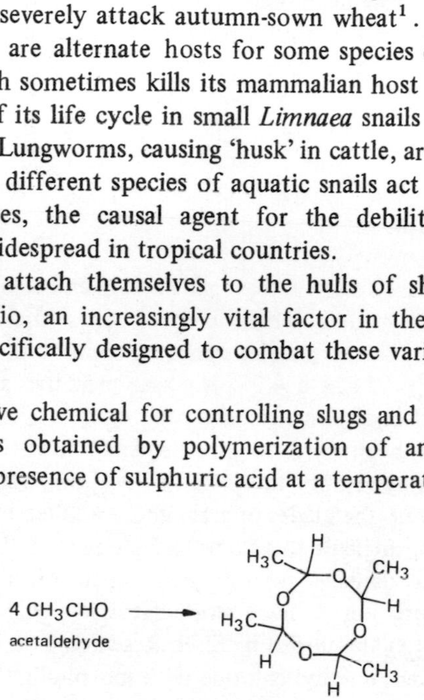

(1)

Metaldehyde functions as a specific attractant and toxicant for slugs and snails and is especially effective against slugs. It is usually formulated as baits containing 2.5–4% of the active ingredient in bran[2]. Meta is useful to control attacks by the field slug *Agriolimax reticulatus* on autumn-sown wheat at a rate of 900 g of the active ingredient in 31 kg of bran/ha. The best results are achieved when it is broadcast on the soil during a moist warm evening a few days after the end of a dry spell of weather[3]. Metaldehyde is highly inflammable, burning with a non-smoky flame, which accounts for its use as a solid fuel which is very convenient for picnic stoves, and the observation that meta was lethal to slugs may have originated from its accidental spillage at a picnic party[1]. Meta is toxic to slugs and snails both by ingestion and by absorption by the foot of the mollusc. The chemical causes an increase in the secretion of slime causing immobilization and eventual death by loss of water (desiccation). Molluscicidal activity is specific to metaldehyde (1); the trimer, paraldehyde with a six-membered ring, is not active and neither is the monomer acetaldehyde[1].

Both dinitro-*o*-cresol (DNOC) and dinitro-*o*-cyclohexylphenol (dinex) (Chapter 8, p. 141) have been observed to reduce the damage inflicted by slugs and snails when they were used as selective herbicides against weeds in cereals and dead slugs and snails have been found after their use. They are also toxic to aquatic snails at 3—5 p.p.m., but have been little used as molluscicides[4,5].

A number of carbamate insecticides (see Chapter 6, p. 98) such as Zectran and methiocarb[2] are very effective against snails when formulated as baits[4]. Methiocarb (2) is obtained by reaction of 4-methylthio-3,5-xylenol and methyl isocyante[2]:

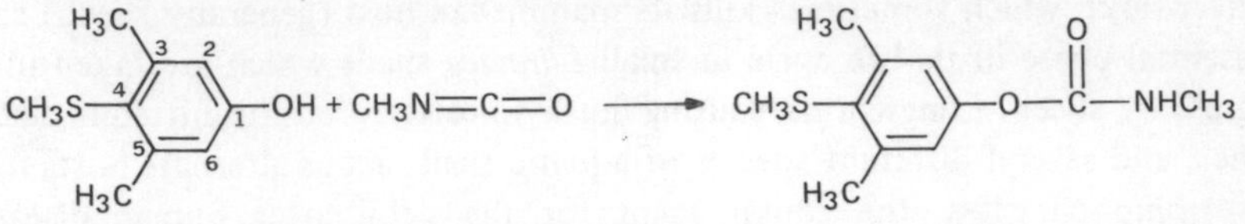

4-methylthio-3,5-xylenol   (2)

Methiocarb (2) is appreciably more active against slugs and snails than metaldehyde. It is formulated as 4% pellets which control slugs and snails at a rate of approximately 5 kg/ha. The organophosphorus compound azinphos-methyl or Gusathion (Chapter 6, p. 75) as a 4% spray was effective against the European brown snail on citrus, but not against slugs[5].

Copper sulphate has been used to kill the snail liver-fluke vectors. Copper sulphate solutions are sprayed onto the grass in meadows where sheep are grazing. The snails tend to climb up the blades of grass and are killed by direct contact with the spray. It is, however, difficult to achieve an adequate kill of the snails without also causing substantial damage to the grass crop on account of the high phytotoxicity of the cupric ion. A much more effective chemical for this purpose is trifenmorph or Frescon (3), developed by Shell Research Ltd. (1966), and obtained by condensation of triphenylmethyl chloride with morpholine[2]:

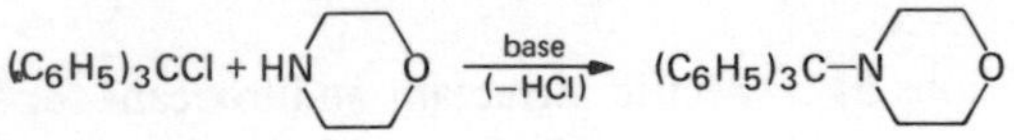

triphenylmethyl
chloride   (3)

Copper sulphate and Frescon (3) are uncommon molluscicides since they are also active against aquatic snails. Frescon (3) is lethal at 1—2 p.p.m., while copper sulphate at 3—4 p.p.m. kills all aquatic snails within 24 hours. Copper sulphate is the cheapest aquatic molluscicide but suffers from high toxicity to fish and algae; it is also easily precipitated by alkaline water and adsorbed by clay particles in the mud. Organic molluscicides, like Frescon, are not adsorbed by clay particles but are adsorbed by organic constituents in the mud. This is a serious problem in their use to control the aquatic snail vectors of *Schistosoma* as these occur in relatively stagnant waters containing large amounts of mud.

The organic molluscicides do, however, have the advantage of being less phytotoxic than copper sulphate or sodium pentachlorophenolate, but it is often difficult to achieve effective distribution of organic compounds in the slow-moving water.

Another important, and widely used, organic molluscicide is niclosamide or Bayluscide (4), introduced by the German Baeyer Co. (1959)[2] and prepared by condensation of 5-chlorosalicylic acid and 2-chloro-4-nitroaniline:

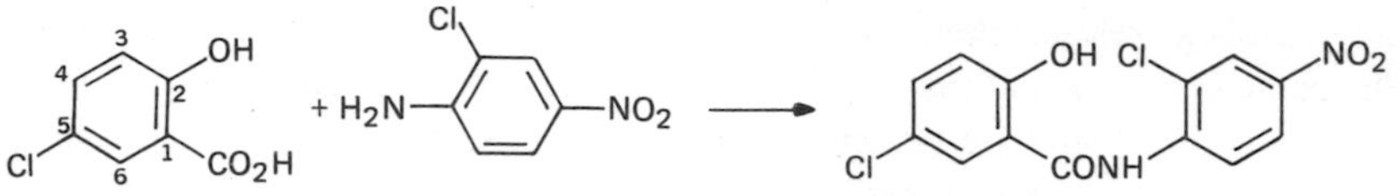

5-chlorosalicylic acid   2-chloro-4-nitroaniline   (4)

Bayluscide (4) is a weakly acidic compound, and is usually formulated as the water-soluble ethanolamine salt[5]. It has been extensively applied in Egypt for control of aquatic snail vectors of the bilharzia disease. Bayluscide is lethal to the snails at 0.3–1.0 p.p.m.[2,5], and has the advantage of low mammalian toxicity: LD_{50} (oral) to rats > 5000 mg/kg.

Special antifouling paints have been developed to combat the growth of barnacles on ships. These often contain such comparatively insoluble copper compounds as cuprous oxide and copper dimethyldithiocarbamate or trialkyltin compounds, e.g. tributyltin acetate. Many of the trialkyltins are extremely toxic to aquatic snails and their eggs at doses of 0.1–0.4 p.p.m. but suffer from high toxicities to fish, mammals, and other aquatic life[1].

The use of chemical molluscicides is only one route to the control of bilharzia which basically arises from the insanitary habits of primitive peoples. For instance, if people could be persuaded not to drink, wash, and excrete in the same water, and irrigation canals were kept flowing freely with little aquatic vegetation, the incidence of the disease could be greatly reduced.

There has also been considerable progress in direct chemical attack on the disease organism when it is actually in the human bloodstream.

References

1. Hartley, G. S. and West, T. F., *Chemicals for Pest Control*, Pergamon Press, Oxford, 1969, p. 106.
2. Pesticidal Manual, 4th edn., British Crop Protection Council, 1974, p. 103.
3. Woods, A., *Pest Control*, McGraw-Hill, London, 1974, p. 103.
4. Fletcher, W. W., *The Pest War*, Blackwell, Oxford, 1974, p. 59.
5. Metcalf, R. L., 'Chemistry and biology of pesticides' in *Pesticides in the Environment* (Ed. White-Stevens, R.), Vol. 1, Part 1, Dekker, New York, 1971, p. 124.

Chapter 13
Some Novel Methods of Insect Control

Many organisms release chemical substances that specifically affect other members of the same species some distance away from the point of release of the chemical. These substances therefore resemble hormones. Consequently such chemicals were originally termed ectohormones, but the modern name is *pheromones*[1 (a)] derived from the Greek words meaning to 'carry' and 'excite'. For chemical insect control the most valuable of these pheromones are the insect sex pheromones or attractants which may be regarded as chemicals which directly facilitate mating, either by attracting another insect from a considerable distance or by inducing the performance of some close-range behaviour concerned with 'courtship'. The majority of sex pheromones are evolved by one sex only although in some species both sexes may have the ability to produce them[2,3]. The sex pheromones of moths have been extensively studied, largely because of early studies by collectors and their economic importance as pests of many crops. The female gypsy moth *Porthetria dispar* releases a chemical attractant which can lure males to her from more than 2 miles away[3]. This gives an indication of both the high potency of the chemical and the sensitivity of the antennae of the male gypsy moth as a receiving apparatus. The threshold amount for response is about 50 molecules or 10^{-14} μg[2]. Recently there has been increased study of sex pheromones because the development of modern analytical techniques, such as chromatography, nuclear magnetic resonance, and mass spectroscopy has enabled some of the pheromones to be identified although only minute amounts were available. Also there have been increasing demands for more selective chemicals for use in pest control owing to a greater awareness of the dangers of environmental pollution (Chapter 14).

Jacobson (1961) considered that the sex pheromone of the female gypsy moth was 10-acetoxy-*cis*-hexadenol or gyptol $(1; n = 5)$[4 (a),5]:

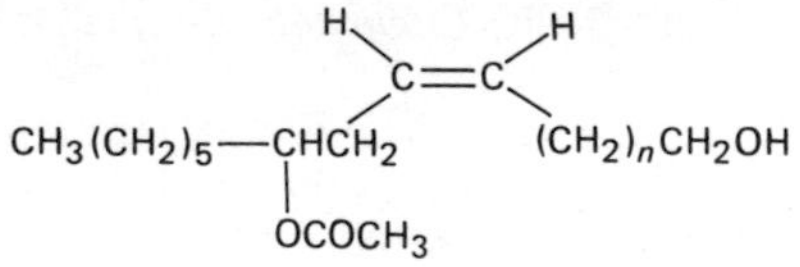

$$
\begin{array}{c}
\underset{\text{H}}{}\quad\underset{\text{H}}{}\\
\text{C}=\text{C}\\
CH_3(CH_2)_5-\underset{\underset{\text{OCOCH}_3}{|}}{CHCH_2}\qquad (CH_2)_n CH_2OH
\end{array}
$$

(1)

A synthetic homologue $(1; n = 7)$ containing two more methylene groups, synthesized from ricinoleic acid (available from castor oil), was found to retain the attractant activity of gyptol (1) and was marketed as 'gyplure'.

Gyptol ($1; n = 5$) is definitely released by the female gypsy moth at the same time as the sex pheromone but studies revealed that neither pure gyptol (1) nor 'gyplure' really have any sex attractant activity towards the male gypsy moths. The variation in the results obtained in tests with these compounds was due to their contamination with trace amounts of active substances. This showed that the true pheromone must have remarkably high attractant properties and it has now been identified as cis-7,8-epoxy-2-methyloctadecane (2) called Disparlure[1] [(a)]:

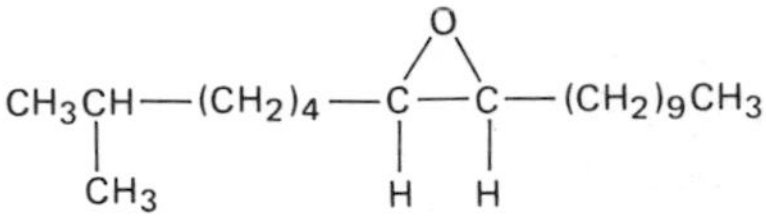

(2)

Other studies[2] showed that the sex pheromone of the natural silkworm moth *Bombyx mori* was 10-*trans* 12-*cis*-hexadecadienol or bombykol (3):

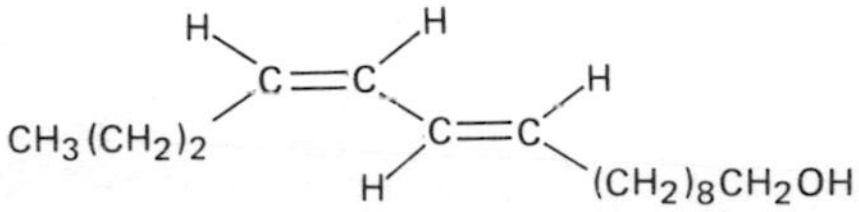

(3)

The attractant property is often critically dependent on stereochemistry; thus of the 4 possible geometric isomers for 10,12-hexadecadienol only the *trans cis* isomer (3) is identical to the natural sex attractant and only this specific isomer has any appreciable activity. The highly sensitive sense organs of the male silk worm moth can detect a concentration of only 10^{-16} g/cm^3 of bombykol emitted by the female.

Other insects, while not attracting members of the species from as great a distance to moths, also use sex pheromones; some examples of known chemical attractants are shown below:

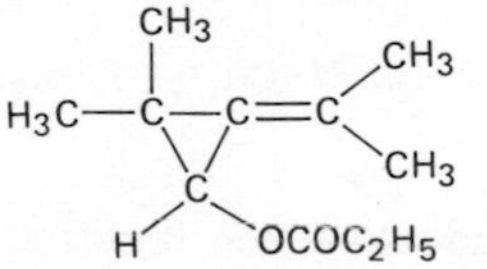

(4)

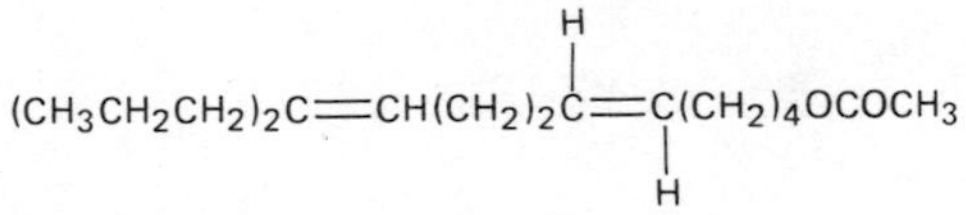

(5)

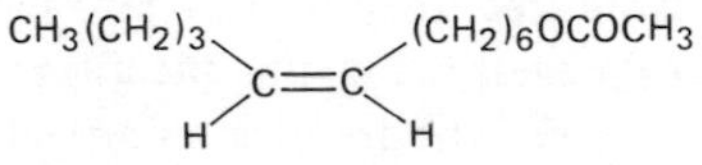

(6)

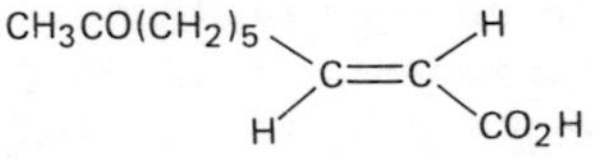

(7)

Compounds (4, 5, 6, 7) have been isolated as the chemical sex pheromones of the American cockroach, pink bollworm, cabbage loper, and honeybee queen respectively[1],[4 (a)]. Generally, attractant activity appears to be fairly closely related to structure; thus small deviations often cause considerable loss in activity. Thus whether the double bonds are *cis* or *trans* and the number of carbon atoms in the chain both appear critical. These molecules are relatively small compared with many natural products; of course their effective action demands both reasonable volatility and specificity in attractant action.

Once chemists have identified a natural sex pheromone they try to synthesize an analogue that is cheaper than the natural compound but retains its activity. The search for synthetic chemical lures has produced several interesting compounds whose function, if any, in nature is obscure; for instance methyl eugenol (8) which attracts the oriental fruit fly and also functions as a feeding stimulant, while siglure (9), medlure (10) and trimedlure (11) are all synthetic attractants for the Mediterranean fruit fly[6]:

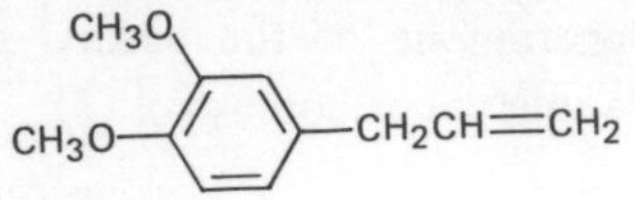

(8) 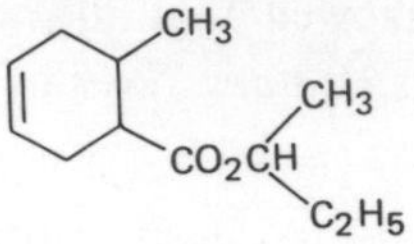   (9)

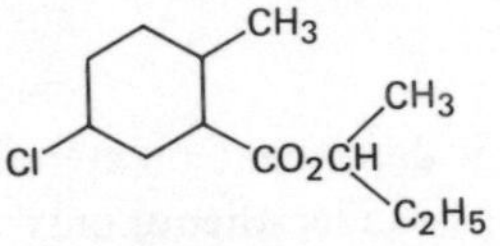

(10) 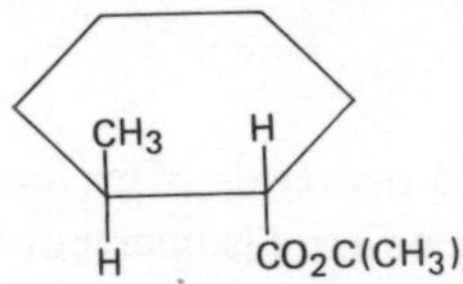   (11)

There are a few examples of synthetic attractants which are effective for insects other than fruit flies. Butyl sorbate (12) is an effective lure for the European cockchafer and methyl linolenate (13) for bark beetles:

$$CH_3CH=CH-CH=CHCO_2C_4H_9$$

(12)

$$CH_3CH_2CH=CH-CH_2-CH=CH-CH_2-CH=CH(CH_2)_7CO_2CH_3$$

(13)

Such synthetic attractants are related to natural products which function as food or oviposition lures, but their precise mechanism of attraction is unknown. It appears that the chemical released by male bark beetles is both a sex and food attractant. When the male beetles invade a suitable tree, they burrow into it and feed on the inner tissues, then they produce faecal pellets containing the attractant which draws both male and female beetles to the tree. The females on arrival are stimulated to join the males in the holes and passages they have made in the tree, while the males immediately start to construct their own galleries. The attractant thus leads to both food supply and mates and provides an extremely effective survival mechanism[3]. Most attractants, unfortunately only affect the adults of the species and immature insects do not appear to be attracted by such food or oviposition lures and are unable to detect their food plants when they are more than a few centimetres away[4(a)].

The use of chemical lures in public health control does not appear promising since insects, bedbugs, fleas, and even horseflies do not seem to possess highly developed olfactory organs. On the other hand, mosquitoes are attracted by chemicals evolved by the human body — blood, sweat, and urine all stimulate yellow fever mosquitoes, and their attractiveness is enhanced by moisture and carbon dioxide[1]. Humans secrete lactic acid in their sweat and its presence probably accounts for the appeal of sweat to the mosquitoes[3].

Oviposition lures are substances that attract gravid female insects of a species and induce them to lay eggs. For example, ammonia and ammonium carbonate exert a powerful attraction for gravid houseflies and green bottle flies respectively. In general, food lures are not as potent as sex lures. Decomposing fruit, fermenting sugar solutions, and digested protein substances have been widely used in traps for a number of insect pests. Such chemically ill-defined materials are usually unspecific and very variable in their effectiveness. However they are useful in preliminary survey studies, and protein hydrolysate — insecticide preparations are still used for the control of some species of fruit fly, although they have been largely replaced by specific chemical lures, e.g. (8–11) (p. 196).

Repellents

Chemical repellents can be either the vapour or contact type, and both must induce the insect pest to move away from them, and they must be acceptable to the host, particularly if the host is a man so that their application causes no discomfort. Research in chemical repellents over the last 30 years or so has indeed been largely concentrated on the production of synthetic chemicals to protect man from insect attack[4(b)]. The advent of World War II caused the United States of America to embark on a major programme of screening chemicals for repellent properties. The discovery of new repellents was vital for the success of military operations against the Japanese in the Far East; indeed at one period a soldier had a 90% chance of being attacked by a tropical disease before he encountered the real enemy. By 1945 some 7000 compounds had been screened by the United States Department of Agriculture and in 1952 the number had been increased to 11,000. The initial tests were for insecticidal and repellent properties against body lice, mosquitoes, and chiggers, and later houseflies, ticks, and fleas were included[3].

From early times a number of herbal extracts had been recommended as insect repellents; in particular citronella oil containing various terpenes such as geraniol, citronellol, and borneol. This deters mosquitoes from alighting or coming near to objects coated with the oil but citronella oil is too volatile and short lived to be an effective repellent[2]. An early fly repellent was dimethyl phthalate (14) (1929) which became an important constituent of skin preparations, but it was not effective against all species and during the War other repellents were discovered. Two of these were 2,2-dimethyl-6-butylcarboxy-2,3-dihydropyran-4-one or indalone (15) and 2-ethylhexane-1,3-diol or Rutger's 612 (16) which were widely employed together with dimethyl phthalate to give a broader spectrum of repellent activity.

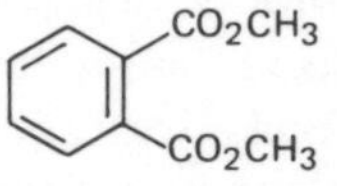

(14) 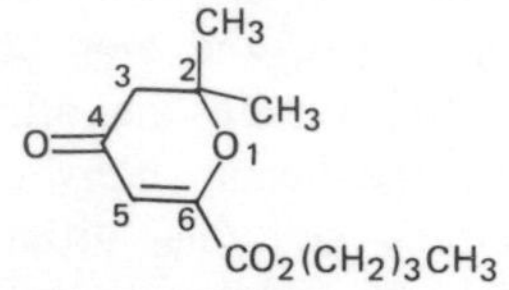   (15)

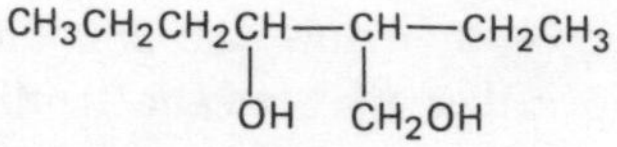

(16) 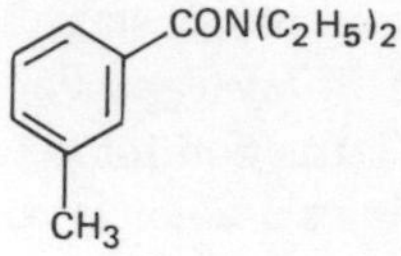   (17)

These mixtures have, however, been largely superseded by N,N-diethyl-m-toluamide (17) or deet which was introduced in 1955. Deet (17) is the most widely used chemical repellent with a broad spectrum of activity against mosquitoes, flies, chiggers, and other biting insects. These chemicals appear to act by cancelling out the attractive stimuli emitted by the animal body at a distance; for instance they prevented mosquitoes detetecting a hot wet surface. Similarly the ancient practice of dragging elderberry twigs over germinating turnip seed to deter attack by flea beetles may be due to the smell of the elderberry masking that of the mustard oils evolved during germination of the seed rather than to a direct repellent effect on the beetles.

The diseases transmitted by mites are important in the tropics and the demands of tropical warfare led to an intensive search being made for repellents against the mite vectors of scrub typhus. Impregnation of clothing, especially socks, with dimethyl or dibutyl phthalate or benzyl benzoate (18) controlled the mites, and against certain species of American ticks, Rutgers 612 (16), benzyl benzoate (18), and N-butylacetanilide (19) were effective repellents:

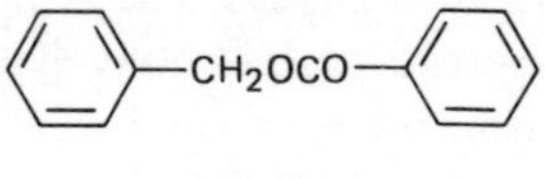

(18) 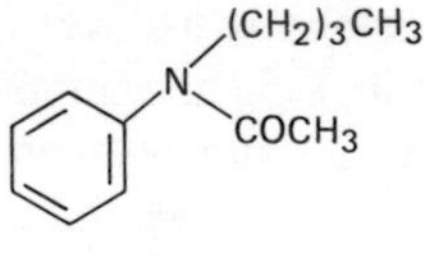   (19)

Recently 2-hydroxy-n-octylsulphide has been claimed to be an effective repellent for cockroaches[1].

Chemical repellents are generally formulated as oils, creams, or gels for hand application or as aerosol packs. Pyrethrum preparations possess considerable repellent properties as well as being insecticidal and when a room is sprayed with a kerosene solution of pyrethrum flies tend to keep away from the treated parts of the house for a considerable time. There is a large potential market for insect repellents for the treatment of livestock against attack of ectoparasites but the present chemicals have to be applied too frequently for convenience or economic benefit. There has recently been considerable emphasis on the introduction of new formulations to make existing chemical repellents more effective and easier to apply. One interesting area is the search for an oral repellent against such insect pests as mosquitoes and houseflies[4(b)].

Very little is known about the mode of action of chemical repellents or the relationship between repellent activity and chemical structure. Repellents have been little used in modern agriculture to deter pests from feeding on crops, although naphthalene has been applied on a small scale in gardens to mask the attractive smell of carrots to the female carrot fly.

Antifeeding Compounds

These are substances which are not necessarily food repellents but cancel out the signal to the appropriate organ in the insect to initiate feeding on the host. In the presence of the antifeeding substance the insect may starve to death while remaining on the host plant[2,4(c)] because it makes the host plant distasteful to the insect so feeding is inhibited. The first antifeeding substance used in agriculture was ZIP, which is still applied to the bark of trees to deter rodents and deer from eating the bark and so killing the tree[3]. 4-(Dimethyltriazeno)acetanilide (20) inhibited the feeding of southern army worms and Mexican bean-beetle larvae when applied at similar dosage rates to commercial insecticides, but the chemical did not affect aphids or mites; it was non-toxic and so did not harm natural predators[2]. The chemical (20) appeared effective when the insect population was comparatively low but lost its value in the presence of a large population of the pest.

$$(CH_3)_2N-N{=}N{-}\langle\ \rangle{-}NHCOCH_3 \qquad R-N{=}N-R'$$

$$(20) \qquad\qquad\qquad\qquad (21)$$

Other substituted diazenes (21) appear to inhibit blood-sucking insects, such as mosquitoes and tse-tse flies from feeding on their hosts[7]. Methyl phenyldiazene-carboxylate or 'azoester' (21; $R = C_6H_5$, $R' = CO_2CH_3$) was effective as an antifeeding agent; the activity has been attributed to the chemical oxidizing the naturally occurring tripeptide glutathione to the corresponding disulphide.

Chemosterilants

The basic idea behind this method of control is the introduction of a large number of sterilized males of the pest species into the area where the pest is to be controlled. The normal female population of the pest is then overwhelmed by the sterile males so that the majority of the matings are not fruitful. If the population of the sterile males can be maintained over several generations and there is no mass immigration of healthy males into the locality, the pest numbers in the area will gradually decline and eventually dissappear [1(b),4(d)]. For success it is vital that the natural population of males is heavily outnumbered by the sterile males from the beginning of the experiment thus ensuring an immediate downward trend in the pest population. This can be achieved either by reducing the pest numbers by application of a suitable insecticide, or choosing a period when the natural population is at a low level.

The idea of sterilization of male insects as a means of pest control was first proposed by Knipling in 1937, but was not practical at the time owing to lack of efficient means for the mass sterilization of the pest insects. Knipling however developed mathematical models showing how the method could eradicate pests. The technique has the advantages of being specific and much more economic than normal chemical control by insecticides and it does not lead to environmental pollution[8].

If the method is to be effective, several criteria must be satisfied: (a) the males of the pest species must be very mobile, (b) when sterilized they must not lose their sexual urge to mate, (c) the female must be satisfied with the act of mating rather than fertility[2,4(d)]. The method is completely specific because the sterile male only mates with females of his own species, and it can be extremely effective as illustrated by its use in the virtual elimination of the screw worm from the United States of America. The larval screw worm is a parasite of cattle and sheep in the American southern states and in various South American countries. It was eradicated by release of sterilized male screw worm flies first in Florida and then in the south-west of the United States. By 1964 outbreaks in Texas were only due to immigration of fertilized females from Mexico and a barrier has been established stretching along the Mexican border patrolled by sterile male screw worm flies to cut down the danger of reinfestation[1(b),2].

Other pests for which the sterilization technique looks promising include Mediterranean fruit flies, melon flies, sheep blowflies, and horseflies. Thus in 1976 more than 10^7 sterile male Mediterranean fruit flies were released over an area of 75 sq miles near Los Angeles to halt an invasion of fruit flies. The cockchafer was eliminated within a trial area by this technique and it looks promising for control of the codling moth, but against mosquitoes the method was unsuccessful.

Artificially reared insects can be sterilized by exposure to X-rays or gamma rays before release in the target area. Also there are several chemicals known to produce sterility in insects, which are termed chemosterilants. These are essential for the sterilization of natural pest insects in the field since the chemosterilant can be incorporated in a suitable food bait. Since 1947 the United States has embarked on an intensive search for chemosterilants and some 200 chemicals are known to sterilize certain insect species. The chief targets of the tests were houseflies and mosquitoes. Chemosterilants may be subdivided into alkylating agents, antimetabolites, and miscellaneous substances.

The alkylating chemosterilants are very reactive chemicals which replace a hydrogen atom by an alkyl group. Some of the most promising members contain aziridinyl groups in which the nitrogen atom is attached to electron-withdrawing groups, such as P=O, C=O, CN, SO, or SO_2. Important examples are apholate (22), tepa   (23; X = O, R = H),   thio-tepa   (23; X = S, R = H),   and   metepa (23; X = O, R = CH$_3$) and the diurea (24)[2,9]:

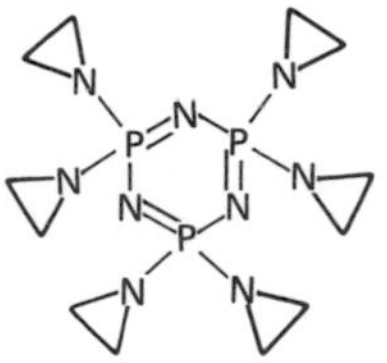

(22) 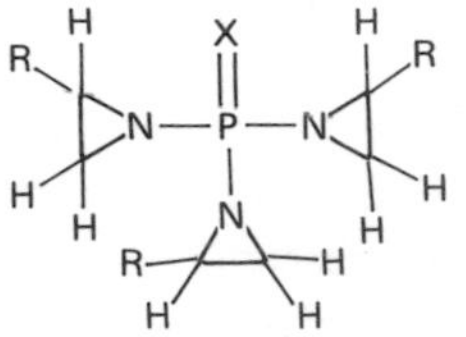   (23) 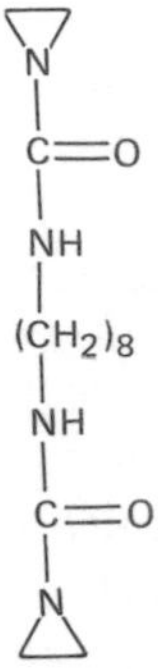   (24)

These chemicals are currently the most promising chemosterilants and effect sterilization at doses lower than those required for general toxicity and, depending on the species of insect and the dosage used, they may act by the insect eggs not developing, or not hatching, or hatching into larvae which die before attaining maturity[10-12]. Other alkylating agents, which are promising chemosterilants, include analogues of nitrogen mustard (25), named because of the similarity to the World War I mustard gas, and sulphonic acid esters, e.g. bisulphan (26):

$(ClCH_2CH_2)_2N\text{—}CH_3$ $CH_3SO_2O(CH_2)_4OSO_2CH_3$

(25)   (26)

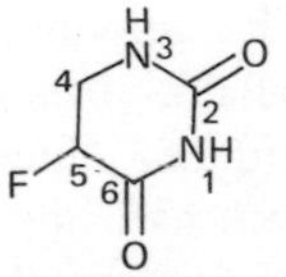

(27)

Chemosterilants are potentially dangerous chemicals and the majority have high mammalian toxicity; they are easily absorbed through the skin and are potentially mutagenic. The mutagenicity could increase the danger of the development of a resistant strain of the treated pest insect.

Antimetabolites are chemicals that mimic natural biologically active metabolites, so they may replace them in a biochemical process with consequent alteration or inhibition. Some of the antimetabolites are chemosterilants and several owe their action to functioning as analogues of the purine and pyrimidine bases occurring in nucleic acids (Chapter 3, p. 36). An example is 5-fluorouracil (27) which can replace uracil in RNA so disrupting its normal function.

The miscellaneous group of chemosterilants covers a wide variety of compounds, including antibiotics like mitomycin and cycloheximide, the alkaloid colchicine, hexamethylphosphoric triamide, triphenyltin, urea, thiourea, and derivatives of s-triazine[7,9].

Hormones and Growth Inhibitors

The majority of existing synthetic insecticides suffer from the disadvantage of causing both environmental contamination and the development of resistant strains of insects. Insecticides that are either insect hormones or mimic their action have been termed 'third-generation pesticides'[1(c)]. Such pesticides would be unlikely to induce resistance in the target pest, and would be selective and therefore harmless to natural predators.

The juvenile hormone released from the *corpora allata* glands in the insects head, together with the moulting hormones or ecdysones circulating in the insects' blood, play vital roles in the insects' growth, development, and reproduction. As insects carry their skeletons or cuticles on the outside of their bodies, the cuticles must be periodically shed (the process of moulting) in order that they can grow larger. Larvae moulting is controlled by these two classes of hormones: ecdysones initiate the deposition of new layers of cuticle and the resorption of old cuticle by the epidermal cells and are therefore essential for moulting. The amount of juvenile hormone present controls the nature of the cuticle laid down — in the larval stages when a large amount of juvenile hormone is available further juvenile cuticle is formed; if it was not present the insect would mature prematurely. On the other hand, when the larva reaches the optimum size, feeding ceases and the insect moults to the pupal stage, when almost none of the juvenile hormone must be present, so that mature or adult cuticle is deposited.

With those insects that pass through a pupal stage, an intermediate quantity of juvenile hormone regulates the deposition of the pupal cuticle, so this hormone perpetuates immature growth and development in metamorphosis, and when it is absent allows maturation. In adult insects, the juvenile hormone controls the development of the ovaries[1(c)]. The amount and timing of the production of juvenile hormone is vital. If it is available at the wrong period or in too large a dose gross abnormalities of growth are induced which generally kill the insect. Further, its presence in insect eggs will prevent hatching and normal development[1(c)]. So if insects are treated with an excess of the juvenile hormone at an early stage in their development their life cycle is distorted and they will remain in a juvenile (larval) stage and will not change *via* pupation into the adult insect[13]. The juvenile hormone may act as a coenzyme for those enzymes controlling larval development, or may alter permeability so that these enzymes are more effective or act directly on the nuclei of epidermal cells.

It is almost 20 years since an active preparation of an insect juvenile hormone was isolated from extracts of the male silk moth (*Hyalophora cecropia*) and its high ability to block insect metamorphosis suggested that the active principle could well have potential as a novel type of insecticide[14]. In 1965 the structure (28) of the juvenile hormone was elucidated[15].

(28)

The failure of efforts to rear specimens of the European bug *Pyrrhocoris apterus* in America were due to the larva not developing properly and this was eventually traced to the paper towelling used to line the cages. The paper was manufactured from balsam fir wood which is not used in Europe. The paper factor or juvabione inhibiting the proper growth and egg hatching of the bug was isolated and shown to have the structure (29) analogous to that of the juvenile hormone (28)[1 (c)].

(29)   (30)

Juvabione (29) also prevented the development of the red cotton bug, an important pest of cotton, which belongs to the same family as the European bug but it did not affect other families of bugs so juvabione appears to exhibit considerable specificity[1 (c)]. The reason for the presence of juvabione in balsam fir is obscure since the European or other related bugs are not found on this tree, though possibly they did once attack it. The isolation of the juvenile hormone mimic from the balsam fir has stimulated the search for other mimics in different plant species. The results have been rather disappointing — from examination of 60 plants only two of them gave extracts showing significant juvenile hormone activity against *Pyrrhocoris*[1 (c)].

The terpene alcohol farnesol (30) and the related aldehyde were active, and later work[1 (a)] showed that compounds containing the methylenedioxy-phenyl group often show considerable activity, and are important insecticide synergists (see Chapter 4, p. 47). These studies initiated a search for new synthetic compounds that can act as juvenile hormone mimics providing 'third-generation insecticides' (Plates 27 and 28). A significant advance was the registration in America of methoprene (31) for control of floodwater mosquitoes and hornflies.

(31)

These compounds act specifically on the endocrine system of insects and this different mode of action enables them to be effective against insects that have developed resistance to conventional insecticides. Methoprene has low mammalian toxicity and shows fairly specific activity against *Diptera* and has low toxicity to other insect species. Thus in its normal use against mosquito larvae, it has little effect on non-target insects such as damselflies, mayflies, and water beetles[3].

Similarly a recently discovered aryl terpenoid mimic (2-ethoxy-9-*p*-iso-propylphenyl-2,6-dimethylnonane) is the most effective juvenile hormone mimic against livestock fly pests, but is relatively inactive against other species of insects[14]. However juvenile hormone mimics do have disadvantages; thus a significant number of insects resistant to conventional insecticides display cross-resistance to juvenile hormone mimics. Such resistance probably arises from the ability of the tolerant insects to degrade enzymically, not only the conventional insecticides, but also the juvenile hormone mimics. The mimics to not disrupt the normal development of insect larvae and do not have true larvicidal activity so their use is restricted to those insects that are pests only in the adult phase. Further, if they are to be effective, they must be applied at the right period in the insects' life cycle so the compounds need fairly high environmental persistence for good activity against field populations of insects of varying ages, but unfortunately most of the juvenile hormone mimics break down fairly rapidly in the field. Their use is therefore confined mainly to the dipterous pests of public and animal health; juvenile hormone mimics have been used against mosquitoes. Here mimics, e.g. methoprene, have the advantages of low toxicity to mammals and non-target insects and are effective against resistant strains of mosquitoes. However the poor persistence of the compounds is a disadvantage; to obtain good control of Californian flood mosquitoes it is necessary to use an encapsulated slow-release formulation of methoprene (Chapter 2, p 17).

Juvenile hormone mimics may also be useful as chemosterilants (p. 200), thus a dose of 1 μg or less of methyl farnesoate causes a female European bug *(Pyrrhocoris apterus)* to lay sterile eggs for the rest of her life, and other insects could be controlled in a similar way, e.g. the human louse and the yellow fever mosquito[1(c)].

Insects and other arthropods have tough cuticles or exoskeletons which support their internal organs and muscles, so in order to grow, the cuticles have to be periodically discarded. This process known as moulting or 'ecdysis' is therefore essential for insect growth. It is controlled by steroidal hormones known as ecdysones – the two most important are α- and β-ecdysone (**32**; R' = CH$_3$ or OH respectively) which were isolated from silkworm *(Bombyx mori)* pupae[16].

(32)

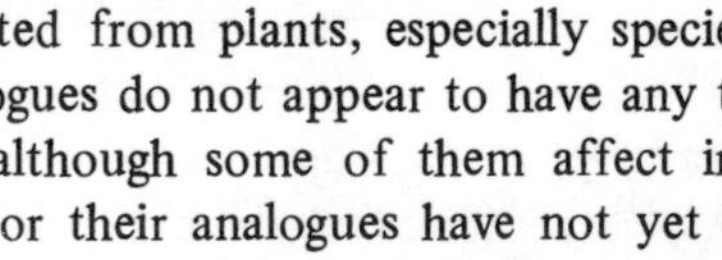

(33)

Closely related steroids have been isolated from plants, especially species of conifers and ferns, but these ecdysone analogues do not appear to have any toxic action on insects feeding on the plants, although some of them affect insect metamorphosis[1(c),16]. Moulting hormones or their analogues have not yet been commercially exploited.

Although the scope of the juvenile hormone mimics has not been nearly as great as was originally envisaged, their development has led to the discovery of other synthetic compounds which act by disruption of insect growth. Such compounds are not juvenile hormone mimics and consequently do not have the disadvantage of only being effective at maturation moults[14]. An example is diflubenzuron (33) – this is not a juvenile hormone mimic and it disrupts the larval as well as the muturative stages. Diflubenzuron (33) is highly active against mosquitoes, livestock fly pests, and many larval crop pests – the spectrum of insecticidal action is much wider than that of juvenile hormone mimics. It is reasonably persistent in its action and has very low toxicity to fish and birds[14,17].

The biochemical basis for the activity of diflubenzuron (33) appears to depend on its inhibition of the normal process of insect cuticle formation. The symptoms of poisoning by this compound are death of treated larvae due to their inability to shed the old cuticle at moulting time, and the distortions of any newly formed larvae suggest that the compound impairs the mechanical properties of the new cuticle. Electron micrographs of the newly formed cuticle of cabbage white caterpillars which have been treated with diflubenzuron show that the cuticle lacks the lamellar structure of the untreated insect cuticle, and the compound apparently interferes with the final stage of chitin synthesis, namely the polymerization of N-acetylglucosamine[14]. Diflubenzuron and related phenylureas are not the only non-juvenile hormone mimics to have shown activity as insect growth regulatory insecticides; for instance, 2,6-di-t-butyl-4-(dimethylbenzyl)phenol and 1-buten-3-yl N-(p-chlorophenyl)carbamate specifically disrupt moults of mosquito larvae[14]. Such insect growth regulators appear to possess considerable potential as 'third-generation insecticides' due to their selectivity and lack of toxicity to non-target organisms[1(c),18].

Natural insecticides can also be derived from insects themselves such as spiders, snakes, beetles, ants, bees, and wasps[19,20]. Some of these toxins and venoms are highly potent and selective but the majority are only effective by injection and are not therefore of practical use as insecticides. However arthropod venoms and toxins can serve as useful models of insecticidal action and may have novel modes of toxicity, and assist in the discovery of new synthetic chemicals with unique modes of action[19].

Microbial Insecticides

Insect pests can be controlled by bacterial, fungal, and viral infections, and entomologists have long known that insect populations can be seriously affected by outbreaks of infectious diseases which was first observed with silkworm and honeybee colonies[1(d),4(e)]. Some *Bacillus* species attack chafer grubs turning their blood into a milky fluid and spore powder preparations of these bacteria have been successfully applied in America since the early 1940s for control of the chafer in grassland which has kept the beetle population at low levels.

Bacillus thuringiensis refers to a group of bacterial isolates which have been studied for nearly a century. *B. thuringiensis* was first isolated in 1915 from diseased flour moth larvae and later several varieties were obtained from different species of infected caterpillars[1(d),9,21]. These bacteria grouped together as *B. thuringiensis* all form toxic protein crystals at the time of sporulation. The crystals are very toxic to certain insects and can be used as insecticides against many *Lepidoptera* and they do not appear to be toxic to other organisms, apart from earthworms. The toxic crystal (endotoxin) is chiefly responsible for the paralysis and ultimate death of infected caterpillars[22], although at least five toxic entities have been isolated from *B. thuringiensis*[9].

The toxic protein crystal is the most studied and important of the *B. thuringiensis* toxins; it dissolves in the alkaline gut of caterpillars and other susceptible insects and the protein solution is digested by enzymes in the gut to release one or more toxins[22]. The mode of action of these is not understood, but the first symptom is paralysis of the gut followed by changes in the gut wall affecting its permeability and allowing escape of the alkaline contents with resultant general bodily paralysis and death of the caterpillar. A wettable powder formulation of *B. thuringiensis* called Thuricide is used as a field insecticide and is active against a wide range of lepidopterous larvae, including many pest species. Such commercial spore/crystal mixtures are mainly used against lepidopterous larvae on leaf crops such as cabbage and tobacco. Thuricide appears harmless to mammals, birds, fishes, and most insects apart from *Lepidoptera*[22].

With certain strains of *B. thuringiensis* a second toxin known as β-exotoxin can be isolated from their cultures. β-Exotoxin is thermally stable and possesses a much wider spectrum of insecticidal activity affecting the larval stages of many species of insects (e.g. fleas) and also showing toxicity to a range of mammals[22]. The toxicity only becomes apparent at moulting or metamorphosis; exotoxin is a high molecular weight adenine nucleotide containing an unusual sugar allomucic acid, and there is some evidence that it acts by interference with nucleic acid metabolism and protein synthesis. In future it may be possible to develop new strains of the bacillus with more useful insecticidal potency[21].

Viruses

Insects can be associated with viruses either as hosts or as vectors. Certain viruses are parasitic and pathogenic to insects and such viruses are enclosed within protein crystals, capsules, or membranes[1(d)]. Virus diseases in insects have long been known; the first record (1527) was of the jaundice disease of silkworms.

Insect larvae, especially those of the *Lepidoptera*, are often attacked by viral diseases described as polyhedral inclusion viruses. The viruses are embedded in a protein matrix which is generally polyhedral in shape. These viruses, known as baculoviruses, are the most promising insecticides and have been successfully sprayed onto forest trees[23,24] to control sawflies. However experiments with viruses of *Lepidoptera* have not been so effective, but a polyhedral virus of the alfalfa caterpillar shows promise against this pest in America. Viruses have also been applied successfully against the cabbage looper, the cotton leaf worm, and the forest tent caterpillar[24]. Some diseases are caused by non-inclusion viruses, one of which occurs in natural populations of the *Drosophila* species which makes the flies very susceptible to even small concentrations of carbon dioxide. A few viruses are known to infect mites; thus the citrus red mite rapidly succumbs to a paralysing non-inclusion virus. This is unstable so it cannot be applied as a spray and is only effectively transmitted by the release of infected males[1][(d)].

Virus diseases of higher animals have been used in certain cases; the most outstandingly successful was the control of the rabbit population in Britain by the myxoma virus causing myxomatosis[23], which was spread by releasing a few rabbits inoculated with the disease. The effectiveness of the virus is now threatened by the emergence of rabbits who have acquired resistance to the disease, and attempts to control rats by a virus disease have been largely unsuccessful (Chapter 10, p. 177).

Fungi

Several insect—fungus associations are pathogenic and hence microbial control of certain insects can be achieved by fungi. The early history of the discovery of fungi parasitic on insects has been discussed by Steinhaus (1956)[23] and led to the elucidation of the principles of the fungal control of insects. The muscardine disease of silkworms was shown (1835) to be caused by a fungus *Beauveria bassiana,* and this organism also attacks a number of insect pests including the codling moth and the European corn borer. The Russian Metchnikoff first tried (1879) to control wheat cockchafers by application of a preparation of the green muscardine fungus (isolated from diseased wheat cockchafers and grown on sterilized beer mash). This fungus is pathogenic to many insect species, including mosquitoes which are susceptible in the larval stage, and could be a useful weapon in vector control.

The muscardine fungi have considerable potential as microbial insecticides, because they can be easily raised on artificial cultures, are pathogenic to a number of insect pests, and appear generally harmless to vertebrates. In contrast, the related *Aspergillus* species of fungi can result in severe diseases in vertebrates so they are unlikely to be of any value for insect control.

Species of the aquatic fungus *Coelomyces* are chiefly pathogenic to mosquito larvae and have been used in field trials for control of mosquitoes; the fungus appears quite promising, but it is difficult to obtain and so far none of the species have been grown successfully on artificial media. Entomogenous fungi are sometimes of importance in the natural control of insect populations[11]. Some members of the entomophthorales are common pathogens of houseflies and in damp weather in the autumn dead flies are often seen on windows with their bodies surrounded by a halo of discharged spores. This genus of fungi attacks many kinds of insects and some members appear to be the only important pathogens of aphids. They have been applied to control, for instance, the European apple sucker in Canada, the spotted alfalfa aphid in California, and aphids on potatoes in Maine[1][(d)].

Nematodes or eelworms are difficult to control chemically (Chapter 11) and it is therefore interesting that several species of fungi are predacious on eelworms which they trap by adhesive snares and hyphal rings. When the nematode enters the ring it contracts, trapping it fast, and then the fungal hyphae penetrate the body wall of the nematode and extract the contents. Such predacious fungi have been used in

208

England, France, and Hawaii against nematodes when different strains of the fungi were discovered to show varying degrees of aggressiveness towards nematodes[1(d)]. It could therefore be advantageous to inoculate the soil with aggressive strains of the fungus, although this characteristic may not be permanent. A general problem is the difficulty of obtaining widespread growth of the pathogenic fungus artificially, before the insect pest has attained almost epidemic proportions[23].

References

1. Woods, A., *Pest Control*, McGraw-Hill, London, 1974, (a) p. 280, (b) p. 249, (c) p. 88, (d) p. 209.
2. Hartley, G. S. and West, T. F., *Chemicals for Pest Control*, Pergamon Press, Oxford, 1969, p. 98.
3. Fletcher, W. W., *The Pest War*, Blackwell, Oxford, 1974, p. 137.
4. *Pest Control: Biological, Physical and Selected Chemical Methods* (Eds. Kilgore, W. W. and Doutt, R. L.), Academic Press, New York, 1967, (a) p. 241, (b) p. 267, (c) p. 287, (d) p. 197, (e) p. 31.
5. Shorey, H. H., Gaston, L. K., and Jefferson, R. N., *Advan. Pest Control Res.*, **8**, 57 (1968).
6. Green, N., Beroza, M., and Hall, S. A, *Advan. Pest Control Res.*, **3**, 129 (1960).
7. Galun, R., Kesower, E. M., and Kesower, N. S., *Nature (London)*, **224**, 181 (1969).
8. Knipling, E. F., *Principles of Insect Chemosterilization* (Eds. Labrecque, G. C. and Smith, C. N.), Appleton-Crofts, New York, 1968, p. 8.
9. Eto, M., *Organophosphorus Pesticides: Organic and Biological Chemistry*, CRC Press, Cleveland, Ohio, U.S.A., 1974, p. 304.
10. Bushland, R. C., *Advan. Pest Control Res.*, **3**, 1 (1960).
11. Bořkovec, A. B., *Advan. Pest Control Res.*, **7**, 1 (1966).
12. Bořkovec, A. B., 'Chemosterilants for male insects' in *Insecticides* (Ed. Tahori, A. S.), Gordon and Breach, New York, 1972, p. 469.
13. Kay, I. T., Snell, B. K., and Tomlin, C. D. S., 'Chemicals for agriculture in *Basic Organic Chemistry*, Part 5, Wiley, London, 1975, p. 430.
14. Ruscoe, C. N. E., *Proc. 8th Brit. Insectic. and Fungic. Conf.*, Brighton, **3**, 927 (1975).
15. Roller, H. and Bjercke, J. S., *Life Sciences*, **4**, 1617 (1965).
16. Horn, D. H. S., 'The ecdysones' in *Naturally-Occurring Insecticides* (Eds. Jacobson, M. and Crosby, D. G.), Dekker, New York, 1971, p. 333.
17. Mulla, M. S., Darwazeh, H. A., and Norland, R. L., *J. Econ. Entomol.*, **67** (3), 329 (1974).
18. Bowes, W. S., 'Juvenile hormones' in *Naturally-Occurring Insecticides* (Eds. Jacobson, M. and Crosby, D. G.), Dekker, New York, 1971, p. 307.
19. Beard, R. L., 'Arthropod venoms as insecticides' in *Naturally-Occurring Insecticides* (Eds. Jacobson, M. and Crosby, D. G.), Dekker, New York, 1971, p. 243.
20. Cavill, G. W. K. and Clark, D. V., 'Ant secretions and canthardin' in *Naturally-Occurring Insecticides* (Eds. Jacobson, M. and Crosby, D. G.), Dekker, M., New York, 1971, p. 271.
21. Angus, T. A., 'Bacillus *thuringiensis* as a microbial insecticide' in *Naturally-Occurring Insecticides* (Eds. Jacobson, M. and Crosby, D. G.), Dekker, New York, 1971, p. 463.

22. Norris, J. R., 'The insect toxins from Bacillus *thuringiensis*' in *Insecticides* (Ed. Tahori, A. S.), Gordon and Breach, New York, 1972, p. 119.
23. Martin, H., *The Scientific Principles of Crop Protection*, 6th edn., Arnold, London, 1973, p. 63.
24. Tinsley, T. W., *Proc. 8th Brit. Insectic. and Fungic. Conf.*, Brighton, **3**, 947 (1975).

Chapter 14
Pesticides in the Environment

The increased use of various types of pesticides in the modern world has led to much greater emphasis on the possibility of serious environmental contamination arising from their use. The scientific attack on pests only dates from about the middle of the nineteenth century (see Chapter 1, p. 3) and in 1963 Rachael Carson's book *Silent Spring*[1] made people aware of the potential dangers of pollution from pesticides.

Pollution may be defined as fouling the environment and has many forms ranging from chemical pollution to that arising from dumping of old cars and other refuse about the countryside. Here one is concerned with pollution due to the widespread use of certain chemical pesticides — wherever a pesticide is applied to the foliage or seed of the crop or to the soil there is a possibility that some of the material is persistent and may lead to serious contamination of the ecosystem.

Since the end of World War II there has been a tremendous expansion in the use of chemical herbicides as a substitute for the age-old mechanical means of weed control. This was stimulated by the shortage and high cost of agricultural labour and consequently the rate of increase in the use of herbicides is much greater than that of other types of pesticides.

Early arsenical weed killers, especially soluble sodium arsenite, were very dangerous to all forms of wild life, including man. These herbicides have high mammalian toxicity and toxic residues remain in the soil for a long time. Since 1961, the use of arsenical herbicides has been officially discouraged and they have not been employed as potato haulm destroyers[2]. Arsenic residues from the continued use of arsenic pesticides increased significantly. Thus more than 3500 lb of lead arsenate were added to 1 acre of a commercial orchard over a period of 25 years and the residues were mainly confined to the top 6 inches of soil.

The dinitrophenols DNOC and dinoseb (see Chapter 8, p. 141) introduced as herbicides in the 1930s are highly poisonous to man and other mammals. Their use has damaged many forms of wild life and they have caused human fatalities; all insects and mammals caught in the insecticide spray are immediately killed. DNOC is so toxic that the quantity applied to an acre (about 4.5 kg) would be sufficient to kill approximately 1000 men. However these dinitrophenols are rapidly degraded after contact with plants or the soil and they do not therefore leave toxic residues. Plants killed by these compounds do not seem to harm animals which eat them unless they do so very soon after the application of the spray. Thus when spraying crops such as peas and beans to kill broad-leaved weeds and for potato haulm destruction, a minimum period of only 10 days is needed between treatment and harvest[3]. These dinitro compounds do not cause any long-term environmental pollution and do not accumulate along food chains

The next major advance in herbicides came in 1942 with the discovery of the phenoxyacetic acid herbicides such as 2,4-D and 2,4,5-T. These hormone-type herbicides act as a mimic for the natural plant growth hormone indole acetic acid. This does not affect animals and these compounds have a very low mammalian toxicity. They are not very persistent and break down the soil within a few weeks of application. There is generally little evidence that the very large-scale application of these herbicides has caused serious environmental damage. However since 1969 the United States government has restricted the use of 2,4,5-T on food crops following experiments indicating that it has teratogenic (foetus deforming) effects on rats and mice. This may be due, at least in part, to the presence of some 30 p.p.m. of the highly toxic and teratogenic 2,3,7,8-tetrachlorodibenzo-p-dioxin (TCDD) (1) produced as a by-product during the high-temperature hydrolysis of 1,2,4,5-tetrachlorobenzene to 2,4,5-trichlorophenol (2)[4]:

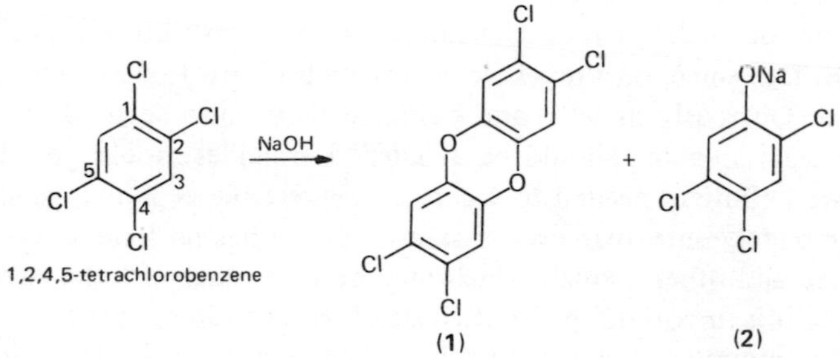

The dioxin (1) is slowly degraded in soil but residues persist for some time in treated plants. The problem seems to have been eliminated by modified manufacturing processes for the phenol (2) that keep the dioxin content <0.1 p.p.m. However there are disturbing indications that pure 2,4,5-T may be teratogenic at high dosages[4], but it is most unlikely to be harmful under normal conditions of herbicide application.

Urea, carbamate, and triazine herbicides (Chapter 8, p. 149) act by interfering with photosynthesis in plants causing them to be starved to death. Animals do not photosynthesize and are therefore not directly affected by these chemicals and, although persistent in their herbicidal action, they have low mammalian toxicity and do not build up along food chains. It is however possible that the triazines may have mutagenic effects[5]. Long-term studies of phenoxyacetic acid, urea, and triazine herbicides gave no indication of a build-up of herbicidal residues in the soil as a result of repeated annual application, nor have they shown any adverse effects on soil creatures such as earthworms, mites, and insects[6].

The application of certain insecticides, acaricides and, more recently, of systemic fungicides has resulted in the appearance of resistant strains of insects and fungi[7]. However in Britain no such effects have been reported in connexion with herbicides, although in other parts of the world some examples of weed resistance are known. Thus a common weed *(Erechtipes hieracifolia)* in sugar plantations has become resistant to 2,4-D and can no longer be effectively controlled by this chemical. It is probable that more examples of resistant weed strains will be continually appearing and this may cause harmful ecological effects — consequently one cannot be absolutely certain about the potential long-term effects of even an apparently innocuous chemical like 2,4-D[8].

Permanent grassland, once a traditional feature of agricultural Britain, has almost disappeared — now grass survives generally as a stage in a crop rotation pattern covering 2–5 years, after which period it is ploughed up. The ploughing almost completely removes the surface-casting earthworms which is unfortunate as they improve soil structure. Now with one application of paraquat all the grass and most of the other weeds can be killed, and since the herbicide is rapidly rendered inactive on contact with the soil, the land can be immediately resown. This technique known as 'chemical' ploughing can also be useful to improve the quality of marginal hilly grassland and there is some evidence that it may inflict less damage to the soil fauna than the traditional mechanical methods of cultivation; for example surface-casting earthworms appear to survive this treatment.

The current use of herbicides does not appear to constitute a major hazard to wild life in Britain since, on the whole, compounds of low toxicity and persistence are employed. Obviously as with any pesticide they must be used correctly and routine herbicidal spraying should be avoided. Herbicides should only be applied when they are definitely needed to secure an appreciable economic gain. One can never be completely sure that even a safe herbicide has no long-term deleterious effects on man and other animals which may be in contact with the chemical over many years, so it is important in the interests of the environment that they are only applied in the minimum quantity to achieve the desired result. Thus those weeds like pansies and spurreys usually appearing in a cereal crop when it is several inches tall need not be killed because they do not seriously compete with the crop plants for light and nutrients and do not interfere with harvesting.

Copper and sulphur fungicidal sprays have been used for a long time (Chapter 7, p. 107). Bordeaux mixture (copper sulphate and slaked lime) is used against vine mildew and potato blight; lime sulphur is used against scab diseases of fruit trees and powdery mildews.

The continual treatment of crops with these fungicides leads to the formation of stable residues containing high concentrations of copper and sulphur which remain in the soil for long periods and affect soil fauna and worms. They may also drain off into ditches and rivers and kill algae and fish. However, in general, fungicidal sprays and washes do not appear to have caused serious harm to wild life. An illustration of the effect of copper fungicides derives from an East Anglian orchard which had been subjected to prolonged treatment with copper fungicides. The surface soil contained 0.2% of copper which seemed to have no deleterious effect on mature apple trees but the soil contained no earthworms so the surface structure of the soil would probably be damaged over a long period. Soil fauna will be similarly affected in many orchards and potato fields. In addition soil fungi which play a vital role in breaking down leaves and other organic matter are killed by copper and sulphur residues so there may be a long-term danger to soil fertility arising from excessive use of copper and sulphur fungicides. Luckily, for many purposes, these are now being replaced by safer organic fungicides like captan.

Organomercury compounds, such as phenylmercury acetate, are widely used as seed dressings to protect cereals from fungus diseases such as smut and bunt. These are remarkably effective in combating attack by the fungal spores. Thus normally 1 oz of an organomercurial compound containing approximately 1% of the metal mixed with a bushel (63 lb) of wheat seed is sufficient to prevent wheat bunt so only about $\frac{1}{10}$ oz of mercury is added per acre. Poultry may suffer when fed on the treated grain but organomercurials (mainly arylmercury compounds) as used in Britain do not appear to present a serious danger to the environment[2,6]. There is no evidence of an appreciable increase in the normal mercury concentration in the soil, although arylmercury seed dressings have been used in this country for some forty years[6]. Arylmercury derivatives are much less toxic than the alkylmercury compounds such as methylmercury dicyandiamide. Consequently these are now banned in several countries notably Sweden where they have caused significant damage to wild life. Thus dead pheasants containing over 20 p.p.m. of mercury were found and the metal also appeared to have accumulated in pike, hawks, and other predators. However mercury is also widely used in industrial processes in Sweden, so these environmental effects may not be due primarily to the use of mercury seed dressings. There is always a danger of mercury fungicides draining off into rivers and the high level of mercury in some sea fish has recently attracted much attention, but most of the metal almost certainly comes from its application in industrial processes, such as the use of mercuric chloride to inhibit the development of slime in pulp manufacture, rather than from organomercurial seed dressings.

Certain sea fish have been shown[9] to contain methylmercury and the biological methylation of mercury was first proposed by the Japanese to explain the Minamata disaster. Any discharge of metallic mercury, inorganic Hg^{2+} ions, or organic mercury compounds into the environment can be converted to methylmercury which can then be taken up by living organisms, e.g. fish, so every effort must be made to reduce pollution of the environment by mercury in any form.

The careful use of fungicides does not apparently cause much damage to the environment, although more long-term studies are needed to assess the effects of residues of copper and mercury on soil fungi and some of them (e.g. captan and dinocap are toxic to fish)[2,6]. Inorganic copper and sulphur fungicides are, however, being phased out and replaced by organic fungicides although at present there is no viable alternative to the organomercurial seed dressings. Several systemic fungicides like benomyl have recently come on to the market (see Chapter 1 p. 8). These are absorbed and translocated within the plant but no harmful residues are left. They are highly effective and, after several years of commercial use, no ill effects have been observed on animals, birds, or even soil microorganisms, and their introduction should reduce the danger of environmental contamination arising from fungicides[6].

Insects are much more closely related to man and other mammals than either plants or fungi and consequently it is perhaps not surprising that the greatest threat to the environment has come from the large-scale application of synthetic insecticides rather than from fungicides and herbicides. Early insecticides included some highly toxic materials, like arsenic and its compounds, such as Paris Green (copper arsenite) and lead arsenate. The latter is still used on a small scale for control of caterpillars on fruit trees and against earthworms and leather jackets in turf[3]. Hydrogen cyanide was also used as a fumigant. The use of dinitro-o-cresol (DNOC) has already been discussed as a herbicide (see Chapter 8, p. 141) and is also employed for controlling aphids, mites and other insects on fruit trees in winter[3]. Sodium fluoride was formerly used against various domestic pests, such as ants and cockroaches, but it was never used extensively on crops otherwise it could have caused serious damage as does fluorine from brickworks and other industries[2].

Two natural insecticides introduced in the nineteenth century were rotenone from the derris plant and pyrethrum extracted from a species of chrysanthemum (Chapter 4). Both these substances are very safe insecticides and would be employed on a much larger scale except for their relatively high cost. These were the main types of insecticides used before World War II. Obviously such highly poisonous substances as arsenic and cyanide were potentially very dangerous to the environment. However their great toxicity was so well known that great care was exercised in their application so that paradoxically comparatively little environmental damage was caused. They have now been almost completely superseded by newer synthetic organic insecticides. These fall into two main classes — the organophosphorus and organochlorine compounds[2]. Organophosphorus insecticides (Chapter 6, p. 68) were developed after World War II from work originally carried out in a search for nerve gases; not surprisingly therefore the early organophosphorus insecticides were very toxic to man and other mammals[2,5,6]. Parathion was the first member of the group to be widely used as a contact insecticide in agriculture and has a wide spectrum of activity being effective against aphids, caterpillars, spider mites, and eelworms[3]. Other early examples included tetraethyl pyrophosphate (TEPP) and schradan. TEPP was indeed the most toxic material ever to be used on farms and it is estimated that 1 oz of this compound could kill 500 men! All these compounds were very poisonous and their use was viewed with considerable alarm — they have caused several human fatalities and full protective clothing must be worn during application. Any birds or small mammals caught in the spray are quickly killed. However organophosphates have the important advantage, over the organochlorine insecticides, that they are comparatively rapidly biologically and chemically degraded in plants, animals, and soil to non-toxic materials. The rate of breakdown depends on the nature of the organophosphate, the formulation, method of application, climate, and the growing stage of the treated plant crop[10].

A consequence of the degradation of organophosphates is that birds or mammals entering the treated area a few hours after application generally survive and after a given period a treated crop is quite safe for consumption. Organophosphates do not accumulate in the mammalian body or along food chains and, after more than 20 years of application, there is no evidence of chronic effects being produced in the ecosystem[5,6]. Another advantage of this group of compounds is that many of them function as systemic insecticides enabling smaller amounts of the active ingredient to be used more effectively which reduces the harmful effects on natural predators since usually only insects actually eating the crop are killed. Recent research on organophosphates has resulted in the introduction of several valuable insecticides, like malathion and menazon, with very low mammalian toxicities, showing that it is possible to obtain considerable selective toxicity to insects within this group of compounds. It may, therefore, be concluded that provided organophosphorus insecticides are carefully applied, those currently in use are unlikely to do serious harm to the environment[2,6,10].

The organochlorine compounds (Chapter 5) represent the other major group of synthetic insecticides, e.g. dieldrin and endrin. The first and most widely used member was DDT, introduced in 1943, and during the war appeared to be the perfect insecticide combining a wide spectrum of insecticidal activity with a comparatively low acute toxicity, about the same as that of aspirin. Indeed the organochlorine compounds did not appear to have any harmful effects on the environment until the 1950s when some doctors became worried by the appearance of these compounds, especially when they were detected in cows' milk after the animals had fed on foodstuffs which had been treated with organochlorine insecticides. In fact in the United Kingdom DDT has probably not caused much harm to wild life, though the related compounds aldrin and dieldrin have killed many species of birds and mammals feeding on them[2]. In Britain gross pollution of the environment by DDT has been avoided probably more by luck than judgement since the authorities considered they could not afford the massive blanket spraying programmes which were characteristic of the use of the DDT and related compounds in the United States of America. The careful application of DDT at minimum dosage has successfully controlled insects in many countries without obvious harmful effects, but when used in ways now considered unwise, this compound has resulted in much environmental damage[11].

In the United States attempts were made to halt the spread of Dutch elm disease by spraying the trees with high concentrations of DDT ($\simeq$28 kg/ha). This treatment was designed to kill the bark beetles which spread the Dutch elm disease fungus. Unfortunately the disease was not halted and large numbers of birds, especially the American robin, were killed. Apparently the DDT got onto the elm leaves which fell on the ground and were subsequently eaten by worms which concentrated the DDT in their bodies and these contaminated worms were eaten by the robins so that they gradually accumulated a fatal dose of the chemical. A very serious feature of organochlorine insecticides is their ability to become concentrated along food chains causing death to organisms at the end of the chain[11,12]. The classic example of this type of effect in the ecosystem was that of Clear Lake in California. The lake was an important centre for recreation but in the summer had clouds of small *Chaoborus* gnats which caused serious complaints to be made to the authorities, so in order to control the gnats a spraying programme was initiated in 1949 using DDD (a closely related compound to DDT but rather less toxic to fish)[9,12]. The first application was a success — the gnats were almost entirely eliminated and other organisms did not appear to have been seriously harmed. In 1950 the gnats were not present in sufficient numbers to cause complaints although during the next two years their numbers built up to about their original level. In 1954 therefore the lake was resprayed with DDD and it was treated a third and last time in 1957. The second spraying again appeared to give satisfactory control, but by the time of the last spraying the gnat and almost 150 other insect species had developed varying degrees of resistance to the insecticide. In December 1954 large numbers of dead western grebes (fish-eating diving birds) were observed and this caused a public outcry. In fact the first spraying does appear to have seriously reduced the breeding success of the grebes in the lake area; indeed from 1950 to 1961 no young were hatched and reproduction remained largely unsuccessful until 1969, some twenty years after the first introduction of DDD into the area. Initially the treatment appeared highly satisfactory — one single application had lasted over several years and a safe and effective means of removing the gnats appeared to have been discovered. The death of the grebes was a striking illustration of the concentration of a toxicant along a food chain. During the period September 1949–September 1957 some 120,700 lb of DDD were applied to the lake from moving barges so as to give a concentration of DDD in water of 0.020 p.p.m. immediately after each application. This very low concentration of DDD appeared quite safe and indeed after 2 weeks no DDD could be detected within the water. However, later studies showed that organochlorine compounds are extremely dangerous in aqueous environments because, owing to their great chemical stability, low aqueous solubility, and high lipophilicity they become concentrated in the living components of the lake's ecosystem[5,9,12].

The disappearance of the chemical from the water after 2 weeks therefore was not due to its degradation, but rather arose from its transport into living organisms (e.g. plankton, plants, frogs, fishes, birds, etc.). After the final treatment (1957), analysis of the lake's ecosystem showed that the microscopic plankton of the lake contained approximately 5 p.p.m. of DDD (=250 times the original concentration of DDD applied to the lake); in frogs the magnification was 2000 times, in sunfish about 12,000 times, and finally some of the sick grebes contained 1600 p.p.m. of DDD in their fat, a magnification of 80,000 times the original concentration of the chemical.

The greatest danger from the use of organochlorine insecticides seems to occur when water is contaminated, because fish and other aquatic organisms have the capacity to absorb the chemical from the water and concentrate it in their fatty tissues. This is especially true of fishes because, in the course of respiration, they pass a large volume of water through their gills so the lipophilic organochlorine compounds are absorbed from the water into the fish. Also there is some evidence that DDT and similar compounds inhibit oxygen uptake at the gills so causing fish to die from suffocation[5,12]. Obviously in such an aquatic food chain as that operating in Clear Lake the fish-eating grebes at the end of the chain obtained the maximum concentration of the organochlorine insecticide which in many cases constituted a fatal dose[13]. Many similar examples have been documented of the concentration of organochlorine compounds along food chains, such as the study of the Long Island estuary food web[13,14]. Oysters and other shell fish obtain their nutrients by continually filtering the water, so these creatures, like fish, are very sensitive to the presence of small concentrations of organochlorine compounds and similar lipophilic molecules in the water they inhabit.

Other important food-chain studies involve predatory birds, such as peregrine falcons and sparrow-hawks[12]. There were fewer of these species of birds in Britain in 1966 than in 1950 although the population appeared to be reasonably stable up to 1939. During the war, drastic control measures were adopted, but by the early 1950s the population of these birds was almost back to the 1939 figures. However since then the position has become grave; in 1962 nearly all the breeding grounds of the peregrine falcon in the remote parts of the Highlands of Scotland were occupied and more than 40% of these pairs reared young birds successfully. On the other hand, in southern England only some 20 out of the 66 breeding territories used in 1939 were occupied and only four pairs reared young successfully. Examination of peregrine corpses showed that the majority of them contained substantial quantities of organochlorine pesticide residues. In some cases the amounts could have been lethal. The corpses of other birds killed by shooting or on the roads contained rather lower pesticide residues. Eggs which failed to hatch were also found to contain substantial organochlorine residues. Experiments have shown that breeding can be inhibited by sublethal amounts of these compounds, and they may in addition cause sterility so that eggs are not laid. They may also affect bird behaviour so that the eggs are not incubated and may reduce the thickness of the egg shells – consequently more eggs are broken in the nest[5,11]. The sparrow-hawk, until recently a common bird throughout Britain, is now quite rare and only breeds successfully in the remoter parts of the country.

There was also a marked reduction in the breeding success of golden eagles in western Scotland during the period 1963–6. This coincided with a decrease in egg-shell thickness and consequent increase in breakage in the nest. In this area the main food of the eagles was sheep carrion from which they derived high residues of dieldrin which was used as a sheep dip. On the other hand, in eastern Scotland where eagles feed mainly on wild prey breeding success and shell thickness remained normal during this period. After the ban was imposed on the use of dieldrin as a sheep dip in 1966 there was a marked improvement in the breeding and shell thickness of eagles in western Scotland. Both the peregrine falcon and the sparrow-hawk are also showing a slow population recovery after the restriction on the use of organochlorine insecticides[11].

Since about 1948 the cyclodiene group of organochlorine insecticides, for instance aldrin, dieldrin, and endrin, were introduced as soil insecticides. In Britain dieldrin was extensively employed as a seed dressing for wheat against attacks of wheat bulb fly, birds, particularly pigeons, dig up and eat the treated grain. Consequently in the springs of 1956–61 large numbers of dead birds were discovered, particularly in East Anglia. This was a major cereal growing region and the use of dieldrin seed dressings was clearly implicated. The poisoned birds remained toxic and were eaten by predatory birds and by foxes which in turn died by accumulating a lethal dose of the chemical. In the spring of 1959–60 some 1300 dead foxes were found in areas where dieldrin seed dressings had been applied, and there is no doubt that the deaths arose from the use of dieldrin seed dressings on spring-sown wheat[11]. Dieldrin, endrin, and heptachlor are now only permitted as seed dressings for autumn-sown wheat in fields were high populations of wheat bulb fly eggs have been counted. This practice is much safer since birds do not dig up much wheat seed in the autumn as plenty of other food is available at that time of the year. The effects of DDT, and other organochlorine insecticides, in reducing the average thickness of birds' egg shells probably results from their interference with calcium metabolism[6]. Studies of egg shells of predatory birds show a sharp fall in thickness during the 1945–47 period when DDT was introduced into the environment. Experimental tests in which DDT and other organochlorine compounds were fed to ducks, sparrow-hawks, and quail showed that the chemicals reduced both shell thickness and the hatching success rates.

Salmon in Lake Michigan accumulate quite high levels of organochlorine residues in their fatty tissues and lay eggs containing appreciable pesticide residues, and this caused the death of almost 700,000 salmon fry in 1968. The successful breeding of trout has been similarly affected and rising levels of organochlorine compounds have been observed in several commercially important sea fishes like tuna, mackerel, and hake[5,12].

DDT and other organochlorine insecticides are very stable substances which are only slowly biodegraded. The metabolites like DDE often contain almost as many chlorine atoms as the original compound[11,12]. Consequently the whole world has become polluted with DDT and its metabolites. They are found in rainwater, antarctic snow, air, fish and animal fats – human body fat may contain some 10 p.p.m. of DDT. The universality of these compounds is a remarkable tribute to both the stability and mobility of these compounds[5].

Organochlorine insecticides act as general nerve poisons; their greater toxicity to insects is probably largely due to the greater ease of absorption through the insect cuticle as compared with the mammalian skin. They pose a severe threat to our ecosystem because of their great stability[13,14]. Such 'hard' pesticides are relatively resistant to biodegradation. Thus 50% of the DDT sprayed on a field was still present ten years later and tended to be recycled through food chains. Chlorinated hydrocarbons are very mobile and thus they adhere to dust particles and are blown around in the dusts of the world. They are concentrated in the fatty tissues of living organisms and because the aqueous solubility of DDT is only 1.2 parts per billion parts of water, it is quickly translated from the water into living aquatic organisms often with disastrous effects (see the Clear Lake case, p. 215). As a result of the hazards presented to the environment, the use of the persistent organochlorine insecticides is severely restricted in many advanced countries (e.g. Western Germany, France, Holland, and Japan) that can afford the higher costs of alternative compounds, such as malathion and Sevin. The United States of America banned DDT in 1972; in Britain a voluntary restriction was placed (1966) on the use of aldrin and dieldrin as seed dressings for spring-sown wheat, and in sheep dips.

However DDT remains essential for control of malaria-carrying mosquitoes in the Far East as it is the cheapest available insecticide. At the peak of the malaria control campaign during the 1960s some 66,000 tonnes of DDT, 4000 tonnes of dieldrin, and 500 tonnes of γ-HCH were applied annually. As the vectors quickly became resistant to dieldrin little of this chemical is now used, and the amount of DDT has fallen to some 46,000 tonnes each year. The World Health Organization consider that to contain malaria this level of DDT usage will have to be maintained since no competitive insecticides are currently available.

The organochlorine insecticide methoxychlor does not apparently accumulate in fats because unlike the majority of organochlorine compounds, it is biodegraded by oxidative demethylation. The use of methoxychlor therefore will probably not result in long-term environmental pollution and so the use of this compound is often permitted[15].

Organochlorine insecticides owe their activity to their interference with the nervous system (Chapter 5, p. 54), but there is no conclusive evidence that these pesticides at the present levels of exposure cause any long-term harmful effects to man[6,12] though they may reduce the life expectancy of those exposed to these compounds from birth (aged 25 or less). However further studies are needed in view of experiments showing that high doses of DDT increase the incidence of liver cancers in mice[5]; there is also concern that the mutations observed in some treated animals might also occur in humans[13]. DDT stimulates production of the female sex hormone oestrogen and certainly affects the sex hormones of birds and rats; possibly the chemical might initiate hormonal changes in man[5]. It has been suggested[5,12] that there may be some correlation between high DDT levels in the brain, liver, and adipose tissues of terminal cancer patients and those who died from coronary diseases.

DDT and other organochlorine insecticides at concentrations of 10 parts per billion inhibit the photosynthesis of 4 species of phytoplankton[5]. Since all life on the earth depends on photosynthesis this observation appears alarming, but lower concentrations of DDT (1 part per billion) did not affect photosynthesis and the actual concentrations of DDT and its metabolites in the sea are of the order of 1 part per trillion or less so there is little danger of damaging photosynthesis by photoplankton, especially since there are many thousands of different species of plankton[6].

The extensive use of the persistent organochlorine insecticides has led to the appearance of resistant strains of insects (Chapter 5, p. 55). A danger is that larger doses of organochlorine insecticides may be applied to try and control the pest which will result in greater environmental pollution. Luckily other types of insecticides such as organophosphates and carbamates are available for control of pests which have acquired tolerance to organochlorine insecticides.

In addition to the organochlorine insecticides mentioned, some 1200 tonnes of chlorophenols are employed as broad-spectrum biocides against insects, bacteria, fungi, algae, and plants. Their major use is as antirotting agents in non-woollen textiles and timber. Chlorophenols can be microbiologically and photochemically degraded but little is known about the speed of their breakdown under natural conditions, how long they persist in living organisms, or their effects on aquatic life at sublethal concentrations[16].

Polychlorinated biphenyls (PCBs) are widely distributed in the environment due to their various industrial uses as, for instance, electrical insulators, transformer fluids, plasticizers, lubricants, and as constituents of wax polishes and sealing compounds[17]. PCBs are considered to constitute a danger to public health and the new EEC regulations[18] restrict their use to electrical insulators and certain other specified uses where there is little danger of environmental contamination. The widespread application of PCBs probably makes them a greater environmental hazard than the organochlorine insecticides, even in 1971 the world production of PCBs was still 50,000 tonnes[17]. Virtually nothing is known about the effect of PCBs on animal populations; sublethal doses may affect both the survival and reproduction of animals[11].

In any consideration of organochlorine insecticides, as well as their pollution of the environment, the tremendous benefits obtained by the use of these compounds must be remembered. In the Far East there is currently no economically viable alternative to DDT for the control of disease-carrying vectors of typhus, bubonic plague, and malaria. During malaria control campaigns large quantities of DDT were applied to the interior walls of the homes of millions of people at concentrations of 2 g/m^2. To control yellow fever, DDT was introduced into domestic water supplies at 1–5 p.p.m. without causing any observed ill effects. In the war, soldiers had their clothing dusted with DDT to control body lice and employees of industries manufacturing organochlorine insecticides frequently contain as much as 200 p.p.m. of organochlorine residues in their adipose tissues. Careful monitoring of these workers over many years has not however revealed any illnesses that could be attributable to the presence of these residues.

Future research on agricultural pesticides (Chapter 15) will be increasingly directed towards the discovery of chemicals, like menazon, which combine low mammalian toxicity with specific activity against the candidate pest and so do not interfere with natural predators. Such pesticides would be valuable in integrated biological–chemical control programmes, a very promising avenue of future pest control measures, since smaller amounts of chemical pesticides would then be needed to obtain effective control of the pest, lessening the dangers of both environmental pollution and the development of resistance in the pest species.

Several new chemical pesticides act indirectly[19]. Certain systemic fungicides (Chapter 7, p. 126) appear to function by increasing the resistance of the host plant to attack by the pathogen. There are also chemicals which can be used against insect pests because they act as attractants, repellents, chemosterilants, or growth hormone mimics (Chapter 13). Such compounds often have very high and specific activity to the target pest combined with low mammalian toxicity, so their use is unlikely to be associated with environmental hazards.

These developments should therefore lead to a reduction in the environmental pollution by pesticides. This will be assisted by increasingly stringent control measures governing the use of 'hard' pesticides, such as DDT, which do not suffer reasonably rapid degradation when released into the ecosystem.

References

1. Carson, R., *Silent Spring*, Hamish Hamilton, London, 1963.
2. Mellanby, K., *Pesticides and Pollution*, Collins, London, 1970.
3. Approved Products for Farmers and Growers, Ministry of Agriculture, Fisheries, and Food, London, 1978.
4. *Herbicides: Chemistry, Degradation and Mode of Action* (Eds. Kearney, P. C. and Kaufmann, D. D.), 2nd edn., Vol. 1, Dekker, New York, 1975, p. 8.
5. Ehrlich, P. R. and Ehrlich, A. H., *Population, Resources, Environment*, W. H. Freeman and Co., San Francisco, U.S.A., 1970, p. 182.
6. *Pesticides in the Modern World* — A symposium prepared by members of the Co-operative Programme of Agro-Allied Industries with FAO and other United Nations Organisations, Newgate Press, London, 1972.
7. Brown, A. W. A., 'Pest resistance to pesticides' in *Pesticides in the Environment* (Ed. White-Stevens, R.), Vol. 1, Part II, Dekker, New York, 1971, p. 458.
8. Hartley, G. S. and West, T. F., *Chemicals for Pest Control*, Pergamon Press, London, 1969, p. 119.
9. *Chemical Fall Out* (Eds. Miller, M. W. and Berg, G. G.), 2nd edn., Thomas, C. C., Illinois, U.S.A., 1972.
10. Eto, M., *Organophosphorus Pesticides: Organic and Biological Chemistry*, CRC Press, Cleveland, Ohio, U.S.A., 1974, p. 192.
11. *Organochlorine Insecticides: Persistent Organic Pollutants* (Ed. Moriarty, F.), Academic Press, New York, 1975.
12. *Environmental Pollution by Pesticides* (Ed. Edwards, C. A.), Plenum Press, London, 1973.
13. Graham, F., *Since Silent Spring*, Hamish Hamilton, London, 1970.
14. Rudd, R. L., 'Pesticides' in *Environment: Resources, Pollution and Society* (Ed. Murdoch, W. W.), 2nd edn., Sinauer Associates Inc., Sunderland, Mass., U.S.A., 1975, p. 324.
15. Brookes, G. T., 'Pesticides in Britain' in *The Environmental Toxicology of Pesticides*, Academic Press, New York, 1972.
16. *The New Scientist*, 64, 908 (1974).
17. Green, M. B., 'Polychloroaromatics and heteroaromatics of industrial importance' in *Polychloroaromatic Compounds* (Ed. Suschitzky, H.), Plenum Press, London, 1974, p. 403.
18. *Chem and Ind.*, (20), 821 (1974).
19. Fletcher, W. W., *The Pest War*, Blackwell, Oxford, 1974, p. 130.

Chapter 15

Future Developments

The original chemical pesticides were general poisons with non-specific activity; thus early herbicides like sodium chlorate and copper sulphate were total weed killers which could not be effectively applied as selective herbicides. Likewise insecticides such as hydrogen cyanide, lead arsenate, and Paris Green were highly poisonous materials with a wide spectrum of insecticidal and mammalian toxicity.

Similarly, such fungicides as sulphur, Bordeaux mixture and organomercurials tended to be comparatively non-specific in their toxicity towards fungi.

Later work led to the discovery of less poisonous and more selective organic chemical pesticides; illustrative examples were the phenoxyacetic acid selective herbicides, certain organophosphorus insecticides like malathion, and the trichloromethylthio fungicides, e.g. captan.

There is now much greater awareness of the dangers of environmental pollution arising from the widespread application of chemical pesticides (Chapter 14), and candidate chemicals have to pass increasingly stringent tests regarding toxicity, and residue formation before they can be marketed as pesticides in many countries. This has caused research on new pesticides to be increasingly concerned with producing chemicals which are safer and more selective in their action.

The ideal chemical pesticide would have high specific toxicity against the target pest, should not persist longer than necessary to achieve its objective, and would not affect the rest of the ecosystem, so that natural predators and other beneficial insects are unharmed[1].

Some well-known examples of modern pesticides approximating to these criteria are the systemic selective aphicides menazon and pirimicarb, and several of the modern systemic fungicides, such as dimethirimol, which shows high selective activity against cucumber powdery mildew. Remarkably potent specific toxicity against flying insects has also been achieved with some synthetic pyrethroids such as decamethrin (Chapter 4, p. 46). However, the majority of pesticides currently in use fall far short of these ideals.

The increasing emphasis on systemic rather than contact pesticides leads to more effective use of the chemicals and less danger of environmental pollution. In the case of insecticides, systemic compounds will also be more selective in their action since only those insects actually eating the treated crop will be killed. During the last ten years the most significant breakthrough in chemical pesticide research has certainly been the introduction of commercially viable systemic fungicides[2,3] (Chapter 7, p. 127).

These are highly active and specific in their toxicity, but their use has rapidly led to the emergence of resistant fungal strains. The growth of resistance poses increasingly serious problems in the fields of insecticides and systemic fungicides, so there is a continual need for the development of new types of specific pesticides to combat resistant insects and fungi. At present there is little sign of the widespread emergence of weed strains showing tolerance to herbicides, although this may happen in the future. Certainly the development of resistance in common broad-leaved weeds in cereals towards phenoxyacetic acid selective herbicides would pose a very serious threat to world cereal production.

In addition to new classes of chemical pesticides that act directly on the pest, recent developments have included behaviour-controlling chemicals such as plant growth regulators (Chapter 8, p. 166), pheromones, insect growth hormone mimics, and chemosterilants (Chapter 13); these are likely to be the most important areas for future research[4]. Current control of insects relies largely on the use of organochlorine, organophosphate, and carbamate insecticides all of which act on the insect's nervous system. Of the novel chemical insect control agents, the sex pheromones probably achieve the ultimate in both sensitivity and specificity; indeed the remarkable sensitivity of insect sensory receptors suggests that if compounds could be found that would block these sensory organs, they might function as outstandingly valuable insect control agents[1].

Many chemical structures influencing insect behaviour have now been identified. Insect juvenile hormone mimics such as methoprene (Chapter 13, p. 203) show high specific toxicity towards mosquitoes and related insects and act by blocking insect metamorphosis. Diflubenzuron is an insect growth regulator which appears to inhibit chitin synthesis and so interferes with the formation of the insect cuticle[4,5]. Another interesting approach to insect control is the use of chemosterilants. This has been applied successfully against fruit flies and there is a continual search for more effective sterilization agents[6] (Chapter 13, p. 199).

These novel insect behaviour chemicals generally have the advantages of high selective toxicity against the target insect pest and they are less likely to induce insect resistance as compared with conventional insecticides. However within the latter field, there is still considerable scope for achieving both selectivity and high activity. For example in the organophosphates, carbamates, and synthetic pyrethroids remarkable selectivity can be achieved by utilizing differences in the ease of penetration of the toxicant to the site of action of different species and by exploiting variations in the nature and rates of metabolism. Examples showing favourable toxicity include the aphicide pirimicarb, the acaricides chlorbenside and tetradifon, and the pyrethroid bioresmethrin.

One aim in pesticides is to obtain the highest possible activity to the pest with the objective of reducing the rates of application and hopefully costs and the danger of environmental pollution. With insecticides some extremely potent compounds are now available. For instance synergized bioresmethrin has a remarkably high specific toxicity to houseflies (LD_{50} 0.002 mg/kg)[1] as compared with 2 mg/kg for parathion (Chapter 4, p. 44). Generally, as new pesticides have been discovered which attack increasingly specific sites this has been followed by progressive increases in activity.

Corbett[7] has suggested that as the enzyme acetylcholinesterase, the principal site of action of existing synaptic insecticides (e.g. organophosphates and carbamates), is widely distributed in the central nervous system of insects and mammals, it should be possible to select more specific nerve transmitters than acetylcholine as targets. For instance, nervous transmission in the locust may also involve relatively specific amine transmitters which would appear to be a good target for attack by a chemical, leading to the development of a fairly specific pesticide for control of locusts. There is substantial research at present for chemicals that affect specific insect neurotransmitters; this is a promising investigation that could lead to important new selective chemical insecticides.

There is a wide range of commercial herbicides on the market with different modes of action, selectivity, and persistency (Chapter 8). Bipyridylium herbicides, when sprayed onto plants, kill all plant growth which has been covered by the spray, but they are immediately deactivated on contact with soil so they can be used in chemical ploughing. On the other hand, triazine herbicides, like simazine, are very persistent and retain their activity in the soil for nearly a year. Some weeds, e.g. couch grass, are difficult to control, especially without damaging other crops, so there is a continuing need for more selective herbicides and recently several compounds, e.g. barban, have been introduced for selective control of wild oats in cereals; the selective control of such grass weeds is a major research objective.

There is increasing emphasis on synthetic plant growth regulators to improve crop yields and harvesting and further developments in this area can be expected in the future.

With fungicides, it appears rather more difficult to devise new routes to substantially more active agents although the general levels of fungitoxicity available are appreciably less than for other biocides. More systemic fungicides, some with new modes of biochemical action, can be expected to appear on the market. A wide range of systemic fungicides will be needed to counter the threat posed by resistant fungi.

Fungicides, like most pesticides, have been discovered as a result of more or less random screening of large numbers of chemicals; any activity was followed by the synthesis of analogous compounds ('molecular roulette') which hopefully led to a structure—activity pattern for the series of compounds under investigation. If some of the chemicals has really useful levels of activity, an effort was made to elucidate the biochemical mode of action which in turn may lead to the design of more active molecules. In the search for more selective pesticides the correct design of the screening tests and the optimum formulation of the toxicant are both vitally important.

More specific fungicides may be discovered in the future by screening them against systems largely confined to fungi; for instance for their effect on chitin biosynthesis, since chitin is only important in the cell walls of fungi and the cuticles of insects. The systemic fungicides polyoxin D and Kitazin (Chapter 7, p. 136) are known to act by interference with chitin synthesis.

Comparatively few important metabolic differences have been discovered between plants and fungi; obviously the disruption of processes unique to fungi would provide a useful basis for the design of new specific fungicides[1].

Screening tests have perhaps, in the past, concentrated too much in evaluating the direct effect of the chemical on the pest organism while insufficient attention has been paid to the effects of the chemical on the host plant and the environment[8]. Some chemicals owe their pesticidal activity to indirect action, by either increasing the resistance of the crop plant to the invading pathogen or by modifying the environment so as to make it less conducive to the support of the pest. Such pesticides are attractive, since their application would be unlikely to induce pest resistance as compared with chemicals exerting a direct toxic action on the pest organism.

Several chemicals are already known that reduce disease levels without directly affecting the pathogen; the effectiveness of mineral oil in the control of *Sigatoka* is probably due to the oil reducing the sugar concentration in the young leaves. Many compounds are known to contain natural antifungal compounds (phytoalexins), and these natural defence mechanisms can sometimes be triggered off by chemicals. The application of copper and mercury salts stimulates the production of the phytoalexin pisatin in peas, and injection of phenylalanine appears to increase the resistance of apple leaves to scab[9]. Chemicals can also modify the phenolic content or amino acid nutrition of the host plant and this can sometimes increase resistance to the pathogen. Chemicals can also act by physically modifying the host plant by inducing the formation of thicker cuticles or starch-filled cells which form barriers to the invading pathogen. To exploit these indirect plant disease control chemicals, a greater knowledge of the differential biochemistry of host plants is needed[1].

There will probably be further introduction of antifungal antibiotics, several of which like blasticidin and kasugamycin are very effective for systemic control of rice blast fungus[8].

More precise knowledge of the various factors that govern the movement of chemicals in plants may enable the discovery of chemicals that are well translocated in shrubs and trees and might control Dutch elm and other fungus diseases of trees[10]. There is also need for chemicals which after spraying onto the leaves of the crop plant are translocated downwards in the phloem to the roots. Such compounds could prove very effective against root pathogenic fungi and many nematodes which are not well controlled by existing chemicals. The majority of systemic fungicides show little downward movement from the leaves, although the synthetic plant growth regulator daminozide (1) (Chapter 8, p. 168) apparently[4] owes its activity against potato scab to downward translocation.

At present a critical gap in the spectrum of activity of commercial systemic fungicides is the control of diseases caused by *Phycomycetes*, e.g. potato blight. Pyroxychlor (2) however when sprayed on leaves is translocated downwards to the roots where it controls root-infecting *Phycomycetes*[8]. Pyroxychlor (2) formulated as granules successfully controlled foot rot in peas.

$$HO_2C-CH_2CH_2CONH-N(CH_3)_2$$
(1)

(2) 2-methoxy-4-(trichloromethyl)-6-chloropyridine with substituents CCl_3, CH_3O, N, Cl

(3) 2-chloro-6-(trichloromethyl)pyridine with substituents Cl, N, CCl_3

The discovery of pyroxychlor emphasizes the importance of correctly designed screening procedures. The tests must allow full interaction to occur between the host and the toxicant; this unique chemical would probably have been missed if only the seed dressing tests had been used[9]. Furthermore, as increasingly specific pesticides are needed their activity could easily not be discovered unless they are exposed to the sensitive organism in the screening tests.

Soil diseases can be greatly influenced by environmental effects such as those deriving from the solid, liquid, and gaseous phase components of the soil, plant exudates, microbial toxins, and crop residues. Certain diseases can be reduced by alteration of the nutrition of the host plant; for example by stabilizing nitrogen nutrition by addition of nitrification inhibitors like nitrapyrin[9] (3).

More attention needs to be paid to the discovery of spore germination inhibitors since various natural compounds with potent and unequivocal activity have been identified[1]; thus methyl 3,4-dimethoxycinnamate inhibits the germination of bean rust spores.

The present chemical pesticide armoury of some 500 different chemicals is insufficient and new compounds are urgently needed for controlling resistant pests and in other specific areas. There are no commercial antiviral agents available against plant viral diseases. Their development is difficult because it involves interfering selectively with the intimate relationship which exists between the virus and its host, but the discovery of plant viricides is undoubtedly a major target for future research effort[1]. Adequate plant bactericides are also lacking and there is considerable scope for the introduction of entirely new toxicants in this field. The problem appears to be the difficulty of finding materials that are sufficiently persistent bactericides but do not harm the host plant. Streptomycin and other antibiotics (Chapter 7, p. 123) are active against certain plant bacterial diseases, e.g. cherry leaf spot, but they are expensive and there is too fine a margin between the concentration achieving adequate disease control and that causing phytotoxic damage.

The availability of more selective chemical pesticides permits them to be used in conjunction with biological control methods and it seems probable that integrated biological—chemical measures of pest control will become more common in the future. Such procedures as manipulating the ecology of host and pathogen, while reserving chemical treatment until the other measures have at least reduced the severity of the pest attack, allows the pest to be controlled effectively by smaller amounts of the often costly chemical. It would also have the advantage of reducing the danger of environmental pollution. The plant breeder has been very successful in producing new strains of crop plants with genetic disease resistance against sedentary root pathogenic fungi, but so far this approach has been less effective in combating the readily dispersable foliage fungi. The disadvantage is that the plant's resistance is generally not permanent and in time the fungus mutates to a new pathogenic strain which is capable of attacking the formerly resistant variety of the host plant. The breeding of resistant crop plants thus degenerates into a race between the plant breeder and the fungal mutation – a new aspect of the age-old war between pests and man for crop plants.

Although less spectacular than the discovery of new types of chemical pesticides, the importance and need for new formulations of existing pesticides must not be overlooked (Chapter 2). Formulations (e.g. wettable powders, seed dressings, granules, microcapsules) can substantially affect the persistency and selectivity of a given pesticide; in efforts to minimize damage to beneficial insects the timing of treatment, method of application, and formulation are adjusted to match the behaviour of the target pest. This ecological approach is capable of further exploitation and could lead to dramatic improvements in the effectiveness of many pesticide treatments[1]; thus microencapsulation can increase persistence by controlling the rate of release of the pesticide. Experiments indicate that direct spraying of pesticides on pests is probably some 10 times less selective than when the toxicant is taken up as vapour from surface deposits. Formulation offers many ways of altering the relative amounts of the toxicant reaching different organisms, e.g. addition of amine stearates to wettable powder formulations of DDT substantially increased the persistency of the dried deposits on the leaves without decreasing the activity. Selectivity could be attained by coating the DDT particles with hemicelluloses since only phytophagous chewing insects possessing the appropriate enzymes (hemicellases) have the ability to degrade the protective coating. More examination should be made of formulations containing synergists which are currently commercially used only with pyrethroids (Chapter 4) since there appears to be considerable scope for increasing the activity of many pesticides by such methods.

In conclusion, there appears little prospect in the foreseeable future that biological control measures, such as the introduction of resistant crop varieties, cultural control, genetic methods, or the use of natural predators will displace chemical pesticides from their dominant position. However, further research on these and other biological control measures is very necessary, to improve their efficiency and enable them to be increasingly employed in integrated control programmes in conjunction with chemical pesticides. The most productive areas of research probably lie in the fields of behaviour-controlling chemicals, microbial pesticides, plant viricides and bactericides, and systemic fungicides, especially those active against *Phycomycetes*.

References

1. Graham-Bryce, I. J., *Proc. 8th Brit. Insectic. and Fungic. Conf.*, Brighton, **3**, 901 (1975).
2. Cremlyn, R. J., *Internat. Pest Control*, **15** (2), 8 (1973).

3. Erwin, D. C., *Ann. Rev. Phytopath.*, **11**, 389 (1973).
4. Graham-Bryce, I. J., *Chem. and Ind.*, (13), 545 (1976).
5. Ruscoe, C. N. E., *Proc. 8th Brit. Insectic. and Fungic. Conf.*, Brighton, **3**, 927 (1975).
6. Turner, R. B., 'Chemistry of insect chemosterilants' in *Principles of Insect Chemosterilization* (Eds. Labrecque, G. C. and Smith, C. N.), Appleton-Century-Crofts, New York, 1968, p. 159.
7. Corbett, J. R., *Proc. 8th Brit. Insectic. and Fungic. Conf.*, Brighton, **3**, 981 (1975).
8. Cremlyn, R. J., 'The biochemical mode of action of some well-known fungicides' in *Herbicides and Fungicides: Factors Affecting Their Activity* (Ed. McFarlane, N. R.), *Chem. Soc. Special Publication* No. 29, The Chemical Society, London, 1977, p. 22.
9. Sbragia, R. J., *Ann. Rev. Phytopath.*, **13**, 257 (1975).
10. Crowdy, S. H., 'Translocation' in *Systemic Fungicides* (Ed. Marsh, R. W.), Longman, London, 2nd edn., 1977, p. 92.

Subject Index

Bold type indicates the more important references. Trade marks have, as far as possible, been indicated by inverted commas.

Abscisic acid, 167
Acaricides, 1, 18, 51, **56–59**, 65, 69, 71, 74, 75, 76, 78, 91, 92, 98, 102, 117, 141, 211, 223
3-(α-Acetonylbenzyl)-4-hydroxy-coumarin, 8, 102, **179–180**, 181.
Acetylcholine, 33, **34**, 40, 80, 81, 82, 95, 100, 101, 224
Acetylcholinesterase, **34**, 40, **80–83**, 87, 91, 95, **100–101**, 224
Acetyl coenzyme A, 30, 31
Acetyl glucosamine polymer, 37, 125, 205, 225
Adenine, 36, 130, 133, 167, 206
Adenine diphosphate, 31, 32
Adenosine cyclic monophosphate, 167, 168
Adenosine triphosphate, 28, 29, 30, 31, 32, 36, 117, 118, 141, 161, 163
A.D.P., 31, 32
Adsorption onto soil colloids, 7, **22–23**, 156, 160, 188–189, 192
Aerosols, 14, 16, 42, 53, 198
Agrochemicals, definition of, 1
Alachlor, 153
Aldicarb, 59, **98–99**, **100**, 101, 189
Aldrin, 6, 24, **64**, **65**, 76, 102, 217
Algicides, 2, 111, 155
Alkylating agents, as chemosterilants 165, 200–201
2-Alkyl imidazolines, 18, 106
Alkyl isothiocyanates, as insecticides, 5, 52, 188
Alkyl tin derivatives, as fungicides, **111**
Allethrin, **43–44**, 47
Allidochlor, 152–153
'Allisan', as fungicide, 118–119
4-Allylbenzyl chrysanthemate, 44
Alphachloralose, **181**
Aluminium phosphide, 174, 177

Amides, as herbicides, **152–153**, 157
Amidines, as acaricides, **59**
Amino acids, 29, 30, 35, 36, 37, 120, 124, 125, 126, 127, 133, 225
Aminoacyl transfer RNA, 36–37, 124, 125
p-Aminobenzoic acid, 122
5-Amino-4-chloro-2-phenylpyridazin-3-one, 157–158
Aminotriazines, as herbicides, 21, 23, 24, 25, 33, **154–157**, 211 224
3-Amino-1,2,4-triazole, 37, **157**
Amiton, 83
Amitrole, 37, 157
Anabasine, 40
Ancymidol, as plant growth retardant, 169
Aniline, 150
Antibiotics, as systemic fungicides, 7, 8, 19, 37, 121, **122–125** 126, 127, 225, 226
  as chemosterilants, 201
Antifeeding compounds, **199**
Antifouling paints, 193
Antifungal soil microorganisms, 127
Antimetabolites, as chemosterilants, 200–201
Antu, as rodenticide, 178
Apholate, as chemosterilant, 200–201
'Aqualin', 155, 156
Arachnids, 2
Arbusov reaction, 80
Arsenic, 7, 50, 214
Arsenical compounds, as insecticides, 3,5, **50–51**, 214
  as weedkillers, 140, 210
Arthropods, 2
Arylphthalamic acids, as plant growth regulators, 169
Arylureas, as herbicides, 21, 22, 25, 33, **150–152**, 157, 211
Aspirin, 215

239

240